Special Places

Betty I. Roots, editor-in-chief
Donald A. Chant and Conrad E. Heidenreich, editors

Special Places

The Changing Ecosystems of the Toronto Region

UBC Press · Vancouver · Toronto

© Royal Canadian Institute 1999

All rights reserved. No part of this publication may be reproduced, stored in a retrieval system, or transmitted, in any form or by any means, without prior written permission of the publisher, or, in Canada, in the case of photocopying or other reprographic copying, a licence from CANCOPY (Canadian Copyright Licensing Agency), 900 - 6 Adelaide Street East, Toronto, ON M5C 1H6.

Printed in Canada on acid-free paper ∞

ISBN 0-7748-0735-0 (hardcover)
ISBN 0-7748-0736-9 (paperback)

Canadian Cataloguing in Publication Data

Main entry under title:
Special places

  Includes bibliographical references and index.
  ISBN 0-7748-0735-0 (bound). – ISBN 0-7748-0736-9 (pbk.)

  1. Natural history – Ontario – Toronto Region. 2. Ecology – Ontario – Toronto Region. 3. Toronto Region (Ont.) – History. I. Roots, Betty I. (Betty Ida) II. Chant, Donald A., 1928- III. Heidenreich, Conrad E., 1936-

QH106.2.O5S63 1999   508.713'541   C99-910644-9

This book has been published with the help of the Royal Canadian Institute.

UBC Press acknowledges the financial support of the Government of Canada through the Book Publishing Industry Development Program (BPIDP) for our publishing activities.
Canadä

We also gratefully acknowledge the ongoing support to our publishing program from the Canada Council for the Arts and the British Columbia Arts Council.

Set in Garamond and Gill Sans by Artegraphica Design Co.
Printed and bound in Canada by Friesens
Designer: Irma Rodriguez
Copy editor: Barbara Tessman
Proofreader: Darlene Money
Indexer: Patricia Buchanan

UBC Press
University of British Columbia
6344 Memorial Road
Vancouver, BC V6T 1Z2
(604) 822-5959
Fax: 1-800-668-0821
E-mail: info@ubcpress.ubc.ca
www.ubcpress.ubc.ca

Contents

Foreword / vii
   Dr. Roberta L. Bondar

Acknowledgments / ix

Introduction / 3

Part 1: **The Broad Physical Basis**

1. The Physical Setting: A Story of Changing Environments through Time / 11
2. Climate / 33
3. Watersheds / 51

Part 2: **From Wilderness to City**

4. Native Settlement to 1847 / 63
5. Spatial Growth / 77

Part 3: **The Past and Present Natural Environment**

6. Ecology, Ecosystems, and the Greater Toronto Region / 93
7. Vascular Plants / 105
8. Mosses, Liverworts, Hornworts, and Lichens / 117
9. Fungi / 129
10. Invertebrates / 151
11. Insects / 161
12. Fish / 177

Contents

13 Amphibians and Reptiles / 187

14 Mammals / 203

15 Birds / 217

Part 4: *The Special Places*

16 From Acquisition to Restoration: A History of Protecting Toronto's Natural Places / 229

17 Special Places / 243

    Waterfront Ecosystems: Restoring Is Remembering / 245

    The Port Lands: The Significance of the Ordinary / 249

    Scarborough Bluffs / 256

    The Savannahs of High Park / 260

    Oak Ridges Moraine / 266

    Credit River / 271

    Humber Valley / 278

    Don Valley / 283

    Duffins Creek / 286

    Rouge Valley / 290

18 Discussion and Conclusions / 295

    The History of the Royal Canadian Institute / 305

    Afterword / 308
      David Crombie

    References and Additional Reading / 310

    Contributors' Acknowledgments / 327

    Contributors / 329

    Index / 331

Foreword

In the 86 years since *The Natural History of the Toronto Region*, the precursor to the present book, was published, the city and indeed the world have changed almost beyond recognition. What remains recognizable are the 'special places,' the ecosystems of Toronto and the earth. As a scientist, and as an astronaut who has seen the planet from the unique viewpoint of the space shuttle *Discovery*, I know both how glorious and how fragile those ecosystems are.

The wonders of the everyday world we citizens of Toronto enjoy, often without realizing the complexity of our natural environment, are celebrated in *Special Places*. But this book is more than a guide; it is a series of thoughtful and carefully researched essays by Dr. Betty I. Roots and her co-editors, Drs. Donald A. Chant and Conrad E. Heidenreich, and many contributors. They offer us a fascinating spectrum – from the vast perspectives of palaeontology, to the portraits of First Peoples, to close-ups of mosses and lichens. Together, these scholars and educators have produced a wonderful present for the Royal Canadian Institute, Canada's oldest active scientific society, 150 years old in 1999. Torontonians and all the readers of *Special Places* owe the Royal Canadian Institute a debt of gratitude for showing us all how the city has changed and what we must do to care for the special place on the planet that Toronto inhabits.

The past has been recalled; the present is described lovingly; and the future is ours to create. Let us take to heart the advice of the authors of this book: we must guide our activities and our use of the environment so that our impact is minimized, assuring that in another 86 years our children's children will still be able to enjoy Toronto's special places.

Dr. Roberta L. Bondar

Acknowledgments

A work such as this needs the efforts of many people to bring it to fruition. We have been fortunate to have that support.

From its inception through the many stages in the development of the book, Professor Ann P. Zimmerman has been a constant source of encouragement and inspiration, especially to Betty Roots, and to her we extend a very special thanks.

Without the unremitting work of our project manager, Sherry Pettigrew, this book never would have gone to press. Sherry gave of herself unstintingly and through it all maintained her cheerful sense of humour. Thank you, Sherry.

We appreciate the support of Kathy Carter, Executive Secretary of the Royal Canadian Institute throughout.

The maps are a unifying component of the volume. With the exception of the historical maps, they were developed by Conrad Heidenreich and produced by Carolyn King at York University. Further development of the maps was provided by Eric Leinberger. We are greatly indebted to all of them.

We wish to thank our authors most sincerely both for their contributions and their patience in responding to our frequent requests for more information. True to the fine tradition of the RCI they are forgoing royalties. Any proceeds from the sale of this book will go to the RCI Scholarship Fund.

Sidebars were not necessarily written by the authors of the chapters in which they occur. Therefore acknowledgment is made to the following, who researched and wrote various sidebars appearing throughout this book: Don Chant, Bruce Falls, Conrad Heidenreich, Warren Kalbach, John Krug, David McQueen, Ted Munn, Henry Regier, Sherry Pettigrew, Wayne Reeves, Glenn Wiggins, Dudley Williams, Siân Williams, and Ann Zimmerman.

Our grateful thanks to Roberta Bondar and David Crombie for writing the Foreword and Afterword respectively.

We thank Laura Macleod of UBC Press for her enthusiasm and help in the negotiations that led to UBC Press becoming our publisher. We thank Barbara Tessman for her work as copy editor and Irma Rodriguez, the designer, for making such a visually appealing book. Camilla Jenkins, who assumed responsibility for the book for the Press, has given meticulous and patient attention and we thank her warmly for seeing the book through to completion.

We also greatly appreciate all our friends and colleagues who have offered constructive advice and helped in various ways.

We are grateful to Northway-Photomap Inc., and in particular to Rick Teminski, for donating the two aerial images of Metropolitan Toronto that appear in Chapter 5.

Acknowledgments

Financial support was given by the University of Toronto, both by the Provost and the Dean of the Faculty of Arts and Science, and by The Richard Ivey Foundation. To these we offer our sincere thanks.

We are grateful for permission to use the photographs that appear on the following pages: p. 2, night sky, courtesy Terence Dickinson; p. 8, Scarborough Bluffs, courtesy John Westgate; p. 10, gravel pit, courtesy David McQueen; p. 32, clouds, courtesy Betty I. Roots; p. 50, Wychwood Ravine, 1907 (City of Toronto Archives SC244-1246B); p. 60, from *First Settlement,* by William Bartlett, courtesy Conrad E. Heidenreich; p. 62, from *A Forest Scene,* by William Bartlett, courtesy Conrad E. Heidenreich; p. 76, highway cloverleaf, courtesy Betty I. Roots; p. 90, beetle chambers, courtesy Martin Hubbes; forest creek, courtesy Betty I. Roots; p. 104, autumn scene, courtesy Betty I. Roots; p. 116, lichen, courtesy John Krug; p. 128, pleated inky cap, courtesy David Malloch; p. 150, zebra mussels, courtesy W. Gary Sprules; p. 160, monarch butterfly, courtesy Frank Parhizgar; p. 176, lake trout, courtesy Great Lakes Fishery Commission; p. 186, eastern milk snake, courtesy Robert Johnson; p. 202, little brown bat, courtesy M. Brock Fenton; p. 216, peregrine falcon, courtesy Ted Muir; p. 226, Credit River at Cataract, courtesy Robert Morris; p. 228, High Park, 1918, detail, NAC PA071052; p. 242, birches, courtesy David McQueen; p. 294, Rouge River, courtesy John Riley; p. 304, Toronto skyline, courtesy Betty I. Roots.

Special Places

Introduction

*Betty I. Roots, Donald A. Chant, Conrad E. Heidenreich,
Henry R. Regier, Ann P. Zimmerman*

The Royal Canadian Institute (RCI) is the oldest active scientific society in Canada. To commemorate its 150th anniversary, the council of the Institute decided to prepare a successor to its 1913 publication, *The Natural History of the Toronto Region, Ontario, Canada,* edited by Joseph H. Faull, professor of botany at the University of Toronto. At the outset, it is necessary to define the greater Toronto region: it has changed so much in the past 100 years, yet it remains a coherent geographical area bounded east and west by the Rouge and Credit River watersheds and by the Oak Ridges Moraine to the north (Figure 0.1).

Major changes have taken place in the biological sciences since 1913. At that time, limited knowledge and limited understanding of ecosystem thinking created the approach of making biological inventories: lists of the plants and animals in the Toronto area. This was a reasonable starting point. Science, after all, proceeds from observation. Seen through the eyes of modern science, however, it is not enough. We must now attempt to understand how organisms relate to one another, how they form interdependent living systems, and how they relate to their physical and chemical environment. This is the ecosystem approach, and the one selected for this book. The RCI could have chosen other themes – cultural, social, economic, perhaps. But throughout its long history it has been an organization focused on science, and it was appropriate to select as our theme the natural world of the greater Toronto region and the changes that have occurred over the lifetime of the RCI.

This book is an attempt to assess our current state of knowledge through modern eyes and also to assess the changes that have occurred in the natural system of the greater Toronto region during the last 100 years through urbanization, the growth of human population in the area, and the rapid growth of technology. These changes, whether good or bad, are now history. Our larger ambition is that with insight and understanding of what has happened in the past, and is happening right now, we can avoid mistakes as we move into the new millennium and beyond.

The title of this book suggests that ecosystems are special places. Various chapters tell you about specific ecosystems such as High Park or the Rouge valley, the environmental conditions that influence the ecosystems of the greater Toronto region, or more about particular kinds of plants or animals that you might expect to find in many of these systems. But what exactly is an ecosystem? Can a city be an ecosystem? How big are ecosystems? How small can they be? How many species can they hold? What makes them different from one another? And why do we call them 'special'?

The term 'ecosystem' is attributed to a British plant ecologist, A.G. Tansley. In an article written in 1935, Tansley defined an ecosystem as the complex of organisms and physical factors present in a particular environment. Interestingly, ecology, the study of ecosystems, had been defined almost 70 years earlier by Ernest Haeckel, a German scientist. Moreover, in a classic 1887 address, *The Lake as a Microcosm,* Stephen Forbes had spoken of the interconnectedness of aquatic organisms and their relationship to 'the entire system of conditions affecting their prosperity,' so the study of ecosystems clearly was not waiting for Tansley. Nevertheless, almost everybody has heard some variation on the story that the word ecology and, by extension, the word ecosystem derive from *oikos,* a Greek word meaning 'home.'

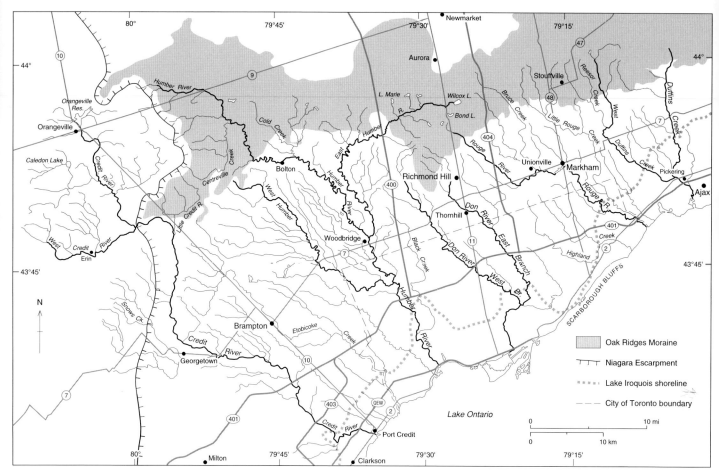

FIGURE 0.1 The greater Toronto region.

So at one level, ecosystems have something to do with organisms and their homes (or their environment). This simple definition is easy to remember, but how useful is it? Does it help us answer the questions about ecosystem size, or diversity of species, or system persistence? The definition of an ecosystem as organisms and their environment is really just a description. Describing the attributes of anything is an important first step in science, but it doesn't provide a basis for asking (and answering) questions about *why* or *how* something has happened, nor does it provide any insight into what might happen in the future.

It is also instructive to ask how we can be more explicit. Consider for a moment the environments that a microbial ecologist might be concerned with and compare them with the environments associated with caribou or grizzly bears or whales or black oak. How large an area needs to be considered to census these quite different organisms adequately? What kinds of information should we collect? Do we need to sample once a week, once a month, or once a year? How long do we need to keep collecting data? No two studies are likely to be designed in exactly the same way, so perhaps we can understand why a very general definition works best – we might not be able to get two ecologists to agree on anything more specific.

As you visit or read about the ecosystems outlined in this book, you will see that different authors work at different scales of time and space. This breadth is characteristic of ecology. Some ecologists focus on organisms, identifying themselves as plant biologists or wildlife ecologists. Others emphasize habitats, studying the ongoing activities in terrestrial or aquatic systems. Some ecologists concentrate on specific contexts such as forest, agricultural, or conservation ecology. Others adopt particular outlooks such as physiological, behavioural, or evolutionary ecology. Each of these researchers makes different assumptions; as a consequence, their working definitions of ecosystems will differ.

Even if we accept Tansley's definition of an ecosystem – local communities of organisms interacting with one another and with the chemical and physical elements making up their environment – what is left out is as intriguing as what is in the definition. What are the local boundaries of the community? Is a hollow of a tree, filled with water, leaves, and a wide assortment of microscopic organisms, an ecosystem? Of course it is. But that tree is part of some larger entity – a pasture, park, or forest – which in turn may be part of an even larger system.

The limits one researcher might set in order to study salamanders may be different from those set by an entomologist who studies beetles, which will differ again from a mycologist interested in fungi. An ornithologist studying the diversity of bird species will have different objectives than a behavioural ecologist interested in how different birds forage for food. So ecosystems can be nested one inside the other, and they can have overlapping 'boundaries,' often defined to suit the convenience of the observer.

If the boundaries of ecosystems are often observer defined, then they must be permeable. In fact, ecosystems are inherently leaky: at a minimum, energy and nutrients move in and out. More likely, individual organisms – for example, spiders, and starlings, and propagules (such as seeds) – move in and out as well. Nevertheless, ecologists generally accept that the organisms within a system interact with other organisms in the same system more than with those in different ecosystems. It is the nature of those interactions, the degree of system openness, and the kinds of relationships among the biotic and abiotic elements of the system, that influence how it operates.

We have still more questions: if ecosystems can be a variety of sizes (the issue of spatial scale), how long do they last or persist (an issue of temporal scale)? At what point has a system changed sufficiently to be considered a new ecosystem? Individual organisms are always dying; new organisms are being born. Yet under natural disturbance regimes, the system usually persists. If our lifespans were 800 rather than 80 years, would we be more cognizant of a slow oscillation between maple dominance and beech dominance in forests in southern Ontario? Are there maple-dominated patches and

Ecosystems

Most of us think of an ecosystem as being some place other than where we are: as a place to visit and, one hopes, to enjoy. Actually, each one of us is an ecosystem: we harbour millions of bacteria, fungal and mould spores, and pollen grains; more than 90 percent of us have hundreds of mites living on us, in our pores and hair follicles (no one knows why about 8 percent of us have none of these tiny creatures). The pillows on our beds have tens of thousands of these mites. If we are unlucky, some of us also are home for lice and fleas.

Our homes are ecosystems as well. We cohabit with dust mites, spiders, flies, millipedes, centipedes, book lice, silverfish, ants, and pseudoscorpions, and with innumerable spores, bacteria, fungi, and moulds, all interacting with one another and with us. Some of our homes have rats and mice, and cockroaches, and there may be squirrels, bats, and raccoons living in our attics.

Perhaps the most interesting nearby ecosystems are those of our own backyards, or the parkette around the corner. These are wildernesses in microcosm. The soil is bursting with life – earthworms, nematodes, fungi, moulds and bacteria, insect larvae, wireworms, cutworms, beetles, and mites (thousands per cubic metre), all amidst a tangle of roots. The herbaceous plants and trees and shrubs above ground provide shelter and food for a myriad of insects and for those that feed on them. The air is alive with flying insects and birds that flit in and out of the system. Perhaps a hungry crow with fierce eyes will pause in its travels (what sights has it seen that we will never know?). There are squirrels, and at night bats and perhaps a raccoon looking for an easy feed. If you are lucky, there is a toad, pop-eyed, squat, roly-poly, and very deliberate, taking its time to decide if you are small enough to eat.

These places are true wildernesses just as much as a remote stream or a primeval forest, and can refresh us and remind us of the magnificence of nature – *if* we pay heed to them.

beech-dominated patches, or are these the respective ends of a continuum of dynamic system states that characterize a beech-maple landscape?

Norway maples appear so aggressive that there are concerns about the continued local existence of some native maple species. What makes Norway maple so successful? Can it be considered as ecologically equivalent to any of our resident maples? What will happen to species that depend on sugar maple or silver maple if these trees are replaced by Norway maple? Purple loosestrife and zebra mussels appear to be crowding out some of their native counterparts. As the changes in species' interactions reverberate through the system, do we recognize a point where one system has changed into another?

Chapter 1 shows us how the landscape of the greater Toronto region has been sculpted by time and by ice, wind, and water. If we went back 12,000 years, would we call the recently exposed rocks, sand, and gravel of the Toronto region an ecosystem? And what about the future? Rapidly increasing concentrations of greenhouse gases may alter the climate more rapidly than it has changed in the past. Will communities of organisms, adapted to natural rates of change, have time to adjust?

And what about other changes wrought by that most domineering of species, we humans? Should we consider the built environment and the organisms within it as an ecosystem? Such an approach *can* be an extremely productive one. Witness the success of the Royal Commission on the Future of the Toronto Waterfront. Its interim report captures some elements from our preceding discussion and implicitly recognizes others as it describes what it means to approach the urban environment as an ecosystem:

> An ecosystem approach understands that humans are part of nature, not separate from it; it recognizes the dynamic nature of the ecosystem – a moving picture rather than a still photograph – that incorporates the concepts of carrying capacity, resilience, and sustainability suggesting that there are limits to human activity; and uses a broad definition of the environment: natural, physical, economic, social, and cultural. The ecosystem approach encompasses both urban and rural activities; is based on natural geographic units, such as watersheds; embraces all levels of activity – local, regional, national, and international. It emphasizes the importance of living species other than humans and of generations other than our own. It is based on an ethic in which progress is measured by the quality, well-being, integrity, and dignity it accords natural, social, and economic systems.

The role of information within evolving reality is at the centre of analysis of all versions of ecosystems. In ecology, for example, gene mutations and chromosomal transformations occur endlessly in the evolutionary process at population levels. During the development of the juvenile organism, the genetic information encoded in the genes, chromosomes, and other genetic material guides the developmental process. Many diseases act by 'misinforming' their host to act in a way that is advantageous to a disease organism. Thus, many, perhaps most, kinds of organisms of all the kingdoms of living things have an ability to learn from experience, which increases their probability of surviving some of the threats that they encounter or of benefiting from some opportunity. Some species, including humans, have evolved ways to transmit learning from one individual or group to another. In the study of ecosystems, we learn how to transmit learning achieved by other species to our species, for example.

In its most encompassing sense, the 'ecosystem approach' is influencing the current evolution of the greater Toronto region. Thousands of people share some intuitive sense of what needs to be done to improve our cultural-natural regional ecosystem. Among the governance institutions that are taking a leadership role in this evolutionary development are the Waterfront Regeneration Trust, the Toronto and Region Conservation Authority, and the Credit Valley Conservation Authority. Each of the three local universities (Toronto, York, Ryerson) has a transdisciplinary program or network of programs of education and research that relate, explicitly or implicitly, to the most encompassing vision of the ecosystem approach. Activist groups such as Pollution Probe, the World Wildlife Fund, and the Canadian Environmental Law Association are leaders with the more encompassing 'eco-approach,' though they may not specifically advertise the fact.

Actions in one geographical location produce ripples and have impacts in every direction. Only with a sense of ecosystems can we attempt to understand the full implications of these impacts: changes in forest cover, for example, affect not only the organisms that live in the forest but also those that use it from time to time, such as migrating birds, and those that live in watersheds fed by water husbanded by forest cover. Changes in bird populations cause changes in the populations of insects on which they feed; and changes in the populations of insects affect populations of their bird predators as well as populations of the plants on which the insects feed and which they pollinate; and so on, and so on.

Organisms cannot be viewed in isolation if we are to understand the workings of the natural world on which we are highly dependent for our own well-being. The chapters that follow first set the stage by describing the physical environment and the changes that have occurred over the years. Then we deal with changes in human population and in our built environment. This discussion is followed by consideration of the organisms that inhabit, and have inhabited, the greater Toronto region and the transformations that have taken place as a consequence of the changes to the physical and built environments. The contributors always attempt to place these alterations in an historical context and relate changes in one group of organisms to consequent modifications in others. Without this emphasis on relationships between these changes and on how those in one location affect others elsewhere in the area of focus, we would simply have been presenting an updated version of the faunistic and floral surveys featured in Faull's 1913 book. The science of biology and our understanding of ecosystems have moved far beyond that point in the intervening 86 years.

This book, then, addresses basic questions: What have we learned from the past? How can we avoid the mistakes of the past with respect to our future use of the natural environment and our impacts on the living system? Inevitably there will be further change and increased impacts on the natural system – we cannot prevent change and freeze the greater Toronto region forever more. But how best can we guide our activities so that these impacts are minimized and so that perhaps we can even remedy some of the harmful impacts of the past that have been a result of creating the urban centre we know as Toronto? It is said that those who do not learn from the past are doomed to repeat the mistakes of the past. The RCI is attempting to do its part to ensure that we have indeed learned from the past.

As a science-based organization, the RCI is disturbed by the poor linkages between what we know and the planning of new development projects and the assessment of their probable impacts. Science can provide answers and predict harmful impacts in advance so that they can be avoided or mitigated, but too often this knowledge and these insights are ignored by decision makers in the greater Toronto region (and elsewhere), with unfortunate results. We hope that this book may tighten these linkages and bring the knowledge of science and decision makers closer together – to the benefit of all those who now and in the future will live in and around the greater Toronto region. We further hope that as the RCI celebrates its 200th anniversary in 2049, there will be less to regret about the way that the natural environment has been treated and that future growth will be more compatible with the integrity of that environment and the emerging concepts of sustainable developments.

Part I
The Broad Physical Basis

The Physical Setting: A Story of Changing Environments through Time

*John A. Westgate, Peter H. von Bitter, Nicholas Eyles,
John H. McAndrews, Vic Timmer, and Ken W.F. Howard*

Landscapes are not immutable; they undergo continuous change through the action of geological forces over millions of years. In the perspective of geological time, the pace of change is slow but inexorable. In contrast, rapid and dramatic changes in the landscape have occurred since settlement of the Toronto region by Europeans – a consequence of human population growth, which, in turn, has stimulated urban expansion as well as industrial and agricultural activities.

This chapter documents the salient physical features of the landscape in the Toronto region and describes the character and timing of the shaping forces involved, interpreted from the landforms, rocks, and sediments. A fascinating but incomplete story of environmental change is unfolded, including the role of human agency in that change.

TORONTO'S DEEP PAST

The place we call Toronto has experienced great and momentous changes in its several-billion-year history – colliding continents, mountains the height of the Rockies, tropical seas, and millions of years of erosion. The record of these remarkable happenings lies beneath our feet, in the rocks, sediments, and soils upon which we live (Figures 1.1, 1.2). Except for a thin veneer of sedimentary rock that has been exposed in river valleys such as those of the Don and Humber, in the cliffs of Lake Ontario and the Niagara Escarpment, and in artificial quarries and road cuts, most of this ancient record (Figure 1.2) is seen only by geologists when they are drilling for water, oil, or some other resource, or by engineers testing for the foundations of buildings.

Our Precambrian Roots

The Precambrian granites, gneisses, and marbles that underlie Toronto constitute our stable 'basement.' These very hard ancient rocks, part of a belt called the Grenville Province, were all heated, altered, and folded one last time about 1.1 billion years ago, during the last Precambrian continental collision. This activity left a wide fault zone that runs almost beneath Toronto, dividing an eastern zone of marble, volcanics, and clastic rocks from a western zone of gneisses. These Precambrian rocks also form a southwest ridge, the Algonquin Arch, which extends all the way across southwestern Ontario and has its axis northwest of Lake Simcoe and southeast of Georgian Bay. This structure influenced the thickness and dip of the younger sediments that draped over it many millions of years later.

Because the rocks of the Grenville Province are of different ages and have been deformed many times, it has been difficult to determine their origin. Nevertheless, many of them were first laid down as sediments in marine and terrestrial environments. Dome-like, layered sedimentary structures called stromatolites, built by microbes, are common in the marble of the Grenville Province, and it is reasonable to visualize the Toronto

The place we call Toronto has experienced colliding continents, mountains the height of the Rockies, tropical seas, and millions of years of erosion.

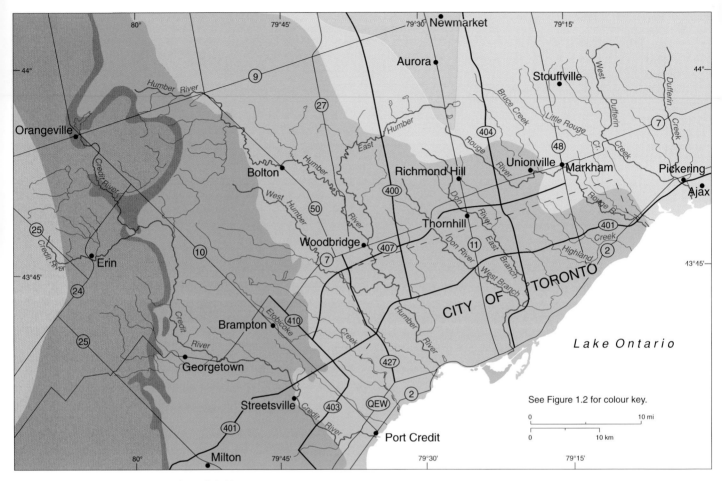

FIGURE 1.1 Bedrock geology of the Toronto region. (After Freeman 1978 and Ontario Geological Survey 1991)

of about 1.3 billion years ago as a sun-lit, shallow ocean in which early microbial life, photosynthetic cyanobacteria, was precipitating and trapping calcium carbonate mud – later to harden into limestone, and finally, by heat and pressure, to be transformed into marble.

Six Hundred Million Years of Wear and Tear

Although the last Precambrian continental collision must have raised mountains the height of the Rockies in Ontario about 1.1 billion years ago, there is nothing more than an irregular erosion surface under the Toronto region. These mountains were ground down by water, wind, and possibly ice over the course of about 600 million years, leaving a rolling plain of relatively low relief where Toronto now stands.

About 600 million years ago, the Sauk Sea, as part of a worldwide rise of sea level, began to build up early North America instead of wearing it down. The sea rose very slowly, advancing about 18 km per million years, and began to deposit a major pile of sediment on the flanks of the ancient continent. It took much of the following Cambrian Period for the sea to 'climb' to today's southern Ontario. The oldest and only records of this sea that exist under the Toronto region may be relatively thin, unnamed sandstones of possible Late Cambrian age that overlie the Precambrian basement. Although these sediments are interpreted as terrestrial – that is, deposited on land – similar rocks in eastern Ontario probably originated in a shallow marine environment (Table 1.1). Environmentally, Late Cambrian Toronto of about 515 million years ago may have been a broad sandy beach covered by quartz-rich sediments from the erosion of the Precambrian mountains.

Toronto in the Tropics

During the Middle Ordovician Period, about 450 million

The Physical Setting

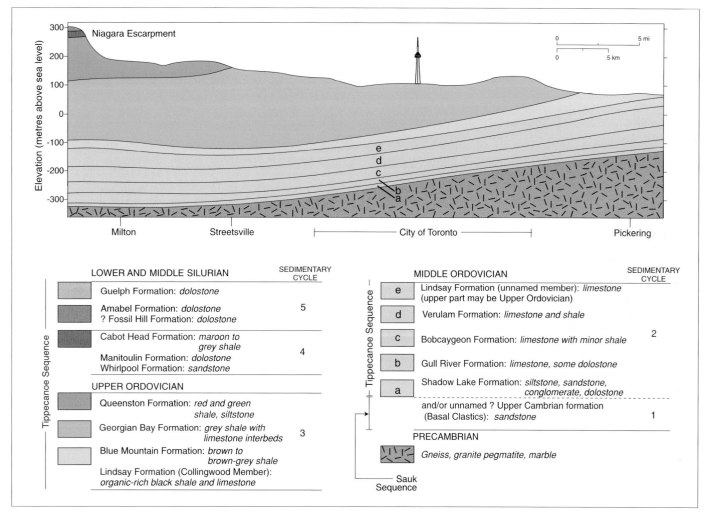

FIGURE 1.2 Schematic east-west cross-section of the bedrock geology under the Toronto region, sub-parallel to the axis of the Algonquin Arch from Pickering to Milton. The scale is the same as Figure 1.1. (Based on information from Freeman 1978 and the Ontario Geological Survey 1991)

years ago, a second major rise of the world's oceans caused a major inland sea, the Tippecanoe Sea, to reach the Toronto region. Toronto was successively many things – a delta, a lagoon, and part of a reef in a tropical sea, for this part of North America was situated just south of the equator at this time. The evidence for this tropical ocean lies deep under the Toronto region in the thick sequence of fossiliferous marine rocks (Figure 1.2, Table 1.1), the lower part of which constitutes a second sedimentary cycle dominated by calcium carbonate-rich rocks, the remains of carbonate-bearing organisms in the sea. The sedimentary rock units are broadly grouped by geologists into a sequence of numbered cycles, as shown in Table 1.1.

Distinct environmental changes began to take place in the Tippecanoe Sea over Toronto by the Late Ordovician, about 440 million years ago. Collision of eastern North America with the proto-Atlantic plate resulted in the Taconic Mountains to the east being pushed skyward; the cycle of erosion started by this event began to cover the Toronto region with muddy and sandy sedimentary deposits. Whereas earlier cycle 2 marine environments had been shallow, sun-lit, relatively clear, and filled with oxygen, the change to muddy-water conditions, sometimes deficient in oxygen, marked a profound and lasting change. It resulted in mostly softer sedimentary rocks such as sandstones, siltstones, and carbonate-rich shales, that now form the lowland

TABLE 1.1

Sedimentary rock units, their characteristics, and ancient sedimentary environments of the Toronto region

Sea/Sequence	Cycle	Series	Stratigraphic unit	Thickness	Rock type	Environment
	Cycle 5	Middle Silurian	Guelph Formation	4-100 m	Dolostone	Marine reef and interreef
			Amabel Formation	up to 38 m	Dolostone	Shallow, high-energy marine shoal to deeper water marine
			? Fossil Hill Formation	up to 24 m	Dolostone	Marine shoal, high to low energy
	Cycle 4	Lower Silurian	Cabot Head Formation	10 to 39 m	Maroon to grey shale	Offshore, deep-water marine to nearshore marine
			Manitoulin Formation	up to 25 m	Dolostone	Offshore, deep-water to nearshore marine
			Whirlpool Formation	up to 9 m	Sandstone	Shallow marine to river environments
Tippecanoe Sea/Sequence	Cycle 3	Upper Ordovician	Queenston Formation	45-335 m	Red and green shale and siltstone	Deltaic, non-marine to nearshore marine; shallow marine, intertidal to supratidal
			Georgian Bay Formation	125-200 m; 100 m average	Grey-blue shale with siltstone and limestone interbeds	Shallow-water marine; high energy storm-affected marine shelf
			Blue Mountain Formation	up to 60 m	Brown to brown-grey shale	Quiet-water marine; below storm wavebase
			Lindsay Formation (Collingwood Member)	10 m maximum	Organic-rich black shale and limestone	Deep, quiet-water, oxygen-deficient marine shelf
			(Unnamed Lower Member)	57 m maximum	Limestone	Shallow marine shelf to marine shoal
			Verulam Formation	32-65 m	Limestone and shale	Open, normal marine shelf or ramp; upper part of shallow marine shoal
			Bobcaygeon Formation	7-87 m	Limestone with minor shale	Shallow, subtidal marine; high energy, open marine
	Cycle 2	Middle Ordovician	Gull River Formation	7.5-136 m	Limestone with some dolostone	Highly saline, supratidal to intertidal marine flat; upper part more open marine subtidal lagoon
			Shallow Lake Formation	normally 2-3 m; 15 m maximum	Siltstone, sandstone, conglomerate, dolostone	Nearshore marine, river and delta environments
Sauk Sea/ Sequence	Cycle 1	? Upper Cambrian	Unnamed Formation (Basal Clastics)	?	Sandstone	Broad sandy beach; shallow marine to nearshore marine?
		Precambrian	?	?	Gneiss, granite pegmatite, marble	Shallow water marine (marble only)?

below the Niagara Escarpment. They can be examined in the Toronto region as outcrops in creeks and rivers, in quarries and building excavations, as well as along the base of the Niagara Escarpment (Figures 1.1, 1.2).

Sedimentation over the Toronto region was not only occurring in ever shallower waters but also contained increasingly more quartz mud and silt. The black, organic-rich carbonates and shales of the Collingwood Member, which is the basal rock unit of the Lindsay Formation, started this new, more clastic third cycle (Table 1.1). Most geologists believe that the Collingwood Member was laid down in a deep, oxygen-deficient marine environment. The succeeding Blue Mountain Formation, consisting of brown to brown-grey shale, was deposited under shallower, more oxygen-rich conditions, but still below the disruptive energy of waves. The environments of the Blue Mountain Formation were followed by yet shallower, wave- and storm-influenced marine environments in which the blue-grey shale and minor siltstone and limestone interbeds of the Georgian Bay Formation were deposited.

Most of Toronto is directly underlain by the Late Ordovician rocks of the Georgian Bay Formation. They crop out along the banks of the Humber and Credit Rivers as well as in the many quarries opened to extract shale for brick manufacture, such as the Toronto Brick Company quarry in the Don valley and the Canada Brick Company quarry in Mississauga.

The progressive shallowing of the sea during the Late Ordovician cycle 3 culminated in the deposition of an enormous sheet of mud, the Queenston Delta, that not only covered Toronto but extended east into New York State and south as far as Pennsylvania. It was in the shifting, often exposed, muddy environments of the Queenston Delta that the red and green shales and siltstones of the Queenston Formation were deposited. These are now seen as outcrops along the rivers and creeks near Oakville, Bronte, Burlington, and Milton. Like the Georgian Bay Formation beneath it, the Queenston is used in brick manufacture.

A significant gap in the sedimentary record follows – the result of a temporary withdrawal of the Tippecanoe Sea caused by a worldwide drop in sea level due to Late Ordovician glaciation in North Africa. In the Early Silurian Period, about 430 million years ago, the Tippecanoe Sea reinvaded the Toronto region and began sedimentary cycle 4. The sandstones, dolostones, and shales of this cycle completed the mud- and silt-dominated Middle Tippecanoe Sequence. These formations – Whirlpool, Manitoulin, and Cabot Head – along with the underlying Queenston Formation, form the less erosion-resistant lower and middle parts of the Niagara Escarpment.

The final story of the Tippecanoe Sea in the Toronto region again involves environmental change, as sedimentary cycle 5, dominated by dolostones, began to be deposited. Although the region was just south of the equator for millions of years, it was not until the Middle Silurian that the muddy shallow seas were replaced by sun-lit, clear, marine environments in which abundant tropical organisms could truly grow and flourish. The thickly bedded carbonates of the Amabel Formation are highly visible in the western part of the Toronto region as the thick 'cap' rock of the Niagara Escarpment. The Amabel Formation consists of relatively hard and resistant, light-coloured dolostone, a magnesium-rich carbonate. Its resistance to erosion, combined with the wearing away of the softer rocks of sedimentary cycles 3 and 4 underneath it, have resulted in the formation of the Niagara Escarpment, the erosional scarp or cuesta that is Ontario's backbone.

Removing a Great Weight

The withdrawal of the Tippecanoe Sea from southern Ontario in the Late Silurian or Early Devonian (about 415 million to 400 million years ago) was followed by yet another advance of an inland sea, the Kaskaskia Sea, during the Middle Devonian, about 380 million years ago. Whether any of the Kaskaskia Sea, or later seas, ever covered the Toronto region is not known. Yet there is evidence that many other sedimentary environments must have existed here, that hundreds, if not thousands, of metres of additional sedimentary rock must formerly have covered the Toronto region. The collision of the supercontinent Gondwanaland with North America resulted in a period of prolonged uplift and erosion that lasted nearly 250 million years. This erosion removed many important pages from the environmental and depositional story of the Toronto region.

Ancient Ecosystems and Their Marine Communities

Complex forms of life were common in the many sedimentary environments that existed millions of years ago in the Toronto region. Because most of these environments were saltwater, the fossils found in the local sedimentary rocks are almost entirely of ancient marine life. Probably the earliest life would have been Precambrian

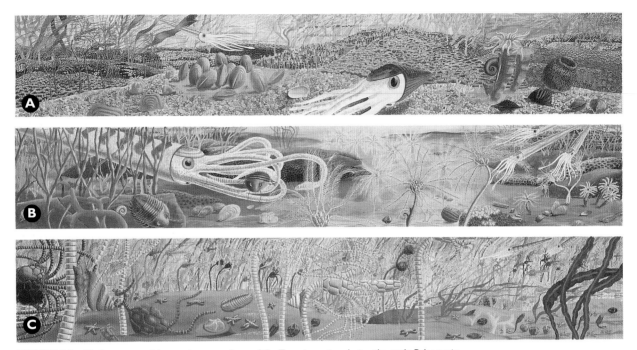

FIGURE 1.3 Marine communities on the ocean floor of the Toronto region during the early Palaeozoic.
(Courtesy Royal Ontario Museum)
A   Crinoid-Bryozoa-Trilobite community: Middle Ordovician Bobcaygeon Formation, 470 million years ago
B   Cephalopod-Trilobite-Bryozoa community: Late Ordovician Georgian Bay Formation, 450 million years ago
C   Coral-Brachiopod reef community: Middle Silurian Fossil Hill Formation, 430 million years ago

single-celled microbes that obtained energy through photosynthesis and thus prepared the earth for larger, multicelled animals by releasing oxygen. Hundreds of millions of years later, when the Sauk Sea covered the Toronto region during the Late Cambrian, not only had multicelled animals evolved, but they had developed hard shells for protection and support. The earliest marine invertebrates in the Toronto region may have been shelled, inarticulate brachiopods. By the time the Tippecanoe Sea covered Toronto about 450 million years ago, during the Middle Ordovician, invertebrate life had become much more diverse (Figure 1.3A). The many Ordovician (Figure 1.3A, B) and Silurian (Figure 1.3C) marine organisms that thrived in the shallow tropical seas of the Toronto region lived in complex communities. Each community in turn was adapted to a particular kind of environment. As these environments changed, individual species either adapted to the changed conditions, moved away, or became extinct.

Fossils may be seen in the sedimentary rocks exposed along rivers and in parks of the Toronto region. Most of these rocks contain well-preserved fossils of the biological communities that lived in long-vanished marine environments. The fossils of the Bobcaygeon Formation (Figure 1.3A), the Georgian Bay Formation (Figure 1.3B), and the Fossil Hill Formation (Figure 1.3C) represent just three of many marine communities that lived and died in Toronto's past.

TORONTO'S ICY PAST

Young, unconsolidated deposits of Pleistocene age – thought to be younger than 150,000 years old – cover the Palaeozoic bedrock throughout most of the Toronto region (Figure 1.4, 1.5). They were deposited during times of marked oscillations in climate: from glacial conditions, when glaciers at least 1 km thick covered Toronto, to interglacial, when climatic conditions were similar to those of today. The thickness of this Pleistocene sedimentary cover is highly variable, because of the presence of numerous bedrock channels. The largest of these buried valleys is the Laurentian Channel (Figure 1.6). This valley was probably a part of an ancient, well-integrated river system that drained what is now the

Great Lakes region (Figure 1.7). Its sedimentary fill is 180 m thick near Bolton. A similar thickness of Pleistocene sediments occurs along the Oak Ridges Moraine (Figure 1.5), whereas along the Niagara Escarpment and parts of the Rouge River valley bedrock is exposed at the surface.

Surface Features

The surface deposits and associated landforms relate to the last phases of glaciation. Most of the southern and eastern parts of the region are covered by deposits laid down by the Lake Ontario ice lobe, which advanced out of the Lake Ontario Basin about 13,000 years ago to abut against the Niagara Escarpment to the west and the Lake Simcoe ice lobe to the north. Till – a massive deposit of stones, sand, silt, and clay (Figure 1.8B) – was smeared onto the ground surface by this glacier. The Halton Till (Figure 1.5) is the most extensive deposit across the Toronto region. In some places, its fluted and drumlinized surface demonstrates the actual direction of ice flow (Figure 1.4). The most prominent landform is the Oak Ridges Moraine, which serves as the divide between the Lake Simcoe and Lake Ontario drainage basins. It is cored with sands and gravels (Figure 1.8C), is at least 80 m thick, and its northern and southern flanks are draped with till of the Lake Simcoe and Lake Ontario ice lobes, respectively. More detail about the Oak Ridges Moraine is given in Chapter 17.

As these glaciers retreated downslope from the Oak Ridges Moraine, lakes formed along their ice margins but were short lived because of continual changes in the position of the melting ice front. Consequently, sediments deposited in them (Figure 1.8D) are thin

A.P. Coleman

Arthur Philemon Coleman was a leader in geological research in the late 1800s and early 1900s. Born in Lachute, Quebec, in 1852, Coleman obtained a master's degree in classics at Victoria University, Cobourg, Ontario, and then studied geology in Germany. He later returned to Victoria University to teach. In 1891, when the university moved to Toronto, he was appointed professor of metallurgy and assaying, and in 1901 became professor of geology until his retirement in 1922.

An avid mountaineer, enthusiastic artist, and scientist, Coleman explored geological sites and recorded his findings in watercolours, which he used to illustrate his books and papers. Many of his written works are both guides to mountain climbing and reports of geological findings.

His first major contribution to geology was his work abroad – in India, Africa, and South America. In these warm climates, to his surprise he found striated rocks and pebbles, evidence of prior glaciation. It was difficult to find a plausible reason for such dramatic climate change in these areas. Although Coleman himself did not believe in the theory of 'continental drift,' it provided a good explanation for many of his findings. In Canada, he made some exciting discoveries in the brick pits of Toronto, finding layers of fossil plants and animals between layers of glacial deposits. These fossils of interglacial organisms forced scientists to rethink their theory that there had been only one period of glaciation and accept that there had been at least two fairly recent glaciations.

Coleman's lifetime contributions include publishing over 100 works, researching glaciation in Canada and abroad, climbing and writing about various mountains, and helping to establish the Museum of Geology in 1912 (one of the five original components of the Royal Ontario Museum). Today, the ROM owns over 300 of Coleman's original watercolours. He was awarded various medals and honours, including awards from the Geological Society of London, the Geological Society of America, and the Royal Society of Canada. He lived a remarkable, full, and healthy life. At the age of eighty years, he travelled to South America and climbed to over 5,000 metres. Before his death in 1939 at the age of 87, he was planning a climbing trip to the Andes.

A.P. Coleman, c. 1910s.
(Royal Ontario Museum Library no. 0000907)

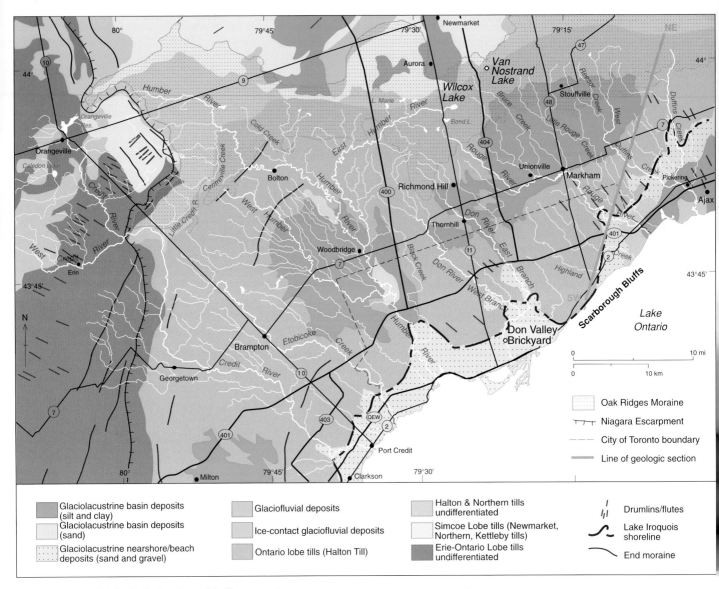

FIGURE 1.4 Surficial geology of the Toronto region.

and discontinuous. The successive ice-frontal positions occupied by the retreating Lake Ontario lobe are shown by the few end moraines, which are ridge-like accumulations of till at the glacier margin (Figure 1.4), but are documented in greater detail by the present drainage pattern (see Figure 3.1, p. 52). Because deglaciation has been so recent, streams still exhibit characteristics imposed on them by the glaciers; there has been insufficient time for erosional processes to produce a more integrated drainage system. The Lake Ontario ice lobe sat in the easternmost part of the Lake Ontario Basin about 12,000 years ago. Ice blocked the St. Lawrence valley, and drainage to the Atlantic was along the Mohawk River in New York State. The lake that ponded in front of the glacier at this time, Glacial Lake Iroquois, existed for a few hundred years – long enough to form an extensive wave-cut terrace and conspicuous bluff (Figures 1.4, 1.9). Beach sands and gravels (Figure 1.8A) cover most of the terrace, but, in the easternmost part of Toronto, erosion has exhumed older till deposits. Subsequently, the lower lake levels that followed from the continued retreat of the glacier energized streams by steepening their gradients and promoted valley development.

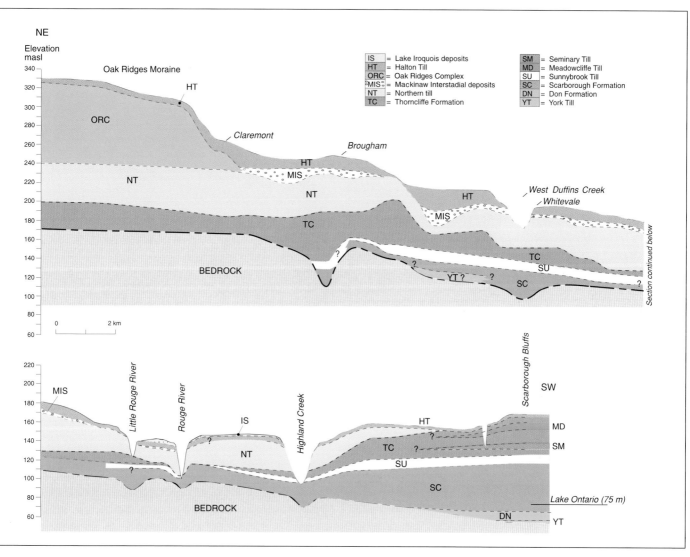

FIGURE 1.5 Simplified geological cross-section of Pleistocene sediments from Oak Ridges Moraine to Lake Ontario. The line of the section shown on Figure 1.4.

In the Deep Freeze with Periodic Defrosting: Big Glaciers and Big Beavers

The earlier story of environmental change beginning with the first occupation of the Toronto region by continent-sized ice sheets is not well known. Fortunately, a few windows afford a glimpse into this past; they are the relatively thick sedimentary sequences exposed along the Scarborough Bluffs, in the Don Valley Brickyard, at Woodbridge, and along the valley walls of the Rouge River and West Duffins Creek (Figure 1.4).

Although we know that large ice sheets covered extensive parts of northern North America well over a million years ago, the earliest evidence for their presence in the Toronto region is about 150,000 years ago, when the York Till was deposited during the penultimate glaciation. This stony, compact, bluish-grey till is best exposed along the Humber River valley and at Woodbridge, where it is more than 5 m thick (Figure 1.10). It is only 1 m thick at the Don Valley Brickyard but can be traced in the subsurface across much of Toronto.

The next glimpse we see is that of a large lake occupying the Lake Ontario Basin, with a river, much larger than the present Don River, flushing sediment into it

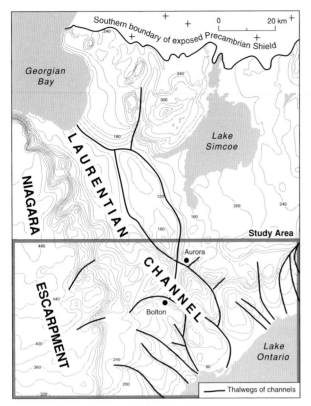

FIGURE 1.6 Bedrock topography and buried bedrock channels in the area between Lakes Ontario and Simcoe and Georgian Bay. The Laurentian Channel connects bedrock basins now occupied by Georgian Bay and Lake Ontario. Contour interval is 20 m.

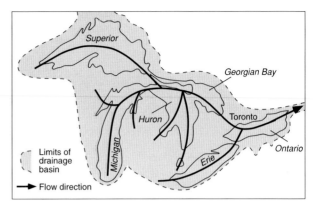

FIGURE 1.7 J.W. Spencer's 1890 reconstruction of the Laurentian River.

from the north, following a route along the Laurentian Channel. Sands and muds were deposited in shallow water along the storm-influenced margin of this lake and can be seen today as a 7 m thick sequence near the base of the north face of the Don Valley Brickyard (Figure 1.11), a site now preserved as an urban park.

The Don Formation at the Don Valley Brickyard is justly famous for its abundant and varied animal and plant fossils, first studied in detail by noted geologist A.P. Coleman. About 500 species of fossils have been recognized and provide a wealth of information on the local environment and regional climate of that time.

Much information about the history of vegetation cover of a region can be obtained from fossil pollen grains preserved in the sediments. Pollen grains for the most part reflect the composition of the upland forest, whereas plant macrofossils, mostly seeds, conifer needles, and wood, are largely derived from local wetland communities. Pollen grains are well preserved in the Don Formation, except for the uppermost metre; concentrations range from 5,000 per gram in sand to 100,000 per gram in clay. These high concentrations mean that representative pollen samples can be obtained from very small amounts of sediment. Tree pollen greatly exceeds herb pollen, and most are of species that live today in southern Ontario (Figure 1.12). We can identify three distinct pollen zones in the layers of the formation, and each can be matched with the pollen 'rain' of one of the three present-day forest regions of Ontario, namely, the deciduous, mixed, and boreal forest.

The lowermost 2 m of the Don Formation is dominated by oak, elm, hickory, beech, and other deciduous trees, indicating a deciduous forest in a warm climate. Sweet gum is also present at this level and suggests a climate warmer than the present, because the northern limit of its range today is the latitude of New York City. The pollen assemblage as a whole indicates a mean July temperature about 2°C higher than that of today. In the 2 m of sediment above this layer, spruce, pine, and hemlock pollens increase, indicating succession to a cool-climate mixed forest. In the uppermost 3 m of the Don Formation, pollen of spruce and pine become dominant at the expense of deciduous trees, with birch and herb pollen, and peat moss spores becoming conspicuous at the top – all indicating succession to boreal forest as the climate continued to cool.

Diatom, ostracod, cladoceran, aquatic mollusc, beetle, and caddisfly assemblages recovered from the Don Formation indicate deposition in shallow water near the mouth of a large river entering a lake in a climate similar to or slightly warmer than the present. The few vertebrates recovered from these sediments include groundhog, two species of deer, bison, large bear, giant beaver (Figure 1.13) and several species of fish.

FIGURE 1.8   Types of sediment exposed in the Toronto region. (John Westgate)
A   Horizontally bedded sands and minor gravel deposited in the shallow water and beach zones of Lake Iroquois about 12,000 years ago. They are about 1.5 m thick and cover loose sands of the Lower Thorncliffe Formation; the contact between these two sedimentary units is defined by the marked change in slope. Cudia Park, Scarborough Bluffs.
B   Till – A massive, stony sediment deposited directly on the ground by glaciers. This is the Northern till, exposed in the north bank of West Duffins Creek, about 3 km west of Pickering.
C   Interbedded sands and stony muds deposited close to ice margin by meltwater streams. Oak Ridges Moraine, just east of Goodwood. Pick is 45 cm long.
D   Horizontally bedded sands and silts deposited in a lake. These sediments occur below the Northern till at same locality as B. Sequence is 2 m thick.

We have no reliable way of dating the Don Formation very precisely. The warm-climate conditions interpreted from fossils in the lower part of the formation as well as the formation's position between glacial deposits argue for a last interglacial age, dated in the deep-sea sediments at 125,000 years before the present.

The pollen spectra of the uppermost part of the Don Formation and lowermost part of the overlying lacustrine clays of the Scarborough Formation at the Don Valley Brickyard are remarkably similar and support a continuity of sedimentation under broadly similar environmental conditions. A major river, following the Laurentian Channel, discharged sediment into a high-level lake, building a large delta, which is best exposed along the Scarborough Bluffs (Figures 1.4, 1.5) where it is almost 50 m thick. Clays predominate in the lower part of this deltaic sequence, known as the Scarborough Formation, whereas sands predominate in the upper part. This upward coarsening of sediment reflects the progressive growth and extension of the delta into the lake, deepwater distal clays being later covered by shallow-water sands of a braided-river system. Reworked organic matter is present throughout the formation, either dispersed in the sediment or as discrete, discontinuous peaty beds.

◀ **FIGURE 1.9**
Lake Iroquois terrace and bluff at Cudia Park, Scarborough Bluffs. The terrace presently stands about 45 m above Lake Ontario and was formed about 12,000 years ago. Lake Iroquois sands are about 1 m thick and cover sands of the Lower Thorncliffe Formation. The dark deposit is the Sunnybrook Drift. (John Westgate)

◀ **FIGURE 1.10** The York Till penetrated by a large ice-wedge cast, Woodbridge, Ontario. (Nancy Williams)

The tree pollen composition of the lower clays of the Scarborough Formation — dominated by pine, spruce, and birch — is very similar throughout its thickness and, with insect remains, indicates a cool-temperate, boreal climate, such as we now experience along the northern shore of Lake Superior. A mean annual temperature of about 2.5°C is indicated, about 5°C below the present level. Spruce, pine, and birch pollen remain abundant in the overlying sands, but the presence of subarctic beetles indicates colder climatic conditions, with a mean annual temperature of about 0°C.

We can find additional evidence for very cold conditions at the time the Scarborough deltaic sands were being deposited if we look at Woodbridge, where large ice-wedge casts penetrate the York Till (Figure 1.10). These periglacial structures are up to 1.5 m wide and 5 m long, very similar in size to ice wedges in the continuous permafrost zone of Arctic Canada, northern Yukon, the Northwest Territories, and Baffin Island — regions with a mean annual air temperature of -6°C or lower. Ice-wedge casts form as a result of sediment filling a wedge-shaped void after ground ice has melted.

The ice-wedge casts at Woodbridge cut through a sandy bed with scattered molluscs and lenses of organic silt that bear pollen indicating a cold climate, and

FIGURE 1.11 North face of Don Valley Brickyard showing 40 m of late Pleistocene sediments. YT, York Till; DF, Don Formation; SC, Scarborough Clays; PRF, Pottery Road Formation; SU, Sunnybrook Drift; LTF, Lower Thorncliffe Formation; IT, Iroquois terrace. (Nicholas Eyles)

then through an underlying stony slope deposit with molluscs that suggest equivalence to the interglacial Don Formation. They then extend into the underlying York Till. The ice-wedge casts are covered by 5 m of layered silt, sand, and peat, which a single radiocarbon dating has determined to be older than 50,000 years. These sediments contain a pine-spruce pollen assemblage, as well as several beetle species that can be found today in northern boreal and tundra environments. This beetle assemblage is similar to that found in the uppermost part of the Scarborough Formation at the Scarborough Bluffs. These observations suggest that the ice wedges and their subsequent casts date from the same period as the upper part of the Scarborough Formation.

Channels cut into the top of the Scarborough Formation (see the Scarborough Bluffs section in Chapter 17, Figures 17.18 and 17.19) and filled with sediments of the Pottery Road Formation record an erosive interval, probably related to a drop in lake level; marked fluctuations of water depths are typical of ice-dammed lakes. A succession of stony muds, sometimes called diamicts, interbedded with sands and laminated silts and clays drape the deltaic Scarborough Formation and are best exposed along the Scarborough Bluffs (Figure 1.14). These deposits most likely originated in a glacial lake, although some authorities prefer a scenario of alternating subglacial and glacial lake conditions. Their age probably ranges from 50,000 to 25,000 years.

The Laurentide Ice Sheet advanced across the Toronto region about 25,000 years ago and deposited the Northern till (Figure 1.8B). The ice sheet completely covered the Great Lakes Basin. The dominant regional ice-flow direction was probably from the northwest (Figure 1.15A) because the Northern till contains blocks of Gowganda

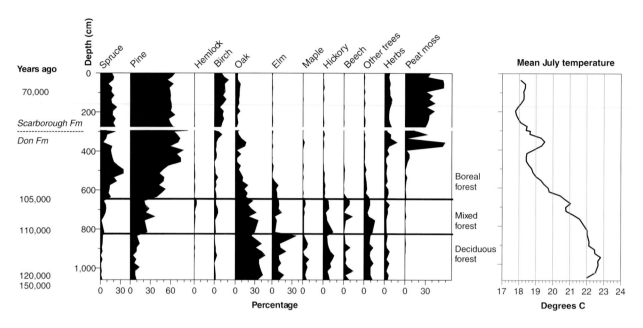

FIGURE 1.12 Stratigraphic diagram of prominent tree pollen in the Don and Scarborough Formations at the Don Valley Brickyard. Percentages are based on counts of 200 tree pollen grains. 'Other trees' include ash, basswood, sweet gum, and black gum. The mean July temperature at Toronto Island Airport is 20.6°C. The time scale is estimated. (Analyst: J.H. McAndrews)

tillite (Figure 1.16), a very distinctive Precambrian rock that is common in tills of the Georgian Bay region. Blocks of Gowganda tillite can be found along the shore of Lake Ontario, but only along those tracts where the Northern till is exposed. We can estimate local ice-flow directions by examining the orientation of two things: the striae, or scratches made on boulders and boulder pavements in till by the moving, debris-charged glacier; and the elongate tillstones. Such measurements on the Northern till in the Scarborough-Markham region again suggest that the ice flowed from the northwest (Figure 1.17).

FIGURE 1.13 Giant beaver (*Castoroides ohioensis*, now extinct), a denizen of Toronto during the last interglacial period, about 125,000 years ago. Length of skeleton is 1.6 m. (Eyles and Williams 1992)

The Laurentide Ice Sheet reached its maximum extent about 18,000 years ago, when its southern margin was in New York State, Pennsylvania, and Ohio. Thereafter, as this ice sheet thinned and retreated, its form became influenced by the topography, and a distinctly lobe-shaped margin developed (Figure 1.15B). Southwestern Ontario was affected by the Huron-Georgian Bay ice lobe and the Toronto region by the Ontario-Erie ice lobe.

Sands, silts, and gravels sandwiched between the Northern and Halton tills (Figure 1.5) record an ice-free interval about 13,500 years ago, when the Laurentide Ice Sheet temporarily withdrew from the Toronto region. A severe periglacial climate prevailed during this Mackinaw Interstadial interval, as shown by the large ice-wedge casts that penetrate the Northern till. Excavations in the Northern till at the Scarborough Town Centre in 1978, for example, revealed an ice-wedge cast of sand, silt, and gravel 2 m wide and over 8 m long. A later re-advance of the Lake Ontario ice lobe deposited the Halton Till, as discussed earlier (Figure 1.15B).

Postglacial Warping and Warming

As the Lake Ontario ice lobe thinned and retreated from the Toronto region about 12,000 years ago, the level of its proglacial lake fell due to the opening of lower

◀ FIGURE 1.14
Tabular, layer-cake form of stony muds (A, Sunnybrook Drift; B, Seminary diamict; C, Meadowcliffe diamict) and interbedded deltaic sands and laminated silts and clays (Thorncliffe Formation). The person (arrow) stands at the contact with the underlying Scarborough Formation delta. (Nicholas Eyles)

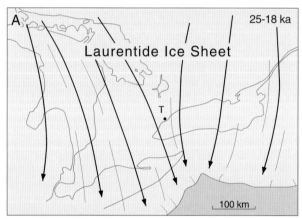

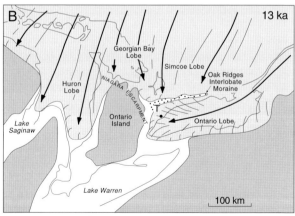

FIGURE 1.15 Regional ice-flow directions: A – at the time of the maximum extent of the Laurentide Ice Sheet, when the Northern till was deposited in the Toronto area, 18,000-25,000 years ago (25-18 ka); B – about 13,000 years ago (13 ka), when the Halton Till and Oak Ridges Moraine were deposited. T is site of Toronto. Dotted area is the Oak Ridges Interlobate Moraine.

drainage outlets to the northeast. When the St. Lawrence Valley outlet opened at the northeastern end of the basin, about 11,500 years ago, the lake level dropped to about 100 m below the modern level, exposing land over large parts of the basin. Subsequently, uplift of the land in response to removal of the glacier's weight – uplift that has been greater in the northeastern part of the basin because the glacier was thicker there – has

FIGURE 1.16 A glacial rock within a glacial rock. Boulder of Gowganda tillite found at easternmost end of Scarborough Bluffs, where Northern till crops out. The faceted, pentagonal form is typical of rocks transported by glaciers. Gowganda tillite formed during a glaciation that affected the area now known as Ontario 2.2 billion years ago. Specimen is 12 cm long. (John Westgate)

The Broad Physical Basis

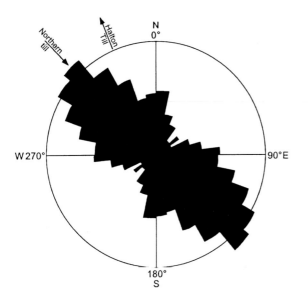

FIGURE 1.17 Alignment of elongate pebbles in Northern till at Scarborough Town Centre. Elongate stones are aligned parallel to direction of ice flow as till is plastered onto the ground. Data are shown as 30° running means based on 100 measurements. The 360° have been divided into 36 10° cells. The number of observations in three adjacent 10° cells (0° to 30°) is averaged and the value placed in the first cell (0° to 10°). Observations in the next three 10° cells (10° to 40°) are averaged and the value placed in the second cell (10° to 20°), and so on.

led to higher lake levels, creating flooded river mouths, large wetlands, and lagoons.

Warm interglacial conditions returned to Toronto about 10,000 years ago. As with the Don Formation, we can get an insight into the vegetation and climatic conditions since deglaciation by examining fossil pollen grains preserved in sediments – in this case, in Wilcox Lake on the Oak Ridges Moraine, near Richmond Hill (Figure 1.4). This small lake formed when a buried ice block melted after retreat of the glacier from the Oak Ridges Moraine about 13,000 years ago. Seven metres of predominantly organic sediment were recovered with the aid of a coring device.

The fossil pollen diagram (Figure 1.18) shows succession from boreal forest, first dominated by spruce and then by jack pine, to a mixed forest of white pine and hemlock with beech and other deciduous trees. This transition occurred about 11,000 years ago. Mixed forest shifted to deciduous forest about 6,000 years ago, when hemlock was almost destroyed by drought that was followed by a pest epidemic. It took a millennium for hemlock to recover.

Forest around the lake was relatively stable until about 600 years ago, when white pine expanded at the expense of beech and other deciduous trees. This succession indicates a 1.5°C decline in mean July temperature,

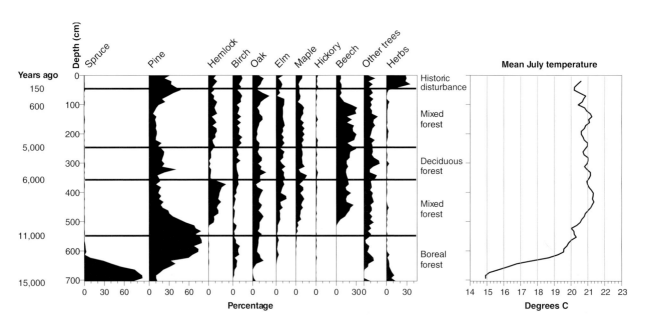

FIGURE 1.18 Prominent fossil pollen grains from a core of sediment lifted from Wilcox Lake on the Oak Ridges Moraine (see Figure 1.4). Mean July temperature at nearby Richmond Hill is 20.8°C. (Analyst: J.H. McAndrews)

which reflects the Little Ice Age, a global cooling event. The last major vegetation change began 150 years ago, with Eurocanadian forest clearance and farming. A half-metre-thick layer of silt was deposited during this period, derived from erosion of soils and surface sediments. This silt contains abundant herb pollen, especially pollen grains of weedy grass and ragweed. A very similar postglacial pollen record has been documented at Van Nostrand Lake, a few kilometres to the northeast of Wilcox Lake.

Future cooling would most likely cause jack pine and then spruce to invade southward, followed by invasion of a continental glacier. Nevertheless, global warming caused by the human-enhanced greenhouse effect will probably delay the onset of the next glaciation.

TORONTO'S VERY RECENT PAST

Environmental change in the Toronto region continued at its geological pace throughout most of postglacial time. Native people lived in harmony with the natural environment, ensured by low population density and low-intensity use of natural resources (see Chapter 4). In contrast, European settlement, which began in earnest in the late eighteenth and early nineteenth centuries, was accompanied by urbanization and unsustainable land-use practices involving 'selective sequential mining' of natural resources. These led to rapid and severe environmental degradation. Three themes illustrate this human-driven environmental change: urban development and associated environmental problems; the impact of population growth and urbanization on groundwaters; and soil degradation.

The Built Landscape

Toronto was designated a city in 1834 and enjoyed rapid growth because of its harbour, easy access to the north, rich farmland, and ability to attract immigrants. Extensive clearance of forest cover and the denuding of watersheds began after 1840 and caused widespread devastation of the habitats of animals and indigenous people. By the 1930s, the effects of deforestation on regional climate and groundwater supplies, especially in the Oak Ridges Moraine, were recognized as a major problem worthy of federal and provincial attention.

Accelerated urban expansion followed the formation of Metropolitan Toronto in 1953, and no break in the pace of urban development can be seen for the near future. The present population of the greater Toronto region is 4.3 million and is expected to increase to 6.5 million by 2021, with most of the growth being directed toward the flanks of the Oak Ridges Moraine, particularly in the Newmarket and Richmond Hill areas. Clearly, there is an urgent need for future urban development to be based on principles that respect the natural environment. The recent formal recognition of ecosystems and watersheds as fundamental units in planning is a step in the right direction.

Groundwater

Copious supplies of good quality water are vital to the growth and development of any large city, and Toronto has benefited considerably from its location on the shores of Lake Ontario. Throughout its history, Toronto has never been short of water, to both use and abuse. Settlers took full advantage of abundant year-round water supplies to establish a thriving lakeshore colony. Access to the hinterland to the north was readily gained via a network of streams fed by springs that issued along the southern flank of the Oak Ridges Moraine. Over time, early pioneers spread into this heavily forested region and began to clear large tracts of land for farming. Shallow wells were found to provide excellent quantities of good quality water throughout the region, particularly along the Oak Ridges Moraine.

As a testimony to the role groundwater has played in the expansive development of Toronto, water well records for the region held by the Ontario Ministry of Environment and Energy now exceed 20,000. The vast majority of rural residents depend on groundwater exclusively. Groundwater is also favoured by farms, golf courses, and rural industries, including the production of bottled water. Municipal users of groundwater include the towns of Milton, Acton, Georgetown, King City, Oak Ridges, Richmond Hill, Aurora, and Stouffville. It must be recognized, however, that surface water from Lake Ontario remains the most convenient source of water for the great majority living in Toronto's urban core.

East of the Niagara Escarpment, the most important aquifers – bodies of rock or sediment that both store and transmit water in economically significant quantities – are sand units found within the layer-cake package of Quaternary sediments. While limestones in the Palaeozoic bedrock can be an important water supply above the escarpment to the west, most of the deep Palaeozoic rocks in the Toronto region are either poorly permeable or produce water that is brackish and

TABLE 1.2

Groundwater-bearing sediments of the Toronto region

Aquifers	Approximate elevation range (m)	Comments
Halton Till aquitard	Variable	Locally drapes the Upper Aquifer but is relatively permeable and allows significant recharge to the underlying aquifer
Upper Aquifer	240-70	Popular shallow aquifer designated by the Province of Ontario as 'hydrogeologically sensitive'; unconfined aquifer, highly permeable with high storage capacity
Northern till aquitard	200-40	Transmits significant quantities of water vertically
Intermediate Aquifer	190-200	Generally favoured for municipal wells; aquifer confined under pressure; highly permeable with moderate storage
Sunnybrook Drift aquitard	Variable	Generally low permeability; known to be moderately permeable where fractured
Lower Aquifer	Variable	Present in deeper bedrock valleys and the Laurentian Channel; highly permeable; confined under pressure; rarely exploited resource

unpleasant to drink. Within the sequence of Quaternary sediments, numerous aquifer units have been identified, fourteen in the Duffins Creek and Rouge River drainage basins alone. Many of the aquifer units are well connected regionally and can be grouped into essentially three aquifer systems separated by less permeable aquitards – bodies of rock or sediment that store water but do not readily transmit it (Table 1.2). A further outcome of this recent work has been to demonstrate that the extensive till sheets, notably the Halton and Northern tills, which were once thought of as being unable to transmit water, and, therefore, of use as host materials for domestic landfills, can transmit water in significant quantities and with it any dissolved contaminants it may contain. Thus, contrary to popular belief, these tills do not provide 'protection' for underlying aquifers. Deep aquifers are simply less affected by contamination introduced at the ground surface because it takes an extremely long time for groundwater to travel from the surface down to the aquifer. There is evidence to suggest that groundwaters in some of the deepest aquifers are thousands of years old and probably entered the system as recharge through the soil zone shortly following retreat of the glaciers.

The most important aquifer in the area is the Upper Aquifer, popularly known as the Oak Ridges Moraine (see also Chapter 17, Oak Ridges Moraine section). The crest of the moraine stands up to 300 m above the surrounding till lowlands and forms the surface water drainage divide between Georgian Bay and Lake Ontario. It also provides baseflow to more than 30 major streams. Well yields in this aquifer are generally excellent and the quality is good. However, the water table is relatively shallow and there is much debate about the potential impact of urbanization and other anthropogenic activities on groundwater quality and supply.

Deeper aquifers in the region are less celebrated than the Upper Aquifer but are just as crucial from the standpoint of water resources. These include the Intermediate and Lower Aquifers, which tend to be preferred for large-scale municipal supply. These flat-lying aquifers are generally thickest and best preserved to the south, notably within the deeper bedrock valleys, the Laurentian Channel being the largest and deepest. Similar high-yielding aquifers are found in valleys of comparable age that cut down through the Niagara Escarpment to enter the Laurentian Channel from the west (Figure 1.6). These aquifers supply the towns of Milton, Acton, and Georgetown. Unfortunately, because so few wells in the region penetrate to bedrock, the hydrogeological nature and areal extent of the deeper 'channelized' aquifers are not well known. They are often described as the greater Toronto region's 'hidden resource' and represent an excellent prospect for exploration and future development.

While Toronto has flourished on the abundance of fresh water supplies, its future growth and prosperity demand that the quality and quantity of this resource be sustained. The public is generally well aware of the need to maintain the quality of surface water, and anti-pollution measures initiated by public pressure during the past twenty years have done much to repair environmental damage caused during the postwar era to Lake Ontario and rivers such as the Don. In contrast, groundwater resources are out of sight, and remain largely out of the public mind. Consequently, their need for protection ranks low on the political agenda, and they remain seriously threatened by human activity.

The sensitivity of groundwater resources to human interference is well documented. Some of the earliest impacts occurred in response to the extensive deforestation of the Oak Ridges Moraine, which allowed soils to freeze during the winter and caused a reduction of natural groundwater recharge. In King Township between 1837 and 1937, over 80 percent of the streams became intermittent because the baseflow had declined, and 17 percent of the wells dried up. In modern times, diminished replenishment of groundwater remains a serious potential concern in heavily urbanized areas, where a large expanse of roofing and impermeable pavement prevents natural soil zone recharge.

Perhaps more critical, however, is the large number of chemical contaminants generated by urbanization and associated industrial activities. The contaminants are released in significant quantities to the subsurface and ultimately degrade groundwater quality. Contaminant audits suggest that road de-icing chemicals, septic systems, leaking underground gasoline tanks, and landfills represent a serious long-term threat to groundwater quality in urbanized areas of south-central Ontario. Chlorinated solvents released from dry cleaners and small manufacturing industries may also constitute a serious problem locally. Studies suggest that because contaminated groundwater tends to move so slowly – typically downward to the water table at 1 m per year and laterally in the aquifer as slowly as 10 m per year – the majority of pollutants released into groundwater during the past thirty years or more of explosive urban growth will not show up in most wells or springs for perhaps hundreds of years.

There is clearly neither room nor time for complacency. With Toronto's rapidly expanding population, the region's groundwater will have a significant economic value and is in dire need of attention. To meet a growing demand, groundwater development is certainly more cost-effective than surface water, which must be raised by pumping from large lakes, and, in many cases, piped over considerable distances. Yet groundwater simply does not attract the high level of political and public concern afforded to streams, rivers, and lakes. Only Halton Region has developed a serious aquifer management plan and only the Oak Ridges Moraine has been recognized as in need of 'protection,' even though it has never been made clear what this term really means. What is quite clear, however, is that southern Ontario, and the greater Toronto region in particular, needs a coordinating agency to regulate, manage, monitor, and protect its valuable groundwater resources. It also needs to recognize that carefully planned conjunctive use of surface and groundwater will be critical if Toronto's growth and development is to be sustained.

Groundwaters in the deepest aquifers are thousands of years old.

Soils

Geological materials deposited on the surface and exposed to the environment are slowly converted to young soils. This is a consequence of chemical and physical weathering, mineral disintegration, and interactions with plants and animals. These processes operate most actively at the soil surface, proceeding progressively downward and eventually forming horizons in the sediment, or parent material. Differences in the kinds of soil are primarily due to differences in parent material, but the effects of vegetation, topography, and age also contribute to variations in soil horizon development. The identity and distribution of major soil types in the Toronto region are shown in Figure 1.19. Detailed descriptions can be found in *The Canadian System of Soil Classification* (1998).

Most of the Toronto region is covered by Luvisols. These soils supported a natural vegetation of deciduous forest, most of which has since been cut down to make way for agriculture. The more weakly developed Brunisols occur on a small forested area of the

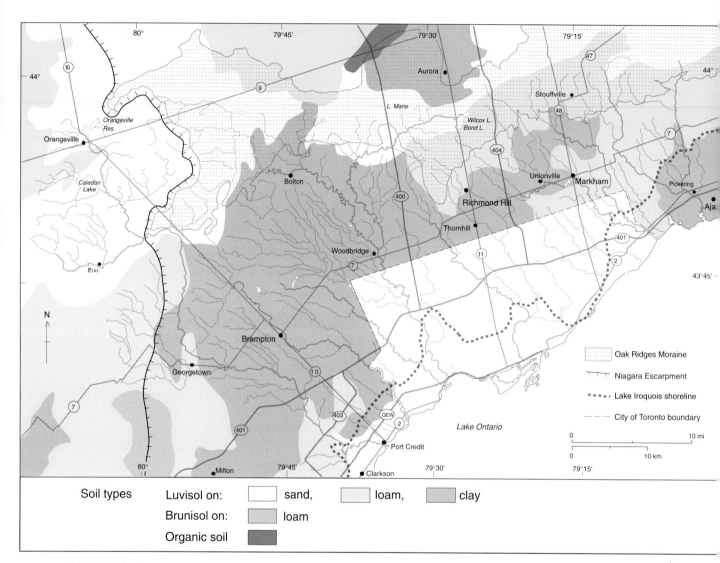

FIGURE 1.19 Dominant soil types in the Toronto region.

escarpment and are derived from thin glacial drift overlying limestone bedrock. Organic soils are found in the Holland Marsh on the northern edge of the region and are composed of decaying plant material formed in poorly drained lowland areas.

The Luvisols of the Oak Ridges Moraine have endured dramatic changes over the last two centuries as a consequence of human activity. Soil stability in connection with changes in forest cover and land use has contributed greatly to the degradation of this ecosystem.

The mixed deciduous forest of southern Ontario contained towering white pine that was prized for lumber by European settlers who came to this area in the early 1800s. The sandy soils supporting these very productive stands were Luvisols that were relatively fertile because of the long-term build-up and incorporation of organic matter in the topsoil and effective nutrient cycling that operated in relatively stable forest systems. This condition can still be observed in the few remnants of 'old growth' forest on the moraine, such as at Joker's Hill.

Unfortunately, much of the forest was 'high graded' by improper harvesting practices that logged only the best trees. Land was also cleared for agriculture, which contributed to large-scale deforestation of the area. Grazing and cropping disrupted natural nutrient cycling and changed once humus-rich horizons to plough layers that were much lower in organic matter and nutrients. The sandy soils were drought prone – a problem that was compounded by the lack of forest cover, which formerly had served as wind breaks and had helped to retain soil moisture.

Much of the Oak Ridges Moraine became unsuitable for sustained farming by the end of the nineteenth century and was abandoned during the Depression in the 1930s. The abandoned farm soils could not support protective vegetation, and as a result wind eroded the topsoil and large areas of 'blowsands' were created. The topsoils were lost, forming newly developed wind deposits of recently exposed parent materials. The large tracts of shifting sands, devoid of vegetation, were called 'the wastelands.' Their occurrence sparked the early formation of an Ontario conservation movement, composed mainly of dedicated farmers and concerned citizens who lobbied local governments for soil conservation and reclamation.

Effective programs of land reclamation and soil stabilization were initiated in the 1920s, mainly by the adoption of aggressive reforestation practices. Politicians and foresters were motivated to create county forests and conservation authorities that would purchase degraded lands for reforestation. Successful techniques involved planting tree seedlings raised at specialized nurseries and using branches to stabilize the surface soil and protect seedlings. Once established, the young trees and the decomposing branches or mulch contributed to the development of new horizons of Regosols, or young soils, on the blowsands. The major species that was planted was red pine, which was best adapted to the sandy, dry site conditions.

CONCLUSIONS

The physical framework of the Toronto region is an amalgam of the consequences of geological events that have operated over a vast period of time. Its rocks and sediments tell a story of colliding continents, large mountain ranges, deltas, tropical seas, continent-sized glaciers, lakes, and animals long since extinct. Huge gaps in our knowledge still exist. Glaciers have waxed and waned across northern North America for over a million years, for example, but our glacial story begins only about 150,000 years ago. Nevertheless, some of those gaps in knowledge will undoubtedly close as new exposures are opened up in the course of the rapid urban growth that now characterizes the greater Toronto region.

The face of Toronto's landscape, wrought over geological time, is now experiencing very rapid change, most of it human driven. Past and present agricultural practices and urban development have degraded the environment, and there is therefore an urgent need for future development to be based on principles that respect the natural world. Recent decisions on urban planning in the greater Toronto region give cause for optimism on this point.

2

Climate

R.E. (Ted) Munn, Morley Thomas, David Yap

The *Natural History of the Toronto Region* (1913) contained a description of the climate of Toronto circa 1913, written by R.F. (later Sir Frederic) Stupart, head of the Meteorological Service of Canada at the time. He based his analysis on observations from a single station, the Observatory on the University of Toronto campus. The meteorological program at that site had commenced in 1840 and included daily observations of temperature, humidity, precipitation, air pressure, sunshine, and cloudiness. These observations were begun by Sir John Henry Lefroy and were regularly published by the Canadian Institute in the mid-nineteenth century.

Since that time, an array of new observing locations and studies have reflected our close concern with the weather. With the growth of the aviation industry in the 1930s, a full observing program was established at Malton Airport (now called Pearson Airport) in 1937 and was supplemented with similar programs over various periods of time at the Downsview, Buttonville, and Toronto Island airports. A full observing program consists of hourly measurements of temperature, humidity, pressure, wind, clouds, and visibility, and daily measurements of maximum and minimum temperature, rainfall, and snowfall. In addition, and particularly since 1951, several volunteer observing stations have been operating throughout the Toronto area, usually limited to daily readings of maximum and minimum temperatures and of precipitation.

Temperature variations – the temperature mesoclimate – across the Toronto urban area were first examined by Middleton and Millar (1936) using an instrumented automobile. On a clear February night, the temperature fell slowly as their car travelled north up Yonge Street, dropped precipitously by about 15°C as the car entered Hogg's Hollow, then rose again farther north (Figure 2.1). Since then, and especially over the past 25 years, several field research programs have been carried out in the Toronto area using instrumented automobiles, balloons, towers, and other methods to examine the urban climate. In 1967, the Province of Ontario passed an Air Pollution Control Act, and a network of stations for monitoring the common air pollutants was established soon after. In 1970, an air pollution index and alert system were introduced.

To explore the subject of Toronto's climate in the twentieth century, we asked the following questions:

R.F. Stupart, 'The Climate of Toronto,' 1913

There are some indications that the climate has changed slightly with the gradual clearing away of the vast American forests.

A period of heat observed during the first four days of July 1911 broke all records, and on six consecutive days the following temperature records were registered: 98, 101, 103, 97, 95, and 92 degrees F [approximately 37, 38, 40, 36, 35, and 33°C].

There is no sleighing at Christmas in more than one year in five, and in many winters the ground is bare during Christmas week. During January and February there is usually sleighing, but towards the end of February the snow melts fast.

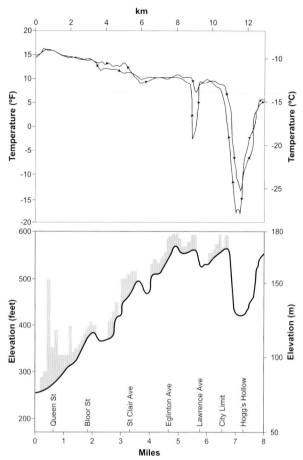

FIGURE 2.1 Air temperature and ground elevation along Yonge Street. Shaded areas represent the height of buildings close to Yonge Street. Air temperatures were measured with a thermometer strapped to the outside of an automobile, at a height of about 0.7 m above street level on a clear night, 22 February 1936, with a weak northerly breeze blowing. Note that the temperature patterns are roughly the same on the northbound and southbound legs. (Middleton and Millar 1936)

(1) Has the climate changed since 1913 at the reference station on the University of Toronto campus? (2) What are the characteristics of the spatial climatic patterns (the mesoclimates) in the Toronto area? (3) What are the very local variations in climate close to the ground (the microclimate)? (4) How will the climate of Toronto change in the twenty-first century?

THE MACROCLIMATE OF TORONTO

Toronto has a continental-type macroclimate, modified to a noticeable extent by the Great Lakes. The region lies under a mid-latitude migratory low-pressure belt that shifts north in summer and south in winter. The individual storms occurring in this belt usually move from west to east and are more frequent and more intense in winter than in summer. Although the best weather often occurs in summer, this season also has the greatest number of thunderstorms, and even tornadoes are not unknown. In the autumn, the Toronto region has, on occasion, suffered from the extra-tropical remnants of Atlantic hurricanes, Hurricane Hazel in 1954 being the most important example.

Weather Systems

The general circulation of the atmosphere is markedly influenced by physical geography. In southern Ontario there is no barrier to prevent cold polar air from sweeping down from the north over Toronto. Nor is there any major obstacle to protect the region from hot dry or warm moist air from the southern United States. However, the Great Lakes do play an important, if somewhat secondary, role in protecting the Toronto region from weather extremes: surface winds, except for those from the northeast sector, must pass over one of the lakes before reaching Toronto. This makes the city somewhat cooler in summer and warmer in winter than the outer suburbs.

Over the continent, the north-south temperature gradient increases in winter, and this is responsible for a greater frequency and intensity of the migratory low-pressure areas and storms. When the Toronto region is under a high-pressure area in winter, cold, clear, and nearly calm days are usually followed by increasing cloudiness and then rain or snowfall. Wind speeds increase from the east as a low pressure area moves toward and over the region. Then, within hours, the wind backs to the northwest, squally showers of rain or snow prevail, and the temperature begins to fall. As the low-pressure area moves away, the weather clears and sunny cold conditions return for a day or two before the cycle repeats. There are variations, of course; pleasant weather may persist for a week or more at a time or, on the other hand, stormy weather may continue for several days.

In the summer, when the storm tracks are usually north of the Great Lakes, Toronto weather is often controlled by either atmospheric stability (temperature inversions) or convective instability. In the latter case, daytime heating of the ground leads to cloud formation, showers, and occasional thunderstorms. On the other hand, when stability and temperature inversions prevail for some time, the humidity rises, pollution is trapped, and air pollution 'episodes' may occur.

In contrast to the marked annual temperature cycle in Toronto, precipitation from winter storms is usually balanced by that from the more local summer convective storms. Winters are stormier than summers, with more cloudy to overcast days, more days with precipitation, and generally higher wind speeds.

Summer

Summer usually begins in late May, when maximum temperatures begin to exceed 20°C and the danger of night-time frost is very slight. July, the warmest month in three-quarters of the years, has average daily maximum and minimum temperatures of 26 and 18°C, respectively. After the summer solstice, the earth continues to gain more heat from the sun than it loses to space, delaying the average warmest period of the year until mid-July in the Toronto region. The highest temperature each summer usually occurs in July or August. The highest temperature ever recorded at the downtown site was 41°C on three successive days, 8, 9, and 10 July 1936. More recently, maximum temperature records were established at several Toronto-area stations in July 1988.

Humidity can be excessive for a few consecutive days two or three times in the course of each summer, but on occasion, when a lake breeze develops, the temperature and discomfort are diminished. July normally has more bright sunshine than any other month; the average is nine hours a day. There is not much difference in monthly precipitation across the summer season, as each month receives about 73 mm on average. The driest month on record was August 1876, when only a trace of rain fell. Thunderstorms occur on 20 to 25 days a year. Most of these occur in summer, on occasion with damaging wind and hail, which is usually local in extent.

Autumn

The transition to the autumn season is usually completed early in September. Daily maximum temperatures fall into the upper teens by mid-month and summer is definitely over. Frosts may occur any time in September and October; the average date of the first frost over the past few decades at the University of Toronto site is 29 October. The end of the growing season, when the mean daily temperature falls below 5 or 6°C, usually occurs by the first week of November, and autumn ends two or three weeks later. October is the driest autumn month, with an average of 63 mm of precipitation, while September and November each receive 76 mm on average. Snow may fall any time in October, but appreciable amounts are not usual until November, when more than 10 cm has fallen in some storms after mid-month. November is usually dull, with considerable cloudiness as the migratory low-pressure systems intensify and move

---

Sir John Henry Lefroy

Capt. John Lefroy, by George T. Berthon, 1853. (Royal Canadian Institute)

Sir John Henry Lefroy was born in 1817 in Crondall, England, the son of the rector of Ashe. At 17 Lefroy joined the Royal Artillery, where he received his scientific training. While still in his early twenties, he was sent to St. Helena in the south Atlantic as director of the Magnetic Observatory. In 1842, he was chosen by the Royal Society to make a survey of the earth's magnetic fields in British North America, which provided important information for accurate surveying and navigation. Two years after his arrival in Canada, he was placed in charge of the Magnetic and Meteorological Observatory of Toronto, where he pioneered the recording of accurate weather observations with reliable modern instruments. He published papers on instruments and record keeping. His daily meteorological records for 6 a.m., 2 p.m., and 10 p.m. at Toronto were published monthly in the *Canadian Journal* of the Canadian Institute from July 1852. Lefroy was elected president of the Institute in 1852 but was recalled to England the following year. After his departure the journal continued to publish the daily records until December 1877. Two places in Canada were named in his honour: the village of Lefroy in Simcoe County and Mount Lefroy near Kicking Horse Pass in the Rocky Mountains.

across the region more frequently. Wind speeds in the autumn are usually much higher than in summer.

Winter

Winter, when on most days the mean temperature is near or below freezing, usually begins by early December. The average Toronto temperature over the December-February period is -3°C. The falling mean temperature curve begins to flatten out in January and the minimum is reached during the first week of February. This lag from the winter solstice is longer than the lag in summer because the Great Lakes have a greater effect on Toronto temperatures in winter than in summer. The coldest day of the year may occur any time from late December to early March. The coldest temperature ever observed at the Observatory was -33°C on 10 January 1859. Several other stations in the Toronto area, with shorter periods of record than that of the Observatory, recorded their lowest temperatures in January 1981.

Cloudiness is at its peak in winter, and half the days have some measurable precipitation. Rainfall is still appreciable, averaging about 23 mm in January. Snowfall peaks in January, with a monthly average of 35 cm. The greatest snowfall recorded in one day was 48 cm on 11 December 1944. Over the entire winter season, Toronto averages 135 cm of snow. The winter with the least snowfall was 1952-3, when only 47 cm fell; on the other hand, 314 cm fell in 1869-70. More recently, there was a record January snowfall of 110 cm in 1999, most of which fell in the first fourteen days of the month.

Spring

Spring usually becomes noticeable some time in March, when, by mid-month, the Toronto temperatures average above freezing. But the date of the last spring frost, averaged over a recent thirty-year period at the Observatory site, is not until 20 April. Maximum temperatures as high as 32°C have been observed in April, but the mean daily maximum that month is but 11.5°C.

Storms are often more intense in March, and wind speeds average higher than in any other month. With more rainfall and less snowfall as the season progresses, the total precipitation for each of March, April, and May averages between 65 and 68 mm.

THE MESOCLIMATES OF TORONTO

The important natural factors influencing mesoclimates in the Toronto region are proximity to Lake Ontario; location relative to the several valleys and ravines that cut through the area; location below or above the old Lake Iroquois shoreline; and distance from the bounding Oak Ridges Moraine to the north and the Niagara Escarpment to the west and northwest (Figure 2.2).

These factors have been increasingly supplemented in the last century and a half by human-induced ones – forests have been cleared, buildings have been constructed, thousands of hectares have been hard-surfaced for roads and parking lots, and in winter (during recent decades) snow has been promptly removed from roads and parking lots. In addition, immense and increasing quantities of heat are produced each year, as well as large amounts of gaseous and particulate pollution, coming from transportation, industry, and space heating. These 'urban' influences have created the Toronto 'heat island' and have been superimposed on the natural factors.

The greatest snowfall recorded in one day in the Toronto region was 48 cm on 11 December 1944.

Lake Ontario has a moderating effect (cooler in summer and milder in winter) on the Toronto climate whenever winds blow from lake to land. The ravines stretching inland from the lake provide preferred paths for lake breezes to move inland, and for cool air to drain down toward the lake at night. In the latter case, the air motion is sufficient to prevent early autumn frosts from developing along the slopes. When the air drains down a hill but has its further progress impeded in some way, however, a frost pocket is likely to develop. This is why the bottom of Hogg's Hollow has low night-time temperatures.

Temperature Patterns

Over the year the Toronto city centre averages 1° to 1.5°C warmer than the surrounding countryside. The tremendous amount of heat released from the city to the atmosphere reduces the rate of cooling, while in the surrounding rural areas, clear skies and often a snow cover enhance the cooling rate. On a winter night, when light winds and clear skies occur, minimum temperatures at Pearson Airport and at the other observing station outside the urban fabric can be as much as 10°C lower than those at the Observatory, although the average difference in January, in all kinds of weather, is

Climate

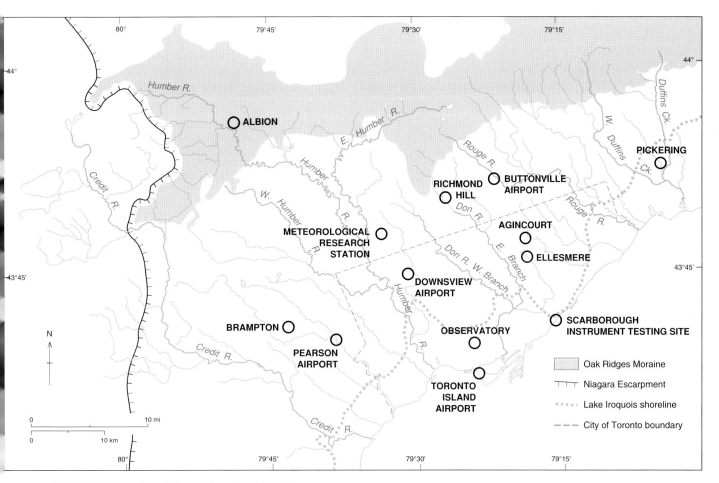

FIGURE 2.2 Locations of the weather observing stations.

only around 2° to 3°C. Immediately along the lakefront, temperatures average about 1°C warmer than at the Observatory.

In July, maximum temperatures average about 26° to 27°C throughout the region, except along the lakefront, where they are a degree or two lower due to frequent afternoon lake breezes. The average night-time minimum temperature in July at the Observatory is 17.6°C, while significantly lower values are reported from the suburbs and outlying areas – 14°C at Pearson, 15° at Richmond Hill, and 16° at Ellesmere (see Figure 2.2 for station locations). When mid-city temperatures are 2° to 3° higher than suburban and country temperatures, the heated city air tends to rise and the cooler country air flows toward the city centre (Findlay and Hirt 1969). Other studies have shown that when surface winds are off-water, the heat island is displaced 5 to 8 km to the northwest of the Observatory in winter but as much as 30 km in summer. With wind blowing from the land to the lake, the heat island is displaced 8 to 13 km to the northeast in summer, while in winter it may be over the lake (Munn, Hirt, and Findlay 1969).

A warmer city centre means heating costs are 10 to 15 percent lower than in the surrounding areas. Close to the lake, and on the island, the annual average cost is about 5 percent more than at the Observatory, due principally to cool springs caused by onshore lake breezes.

In the city core and along the waterfront, on average, there are about 191 frost-free days a year. The frost-free period drops to about 150 days at Pearson Airport and Richmond Hill. At the Observatory, the average frost-free period is from 20 April to 29 October, while at Pearson Airport and Richmond Hill, the period is from 8 May to 5 October.

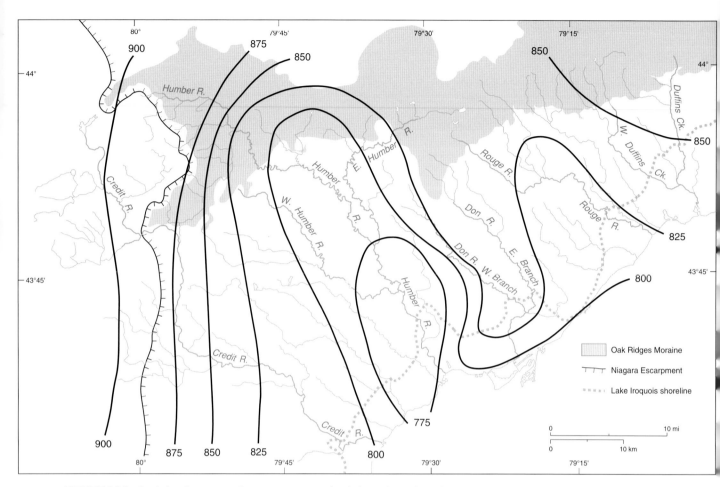

FIGURE 2.3A Isopleths of mean annual precipitation in mm. Isopleths are lines of equal precipitation.

Precipitation

Annual precipitation at the midtown Observatory averages 819 mm. There is a distinct 'rain shadow' (dry area) in the area drained by the Humber and Credit Rivers on the west side of the region, where the average annual precipitation at Pearson Airport is 781 mm. Northeast of the city centre, precipitation increases to average over 850 mm in the upper portions of the Rouge River and Duffins Creek watersheds (Figure 2.3A).

There is also a snow shadow in the Credit and Humber River basins. The average annual snowfall to the west of Yonge Street is less than that to the east: Pearson Airport averages 124 cm, the Observatory 135 cm, and northeast of Highway 7 more than 140 cm. Precipitation that often falls as snow over most of the region falls as rain on the island and along the waterfront, where the average annual snowfall is 111 cm (Figure 2.3B).

Solar Radiation and Sunshine

In the winter months, the Toronto region receives only about one-fifth the radiation received in the high-sun months of June and July. Measurements have shown that, for all months of the year, the eastern suburbs receive a few percent more solar radiation on average than the Toronto Observatory, where pollution and low cloud shield the urban core to some extent (Mateer 1961).

The duration of bright sunshine at Toronto averages less than 100 hours a month during November, December, and January but more than 275 hours in July. Observations have shown that, to the northwest of the city, from 4 to 6 percent more sunshine is received than at Toronto. In general, the mesoclimates of the suburban and rural fringes of Toronto offer brighter conditions than those in the urban core.

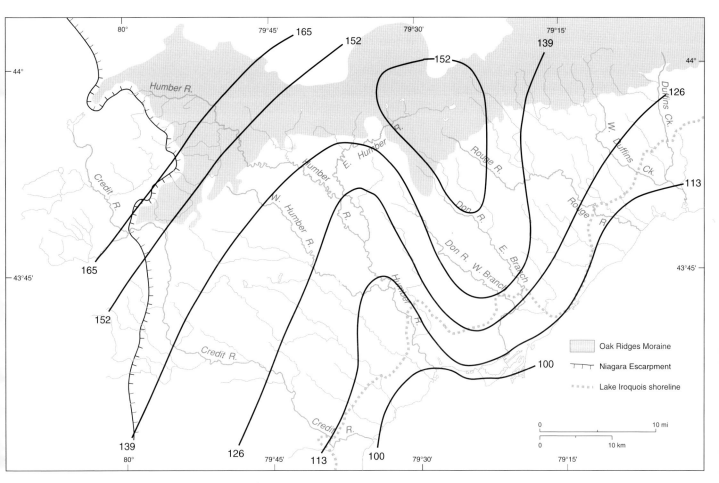

FIGURE 2.3B Isopleths of mean annual snowfall in cm.

Wind

There is so much variability as the migratory pressure systems move over the region that there is no prevailing wind direction at Toronto. At Pearson Airport, the most frequent direction is northwest from March to August and west for most of the rest of the year. At Toronto Island Airport, west is the prevailing direction in winter and east in summer, due to the lake effect. The average annual speeds are 15 km/hour at Pearson Airport and 17 km/hour at the Island Airport. Maximum speeds in March have been as great as 97 km/hour at Pearson and 121 km/hour on the island.

Humidity

In general, urban areas experience relative humidities 2 percent less in winter and 8 percent less in summer than those in the surrounding countryside. Over a 20-year period, the relative humidity at Pearson Airport averaged higher each month than that at the Observatory. The greatest difference in relative humidities, 6 to 8 percent, was in the winter season from November through May, compared to 2 to 5 percent in the other months. As might be expected, humidity values along the lakefront are higher than those inland.

MICROCLIMATES IN TORONTO

Climate-observing stations are located in the centres of grassy plots in order to give representative observations. There is, however, very great local variability around observing stations due to buildings, roads, and trees. These differences are sometimes greater than the differences across an entire network of stations.

Temperatures a few millimetres above a black tarmac surface on a sunny day in summer may climb to 45°C, as compared with screen temperatures of about 25°C, while under a nearby large shade tree the temperature may reach only 20°C.

Architects and engineers have long known about the existence of these small-scale climatic patterns, or microclimates, in cities. Some examples follow.

Changes of temperature and humidity with height: The vertical gradients of temperature and humidity in the lowest 2 m of the atmosphere are greatest on sunny days, when the temperature falls with height, and on clear calm nights, when temperature rises with height. Depending on the radiative properties of the underlying surface, temperature differences can range from 1° to 10°C, while humidity is almost always highest near the surface.

Frost-prone sites: With clear skies and light winds at night, surface cooling is greatest over pavement and dry earth away from walls and trees, which re-radiate heat back to the ground. Natural hollows into which cool air drains are particularly frost-prone, as are bridges and garage roofs, where heat cannot be drawn up from underlying soil.

Immense and increasing quantities of heat are produced each year, as well as large amounts of gaseous and particulate pollution, creating the Toronto 'heat island.'

Shaded areas: Radiative exchanges take place mostly at the top of the tree canopy in the forest climates of the older residential areas of Toronto, giving a surface climate cooler in daytime and warmer at night than over open land. South-facing walls receive more solar radiation than do horizontal surfaces and much more than north facing walls. Small enclosed courtyards receive only reflected sunlight (Verseghy and Munro 1989).

Strong wind effects: Regional airflow parallel to the orientation of an urban canyon channels the wind. Regional airflow at right angles to an urban canyon causes gusts of wind to tumble down from the tops of buildings toward street level. At intersections, these two effects may cause considerable discomfort to pedestrians. Scale-model testing in a wind tunnel, now required before large buildings can be constructed in Toronto, can largely circumvent these problems. In particular, such testing has been done in studies of pedestrian wind fields around Toronto City Hall, Skydome, Simcoe Place, and along Bloor Street from Avenue Road to Church Street (Figure 2.4).

Moderate to strong winds occurring with precipitation produce rain shadows to the lee of buildings. Snow

Mesoclimate and the Siting of Animal Enclosures at the Toronto Zoo

The Toronto Zoo is in a valley 6 km inland from Lake Ontario, just upstream from the confluence of the main branches of the Rouge River. The river has dissected a tableland of unconsolidated sediments, giving a relief of up to 40 m over short distances. The tableland and the valley floor themselves have gentle inclines, but there are extensive forested areas along steep valley slopes beyond the tableland.

In the early 1960s, Raymond Moriyama and Associates, who were at an early stage in designing the Zoo, asked Bruce Findlay and colleagues at the Meteorological Service of Canada two key questions (Higgins and Findlay 1969). What were the mesoclimatic variations across the 6 square km area chosen for the Zoo in the Rouge River valley? Were these variations large enough to be considered significant when deciding where to locate the different species of animals?

In 1967-8, these questions were studied using automobile temperature traverses, foot traverses to examine shade effects, balloon tracking of daytime lake breezes in summer, and daily measurements from two reference observing stations. In daytime, valley temperatures were noticeably warmer in winter and cooler in summer than over the surrounding area but with little variation across the valley. At night, however, the valley was a repository for cold air draining downslope – 0.8°C lower, averaged over a year, and as much as 5°C colder on clear nights with light winds.

In response to this study, sensitive non-indigenous animals were excluded from the low elevations, and extra protection from severe mesoclimates was provided for certain habitats in the tableland.

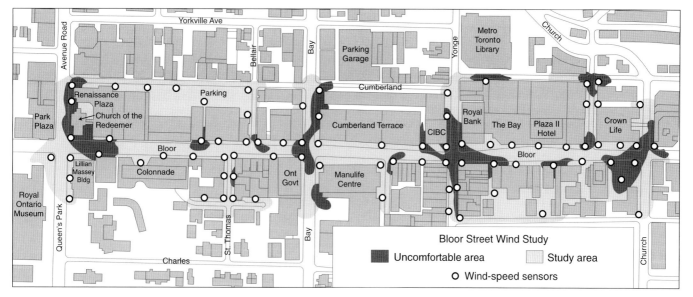

FIGURE 2.4 The wind on Bloor Street is notorious. A wind-tunnel study by RWDI Associates pinpointed the windiest locations and recommended designs for new buildings that would reduce pedestrian discomfort. (William R. Constable, Rowan Williams Davies and Irwin. Reproduced with permission from *Canadian Geographic*, original source.)

drifts also occur near lee corners, where the wind speed drops.

Human-induced variations in areal snow cover: After snow is cleared from city streets, parking lots, and sidewalks, more solar radiation is absorbed at the surface, contributing to the Toronto urban heat island.

A new urban snow surface begins to darken within a day or so, and more solar radiation is absorbed than in the surrounding countryside. This contributes to snow melting and the urban heat island.

Indoor microclimates: In all seasons of the year there are many different controlled climates in homes, offices, shopping malls, sports arenas, and air-conditioned automobiles.

AIR QUALITY

Historical Perspective

The City of Toronto passed its first air pollution control measure, an anti-smoke by-law, in 1907. But it was 1949 before an effective 'control of black smoke' by-law was adopted, and 1957 before the by-law was extended to include all the metropolitan municipalities. By 1968, 95 percent of Toronto's once serious black smoke problem had been eliminated, but concerns remained about other smoke and dust problems and gaseous emissions from combustion and other chemical processes (Ontario Ministry of the Environment 1973).

When the Air Pollution Control Act of 1967 came into effect, the control and prevention of air pollution became the responsibility of the Ontario government, which had previously acted only in an advisory capacity. The act was superseded by the Environment Protection Act in 1971 and new programs were introduced around the same time: control of emissions of all stationary sources of pollution through industrial abatement and fuel conversions; establishment of a network of air pollution monitoring stations; development of an air pollution index and alert system to trigger the curtailment of operations at major sources when necessary; and construction of a detailed air emission inventory and an air quality simulation model to compare the effectiveness of different abatement strategies. Also in 1971, a regulation restricting the sulphur content of fuels sold or used in Metropolitan Toronto came into effect.

These measures proved effective and air quality improved substantially. By the end of 1972, sulphur dioxide in downtown Toronto had decreased from an annual average of 0.10 parts per million in 1967 to 0.03 in 1972, a drop of 70 percent. During the same period, suspended particle loadings had decreased from 0.70 COH (coefficient of haze) units to 0.32, a drop of 54 percent.

Despite the significant decreases in such common air pollutants as sulphur dioxide, carbon monoxide, nitrogen oxides, and total suspended particles, there has been no improvement in ozone, a secondary pollutant whose concentrations exceed the Ontario ambient air quality criterion. Moreover, concern continues over fine particle pollution, which is a factor in respiratory health problems. Public concern is growing over the brown haze frequently visible above the city, a haze due to a mixture of fine particulate matter and nitrogen dioxide in the urban mixed layer up to a height of 1 km or more. The section on watching the sky, pp. 47-9, illustrates one of the effects of changes in air quality over the past 100 years.

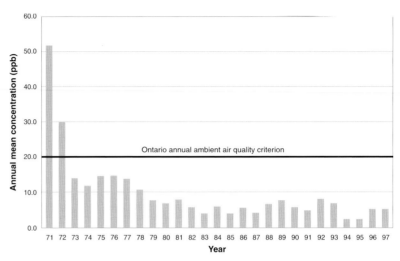

FIGURE 2.5 Sulphur dioxide trend for Toronto downtown, 1971-97.

Air Pollution Index and Alerts

In March 1970, a pollution index/alert system, based on measurements of sulphur dioxide and suspended particles, went into operation in Toronto to issue warnings and prevent the adverse effects of pollution build-up during episodes of stagnant weather. In the first year of operation, the advisory level of 32 was exceeded seventeen times, and the first alert level of 50 was exceeded on two occasions in downtown Toronto. By 1972 there were only two advisory-level occurrences and none at the alert level. No advisory-level warning has been given since 1985.

The maximum air pollution index value ever recorded at the downtown site was 62 in 1975. Based on other earlier measurements, it has been calculated that the highest Toronto index value would have been 155 on 4 December 1962. That particular episode, lasting from 30 November to 4 December, resulted in dense smog over Toronto, which caused suspension of the Grey Cup football game halfway through the match; the game was resumed the following day.

Air Quality Index

Introduced in 1988, the Air Quality Index (AQI) provides real-time public information on air quality at six sites in Toronto. The index is based on measurements of those pollutants that affect health and the environment: sulphur dioxide (Figure 2.5), ozone, nitrogen oxide, total reduced sulphur compounds, carbon monoxide, and suspended particles. The index is calculated on an hourly basis for each pollutant. Levels below 32 indicate that air quality is very good or good, levels between 32 and 49 are described as moderate, between 50 and 99 as poor, and 100 and above as very poor.

A summary of AQI levels in Toronto for 1995 indicates that good to very good air quality occurs most of the time, with little variation across the city. Ozone is the most frequent cause of poor air quality. In the more industrialized areas of Etobicoke and York, suspended particles also contribute to moderate/poor air quality.

Air Quality Advisories for Ground-Level Ozone

In May 1993, a joint air quality advisory program was launched by Environment Canada and the Ontario Ministry of the Environment to inform the public when regional concentrations of ground-level ozone are expected to be high during the May-to-September smog season. A 'Spare the Air' day features an advisory to inform individuals about how they can prevent further deterioration of air quality. Since 1993, the following advisories have been issued for Toronto: 1 advisory for 1 day in 1993; 2 advisories covering 5 days in 1994; 5 covering 8 days in 1995; 2 covering 3 days in 1996; 2 covering 5 days in 1997; and 3 advisories covering 7 days in 1998.

The city's highest air pollution reading, on 4 December 1962, resulted in dense smog over Toronto, shutting down the Grey Cup game halfway through the match.

Considering the rapid growth in Metropolitan Toronto over the last 25 years (the volume of vehicular traffic increased by 250 percent over that period), it is indeed fortunate that air pollution control programs were greatly strengthened in the 1970s. One spectacular success was the effect of a ban on the use of lead in gasoline in the 1970s. Average lead levels in the blood of Toronto children declined significantly in the next decade (Hilborn and Still 1990).

As with climate observing, air pollution monitoring is done at 'representative sites.' But some of the highest pollution concentrations occur at curb-side or directly downwind of short chimneys. The occupational exposures of taxi drivers and traffic wardens may therefore be considerably higher than indicated by readings taken at air pollution monitoring stations.

CLIMATE CHANGE IN TORONTO: HISTORICAL TRENDS

Early Opinions about the Toronto Climate

R.F. Stupart, in his treatment of Toronto's climate in 1913, said little about climate change. A few years later though, in a paper entitled *Is the Climate Changing?*, he acknowledged the geologists' theory of ice ages and claimed that long-term 'climatic variations result from changes in atmospheric circulation due to a varying solar activity' (Stupart 1917). Further, he noted that 'the carbonic acid gas theory finds little favour with the meteorologist,' contrary to today's belief that the release of increasing amounts of carbon dioxide into the atmosphere must produce atmospheric warming. At about this time, one of Stupart's lieutenants, John Patterson, pointed out that 'it is doubtful if long-range forecasting will ever be successful until the fundamental principles underlying the various meteorological phenomena are discovered' (Patterson 1917).

Stupart's report (1917) on his examination of the Toronto climatological record from 1840 to 1916 is worth repeating: 'The precipitation figures give some ground for suspicion that the annual rainfall has diminished somewhat, but they are not conclusive ... In temperature, there has been a manifest tendency towards higher annual mean and mean spring and summer temperature during the past twenty odd years, while any upward winter tendency is less noticeable ... In the early days of the Observatory, [spring] frosts seem to have occurred later and been more severe than in later years, and certainly the official figures conform very exactly with the reports of early settlers.'

The Climate Record

Because of the recognized urban effect, the climate record at the Toronto Observatory is never used to provide statistical evidence of provincial or national climate

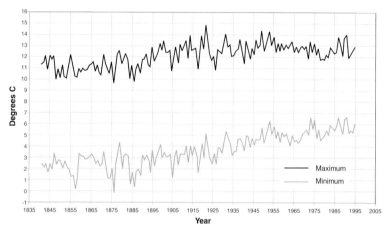

FIGURE 2.6A Annual mean maximum and minimum temperatures for Toronto, 1840-1995.

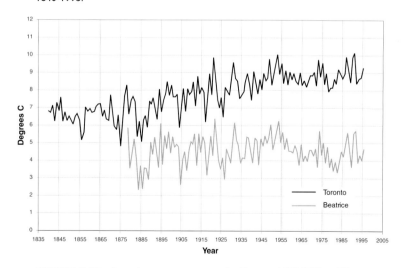

FIGURE 2.6B Annual mean temperatures for Toronto, 1840-1995, and Beatrice, 1878-1995. Beatrice was a rural climatological observing station located 161 km north of Toronto on a farm in Muskoka District.

warming. Regardless of what happened in the rest of the province, Toronto temperatures would have increased in the twentieth century. In fact, minimum temperatures at the Observatory have increased fairly steadily, and now are 3° to 4°C higher than they were in the 1840s (Figure 2.6A). But this is not the case with maximum temperatures, where there appears to have been a change in the 1950s in the upward trend line. Previously, the trend was about parallel with that of minimum temperatures. Since then, the trend line shows slightly cooler maximum temperatures, although the individual 1987 and 1991 values rank with the highest dozen years on record. This trend toward cooler maximum temperatures cannot be attributed to an urban effect, since another study has shown a slight reduction in maximum temperatures over the whole Great Lakes-St. Lawrence region (Skinner and Gullett 1993).

Snowfall and convective rainfall vary so much over short distances that precipitation data cannot be used in climate change studies with the same degree of confidence as can temperature data. Moving means show that at the Toronto Observatory, annual precipitation averaged more than 800 mm in the nineteenth century and then near or slightly below that value until the most recent 25 year period, when precipitation has again increased (Figure 2.7).

Snowfall is one of the more easily recognized indicators of climatic change. At the Observatory over the last century and a half, winter snowfall has averaged 140 to 150 cm, with two major exceptions. There were excessive falls totalling more than 200 cm in winters of the 1880s, and during the 1990s winter snowfall has averaged only 120 cm.

WHAT IS AHEAD?

Most atmospheric scientists believe that changes in *globally averaged* climate are likely to be unprecedented by the end of the twenty-first century (and maybe earlier) because greenhouse gases are increasing. Well-established theory indicates that greenhouse gas accumulation will cause the atmosphere to warm, more at the poles than at the equator, thus disrupting global patterns of winds and ocean currents. There is, however, considerable debate on a number of related questions:

1. Has global climate warming already begun? (The evidence is increasing that climate change has started but the case is not persuasive.)

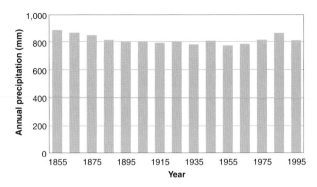

FIGURE 2.7 Mean annual precipitation by decade for the Toronto Observatory, 1855-1995.

2. For any given rate of rise in greenhouse gas concentrations in the next 100 years, how quickly would the climate change? (The oceans have a huge heat capacity that could slow down atmospheric warming.)
3. How would a change in *global* climate affect the climate of southern Ontario, and of Toronto? (Moving from the global to the local scale is a research task of considerable interest around the world.)
4. How will global climate change affect the frequencies of rare climatic events such as floods, droughts, and tornadoes? (This is also a research task of great current interest.)

The bottom line is that climate change could become at the very least inconvenient, and at worst disastrous to the economy and well-being of Toronto and its inhabitants by the end, or perhaps even the middle, of the twenty-first century. Two other possibilities – that the Toronto climate will remain unchanged or that the climate will actually 'improve' (whatever that means) – seem rather remote.

Globally Averaged Climate Change Possibilities for the Latitude of Toronto by the Year 2100

Over the course of the next century, globally averaged climate change has several specific implications for the Toronto region – and other regions in the same latitude. Average temperatures will rise by 1° to 3°C. These increases will be greater at night than in the daytime and greater in the winter than in the summer, although the frequency and intensity of summer heat waves will grow. Precipitation will be affected, with less snow but more rain in the winter. Summer precipitation will become more showery. More evaporation, particularly in summer, will lead to drier soils and greater frequency of drought. Lake levels will decline by as much as 1 m. Finally, stagnating anticyclones will occur more frequently, with the result that high pollution episodes will be more common. Yet there is no consensus on whether the frequencies of rare events will change.

These changes are based on scenarios of continuing increases of greenhouse gas emissions and on computer models of atmospheric processes. If the nations of the world were able to reduce their greenhouse gas emissions sufficiently, climate change might not be so severe. However, United Nations negotiations on implementing the Framework Convention on Climate Change are progressing very slowly, and there is only a remote possibility that emission reduction targets will be met.

In trying to look ahead into the twenty-first century, we must consider several other factors. First is the growing population in the Toronto region. In the 1950s, only small communities existed north of Steeles Avenue, west of Highway 427, and east of the Rouge River, but the urbanization phenomenon now extends almost to Lake Simcoe, to the Niagara Escarpment, and to Pickering. There seems to be no possibility of checking further outward growth, or of preventing elimination of the patches of farmland that still exist within the metropolitan area. The spatial dimensions of the urban heat island and related atmospheric phenomena will therefore grow.

Fifteen thousand years ago, Ontario was covered by a massive ice sheet more than 1 km thick. Palaeoclimatologists assure us that 15,000 years into the future, the greater Toronto region will once again be buried in ice.

A second factor is the growing number of automobiles in the Toronto region. Unless transportation and land-use policies are radically altered, oxidant-causing air pollution emissions can only increase. Of course, new technology may provide temporary relief, but the increasing numbers of automobiles in the coming decades are likely to swamp technological developments and replacements of old by new cars.

The final major factor is that the citizens of Toronto are exhibiting a growing impatience with lack of action on environmental issues. On a positive note, several local groups are actively taking measures to reduce greenhouse gas emissions or are planting trees that take up carbon dioxide from the atmosphere. In 1992 Toronto City Council created the Toronto Atmospheric Fund with the goal of reducing carbon dioxide emissions in Toronto by 20 percent by the year 2005. Its mandate includes educating the public about climate change, promoting energy conservation and efficiency,

Vertical Gardens: Extending the Urban Canopy

According to Professor Roger Hansell of the University of Toronto, the urban heat island is both a symptom and a cause of waste energy. Seen from space, Toronto has the signature of a rock quarry, re-radiating energy with almost no use made of its potential. Even worse, the city imports energy to heat – and cool – the rocks!

Toronto spends more money on hydroelectric energy in the summer to air-condition its buildings than it does in winter to heat them. One way of helping to curb such a consumption of energy is cultivating vegetation canopies, which help to cool the city in summer and reduce air pollution and urban emissions of carbon dioxide and other greenhouse gases to the atmosphere.

Yet vegetation canopies have their limitations: there are far more walls than rooftops re-radiating heat energy in the city. To address this fact, Roger Hansell, Brad Bass, and their students at the University of Toronto are studying the development of vertical gardens for wall surfaces and living 'green screens' for windows, to intercept the sun's energy before it becomes waste heat. Such measures would reduce the costs of cooling buildings at the same time as potentially producing agricultural products and reducing our dependence on energy-expensive electrical supply lines. A variety of movable shutters, trellises with window boxes on adjustable bases, and hanging gardens are being designed.

By the mid-twenty-first century Toronto might well be a rich urban ecosystem, efficiently recycling energy and matter in an aesthetically attractive environment. Hansell and his colleagues envision an evolving architecture that incorporates living green surfaces as attractive as the classical hanging gardens of Babylon.

creating green spaces to absorb carbon dioxide, and supporting community projects related to these goals. At the end of 1998, the fund equity was more than $25 million. The Ten Thousand Trees project, which is involved in planting approximately 10,000 seedlings as part of the reforestation of the Rouge River valley, began in 1990, and now involves 20 organizers and 15,000 volunteers. In 1993, Pollution Probe began the Clean Air Campaign and Commute project, which culminates annually in a special activity day that emphasizes alternate modes of transportation and automobile testing clinics. Since 1996 the rideTOgether program sponsored by the Toronto Environmental Alliance has promoted car pooling and reducing car use in Metro Toronto.

What, then, is the weather outlook for the twenty-first century? In all objectivity, we must admit that it is rather uncertain, especially in terms of timing, but the odds favour a Toronto climate similar to that of Tennessee. Several advantages would accrue from the milder climate. The reduced number of snowstorms would lead to lower costs for snow removal and fewer motor vehicle accidents. Moreover, less snow to be shovelled would result in fewer heart attacks. Another health benefit would be a reduced incidence of broken bones from falls on icy sidewalks. Milder winter weather would lower heating costs and reduce the incidence of respiratory illnesses. The tourist season would begin earlier in the spring and last later in the autumn, although it should be noted that sparse snowfalls would not be appreciated by those in the winter recreation sector. Finally, warmer lake temperatures would allow more swimming and water recreation in the summer.

But there are disadvantages as well. More frequent and more intense summer heat waves would cause an increase in heat-related mortality. Although there would be savings on heating costs in the winter, these would be counterbalanced by greater consumption of energy in summer, through greater use of cooling systems in buildings and motor vehicles. We can expect more frequent and more intense oxidant-type pollution episodes, not only in summer but also in spring and autumn, causing an increase in the incidence of asthma attacks and other respiratory ailments. Lower water levels in Lake Ontario would require shoreline restructuring and additional off-shore dredging. Summer droughts would be more frequent and intense, adversely affecting trees, grass, and gardens. Warmer river water would lead to a decline in successful fish spawning. Warmer water in the lake in the summer of 1998 seemed as well to have been responsible for an algae bloom that affected the water supply. The warmer weather could encourage the northward spread into Ontario of insect pests and vector-borne pathogens such as viral encephalitis. Finally, it is possible that the frequencies of harmful rare events such as tornadoes would increase.

Strategies

Mitigation policies have the goal of reducing the emissions of greenhouse and oxidant-forming gases. Toronto has an active program to reduce its emissions of greenhouse gases, but worldwide action is required if atmospheric concentrations of the gases are to be stabilized. Mitigation policies can be more effective in reducing the oxidant-forming pollutants emitted by automobiles. Here again, however, cooperation is required – in this case with authorities in the United States, as the Toronto region is affected by such emission from as far away as the Ohio River valley.

Adaptation policies have the goal of encouraging people to adjust to a warmer climate with more frequent oxidant episodes. Examples of adaptation strategies include:

- planting species in parks and gardens that are more tolerant of heat, drought, and pests
- providing more shade through more sensitive building design, tree planting, construction of overhangs, and similar measures
- cooling of buildings (in summer) through the use of water from Lake Ontario and of vertical gardens
- emphasizing water conservation through more realistic pricing policies
- improving the efficiency and cost of riding the TTC and GO Transit, and discouraging the use of automobiles
- providing research funds for the development of innovative tools that will help people adapt to climate warming.

A first step toward adaptation would be to develop a system that would provide early warning of climate change. However, climate is so variable that the evidence for climate change may not be statistically significant until it is already too late to take effective action. So perhaps we must wait for effective action until the next generation takes charge. It should be mentioned that children born in the 1990s will be in their fifties by mid-century. Let us hope that by then the point-of-no-return with respect to the climate-warming issue will not yet have passed.

Watch the Stars: A Century of Stargazing from Toronto

Terence Dickinson

Two views of the night sky, using identical exposures with the same camera. The one on the left was taken in Etobicoke looking towards downtown Toronto at night, showing the brightness of the sky from city lights. The one on the right was taken from the author's home north of Kingston, Ontario, where there are few lights. This is what the night sky looked like over Toronto a century ago. (Courtesy Terence Dickinson)

On the evening of 24 September 1892, in a modest lecture hall a short walk from the newly completed provincial parliament buildings at Queen's Park, amateur astronomer George F. Lumsden presented a public lecture entitled 'The Constellations.' After the talk, he escorted his audience outdoors, where he pointed out the stellar configurations above the Toronto skyline. Lumsden later noted that the cool, early autumn evening was perfect for stargazing, and 'the stars shone forth in all their glory.'

By day, Lumsden was a high-level provincial bureaucrat (equivalent to today's deputy minister), though his first love was the stars. During his evening strolls, he was often seen at the centre of a small cluster of people, arm raised to the sky, sharing his interest in astronomy by tracing out the constellations and recounting sky lore.

Fast forward to today, a little more than a century later. The same stars are up there, twinkling in exactly the same constellations that Lumsden pointed out. But unlike the 1890s scene on the Toronto street corner, hardly any of them are visible from anywhere within 25 kms of Queen's Park, never mind from just down the street. Much has changed since Lumsden's day, of course. In the intervening century, Toronto has increased in population by a factor of 20, and the accompanying lights and air pollution have taken their toll. The starry sky used to be the first thing people noticed as they stepped out the door, but now the sky shines with a space-age yellowish brown hue – the air illuminated by the

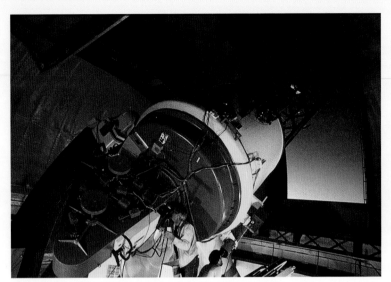

When it was commissioned in 1935, the 1.8 m telescope of the David Dunlap Observatory near Richmond Hill was the second largest in the world. (Courtesy Terence Dickinson)

combined glow from hundreds of thousands of street lamps, parking lot security lights, porch lamps, and so on.

Although the natural beauty of the night has been beaten back by urban lighting, this loss is just part of a larger trend in modern society, where nature in general has been removed, bit by bit, from our daily lives. There is no turning back the clock – not many of us would willingly give up the conveniences that a century of progress has wrought – but there is also no denying that few sights in nature can match the majesty of a star-filled night sky adorned by the silky ribbon of the Milky Way. It still can be enjoyed, though one must travel several hours away from the city's glow to see it.

According to the late Helen Sawyer Hogg, a distinguished University of Toronto astronomer who arrived in the city in 1935 to work at the university's newly completed David Dunlap Observatory in Richmond Hill, the local stargazing conditions were still nearly pristine in the 1930s. She recalled seeing the Milky Way overhead as she stood waiting for a street car outside Eaton's downtown department store. Conditions at the Dunlap Observatory were superb in those days. Toronto was an insignificant glow on the southern horizon. Dr. Hogg and other astronomers were able to conduct world-class research using the giant 1.8 m telescope, then the second largest telescope on the planet.

Once serenely isolated, the Dunlap Observatory is now surrounded by the bustling city of Vaughn. Although a limited program of research continues there, University of Toronto astronomers now travel to remote sites, such as the Canada-France-Hawaii Telescope atop a 4-km-high mountain in Hawaii, to conduct cutting-edge research.

Along with the profound change in night sky conditions in and around Toronto, the past century has delivered an even greater change in our perception of the cosmos. In Lumsden's day, fundamental questions like what caused the sun to shine, or what lay beyond the Milky Way were complete mysteries. The extent and composition of the universe beyond the few million stars visible through a good telescope were simply unknown.

Modern astronomers have gathered compelling evidence that we inhabit one of nine planets that orbit a rather ordinary star, the sun, which is located near the edge of an average-sized galaxy we call the Milky Way. The Milky Way is an immense island of 100 billion stars floating in the abyss of space. Beyond it are billions of other, similar galaxies. The universe does not go on forever though. In the 1920s astronomer Edwin Hubble discovered that the universe is expanding – the galaxies are moving away from each other. Measurements of the velocity of this expansion indicate that it began about 14 billion years ago in what astronomers rather prosaically call the Big Bang.

Time exposure showing the Milky Way, as seen from rural Ontario. Similar detail was visible from Toronto in the 1930s. (Courtesy Terence Dickinson)

Thus, the universe has a finite size and age. Current observations suggest that the universe will continue to expand and eventually the stars will burn out and all will go dark. But that dark end lies hundreds of billions of years in the future.

As a full-time writer of astronomy books for adults and children, I have tried to be a keen observer of people's interest in things cosmic. My first exposure to public fascination in astronomy came in 1960 when I was a teenager and an avid amateur astronomer. I had joined the Royal Astronomical Society of Canada and decided to take my beginner's telescope to a public stargazing event where members of the society set up their telescopes for public viewing in city parks. The venue was High Park. I arrived at dusk to find hundreds of people lined up at each telescope, waiting for a look at Jupiter's cloud belts, Saturn's rings, or the craters of the moon. As I set up my little telescope, a line immediately formed behind me. People took their turn at the eyepiece, asked questions, and were generally enthused by what they saw.

The scene that evening has been repeated at least a hundred times since, as members of the society have continued the public viewing sessions in Toronto parks to this day. When Halley's Comet made its appearance in 1986, traffic jams and crowd control problems accompanied the viewing sessions as thousands of people were eager for a look despite the cold mid-winter temperatures. Similarly, Saturday night public tours of the David Dunlap Observatory, which include a peek through the big telescope if the sky is clear, are often booked solid weeks in advance.

Public enthusiasm for astronomy in Toronto is not limited to telescope viewing. When seats went on sale for a lecture by cosmologist Stephen Hawking at the University of Toronto's Convocation Hall in April 1998, they sold out in the same day. Perhaps a more concrete measure of interest in things astronomical is sales of books, telescopes, and reference material on the subject, all of which have at least quadrupled from what they were a generation ago. Similarly, introductory astronomy courses at York University and the University of Toronto are often full and have to turn away students.

Which brings me back to the business of light pollution and the huge transition from the star-filled Toronto skies of our great-grandparents to the situation today. Actually, almost all the change has taken place since 1960, when the first high-illumination mercury vapour lamps began to be installed on Toronto streets. By the early 1970s the transition was complete. Since then, even more-intense lighting fixtures have been replacing the earlier versions. Moreover, Toronto has been one of the fastest-growing cities in North America in the last half of the twentieth century.

Yet despite the almost complete erosion of our view of the natural beauty of the stars from urban areas, interest in astronomy is paradoxically higher than ever. How to account for it? For one thing, people travel more than they used to, often to vacation spots such as cottage country or a Caribbean resort where the stars shine in stark contrast to the light-washed view at home. A starry night has become something to be cherished – something exotic and mysterious.

Beyond this, though, the bold ideas of modern astronomy – the Big Bang, black holes, the immensity of time and space – likely have an intrinsic appeal in an increasingly secular society perhaps seeking to understand cosmic themes that were once the preserve of traditional religions. The mainstream news media have a role here too. Hardly a week goes by without a report of a new discovery by the Hubble Space Telescope, a distant space probe, or researchers at an observatory here on Earth.

Finally, there is the ultimate question: Are we alone in the universe? The best estimates suggest that there are 10 billion trillion stars in the universe. Many of those stars likely have planets orbiting them. Even if only a tiny fraction of those planets resemble Earth, there could still be billions of opportunities for life as we know it to arise. Then again, we could be alone. We may never know, but try asking that question next time you are standing under a starry night sky and see if a slight chill doesn't run up and down your spine.

Comet Hale-Bopp, seen by an estimated one billion people worldwide in the spring of 1997, was the most spectacular comet visible from Canada since Halley's Comet in 1910. (Courtesy Terence Dickinson)

3

Watersheds

Henry A. Regier, D. Dudley Williams, and Gordon A. Wichert

The current greater Toronto region landscape is young. Chapter 1 tells us that the glaciers melted from this area only about 10,000 years ago. As the ice retreated, it left large deposits of sand and gravel (such as the Oak Ridges Moraine) filled with water in underground reservoirs. Water replenished by rain and snowmelt now leaks from these aquifers all year round, creating the important baseflow of the region's rivers and streams. We refer to the surrounding area from which water drains into a particular river (more than 5 m wide) or stream (less than 5 m in width) as its 'watershed,' although 'catchment' and 'drainage basin' are alternative terms (Gordon, McMahon, and Findlayson 1992).

Watersheds have been associated with the greater Toronto region for millions of years, possibly dating back to when it was a broad sandy beach on the shore of the Sauk Sea in the Late Cambrian (515 million years ago). During the Ordovician (approximately 450 million years ago), the region was a delta running into the tropical, inland Tippecanoe Sea, and, during the Pleistocene, water drained along the huge Laurentian Channel that was part of the ancestral St. Lawrence River system. About 12,000 years ago, ancient Lake Iroquois formed when an ice dam across the St. Lawrence River impounded the waters at a higher level than present Lake Ontario. Lake Iroquois left a shoreline slope, ravines, and bluffs that have been eroding ever since. The region's running waters are responsible for much of this erosion, as streams and rivers are never in equilibrium within their valleys – they are always removing material and transporting and depositing it somewhere downstream. Years later those deposits may again be eroded to be deposited still farther downstream.

Hydrodynamic forces on a rotating planet mean a stream will flow in a straight line only if some strong physical structures constrain it. If a stream is constrained by linear hard banks, whether natural or made by humans, it develops enlarged vertical sinuosities, which may eventually undercut the banks. Constrained by a flat, hardened bottom, a stream will enlarge the horizontal sinuosities to reach laterally beyond the hardened bottom and erode riparian materials. Constraining both the sides and bottom, without creating special energy-dissipating structures, will cause the stream to build up momentum and kinetic energy that manifests itself as intensified erosive power and turbulence downstream of the hardened sections. Sooner or later the self-organizing turbulence of a flowing stream breaks down or circumvents any constraining straight-line structure.

A river's energy spectrum creates a system not unlike a tree – that is, like an oak tree, for example, not a conifer or a palm tree. The branches of the tree represent the networks of small streams that drain the aquifers, typically via springs and seeps. These streams coalesce into larger and larger streams, into rivers, and, eventually, into one major river – the Credit, Humber, or Don – which represents the trunk of the tree (Figure 3.1). As this river reaches its receiving lake, it again branches out – analogous to the tree's roots – as its energy dissipates through wetlands and deltas.

The water in the aquatic system is simply part of a huge circulating cycle – the hydrological cycle, with water moving through air, land, and sea (Figure 3.2). Water, derived from the oceans, evaporates and later condenses to form rain. Where rain falls on land it can: (1) be intercepted by vegetation and evaporate back into the atmosphere; (2) be blocked by an impervious substratum (e.g., rock or clay) at, or close to, the land surface and thence flow downhill to the nearest runoff channel;

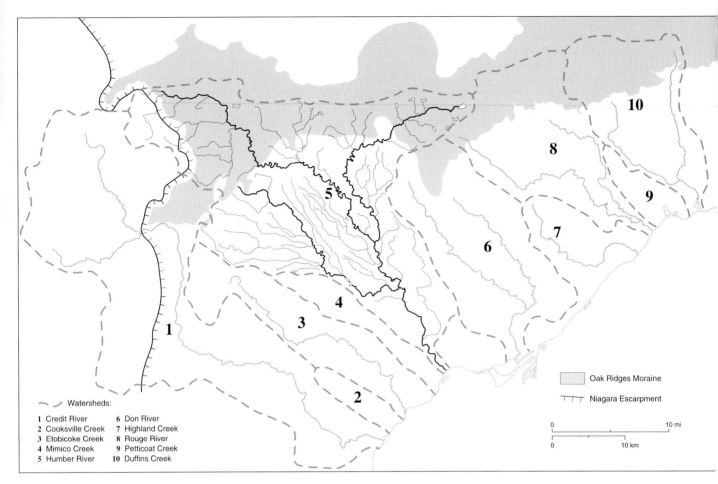

FIGURE 3.1 Watersheds of the greater Toronto region. Tributaries are shown only for the Humber.

(3) penetrate the soil to root depth and thence be transpired through plants back into the atmosphere; (4) percolate through the soil to the groundwater table, from where it contributes to the source of headwater streams (at springs) and the downstream baseflow of rivers. The amount of rain that falls is a feature of a region's climate, whereas partition into the various flow routes depends largely on the details of the local physiography, especially landform and geology. These two factors, then, determine the hydrological nature of the landscape.

The climate of the greater Toronto region shows annual cyclical wet and dry periods. In addition, non-cyclical fluctuations that are wetter than average, drier than average, and of normal wetness all occur within an interval of about 20 years.

Coastal wetland boundaries are defined largely by these periods of extreme wet and dry weather. When an intense onshore wind coupled with a rainstorm occurs in a season of high water level within a period of wet years, the coastal zone as a whole becomes quite water-saturated and both the lakeward and landward boundaries of the wetland move upslope and toward the land. It becomes relatively dry in extended periods of calm weather during a dry season within a period of dry years when boundaries move lakeward. Such high and low episodes effectively determine the land-side and water-side borders of the coastal wetlands over the long term. The size of a coastal wetland ecosystem thus depends directly on the amplitude of water fluctuations in the lake. The biological community of such an ecosystem is also dependent on these kinds of water-level fluctuations or natural disturbances. Damming the St. Lawrence River in the 1950s, with concomitant reduction in lake-level fluctuations, affected all the coastal wetlands of Lake Ontario.

The predominant feature of stream systems results from the dynamic interactions that occur between flowing water, regional geology, physiography, hydrology, and climate (Imhof et al. 1990). Streams and rivers are always changing, yet running-water ecosystems persist. Organisms that live within such a system have evolved in response to a *natural* disturbance regime, exhibiting characteristics and adaptations that suit such dynamic habitats.

OLD-GROWTH FEATURES OF PRISTINE STREAMS, WETLANDS, AND VALLEYS

Ecological processes that proceed quite predictably over thousands of years in settings benign to living things culminate in landscapes that can be termed 'old-growth.' Originally a term used to describe forests, old-growth meant a system with a predominance of trees that were large for the species and the ecological condition of the site, and with unevenly aged trees and a mixed cover varying from patches with multiple canopy layers through to gaps in the canopy that admit sunlight.

An ecosystem that is protected from human interference will, over time, acquire old-growth characteristics and especially numerous relatively large and old organisms that represent many species of trees and shrubs, mammals, birds, reptiles, amphibians, fishes, crustaceans, molluscs, and insects. Different species, especially those that are ecologically dominant, are linked behaviourally to their preferred spatial habitats. High biodiversity and high biomass are the norm. In the old-growth landscape of the Toronto region few of these dominant species posed a threat to humans (that is, there were few predatory, venomous, or toxic forms), and most of them were highly desirable relatives of species that the Northern European settlers had valued in their homelands.

Under old-growth conditions, the shrubs and trees near the streams and wetlands in the valleys of the greater Toronto region were of species that could tolerate 'getting their feet wet,' seasonally or permanently. Willows, alders, white cedars, red maples, and American elms combined to provide a canopy that shaded most of the water surface of small streams. Because the sunlight reached the stream only spottily, there was little photosynthesis by algae or higher aquatic plants in the streams. Most of the organic food for the stream organisms fell into the stream as leaves, branches, and tree trunks, or 'woody debris' generally (Kaushik and Hynes 1971). Some of the aquatic animals also caught terrestrial animals (e.g., trout feeding on bankside insects).

Under fully natural conditions in the Toronto region, baseflow in the streams and rivers was maintained mostly by subsurface flow, through porous sand and gravel and fibrous organic matter. Surface flows occurred with

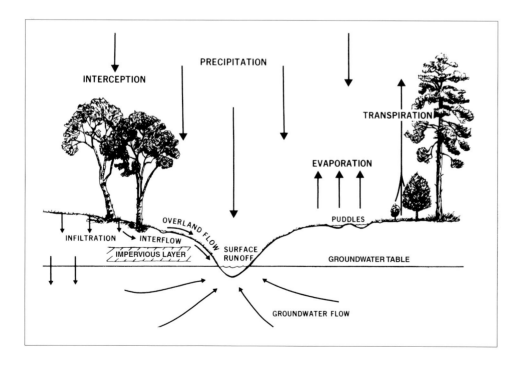

FIGURE 3.2
Local details of the global hydrological cycle.

warm spring rains melting snow over frozen ground, and floods generally occurred in the spring. Melting snow and spring rain produced shallow, temporary pools in areas where the groundwater table was high. These pools supported unique communities of aquatic invertebrate species highly adapted, with fast life cycles and dormant eggs, to living in habitats that dried up by early summer (Wiggins, Mackay, and Smith 1980). Drainage of land for urban and agricultural purposes, however, has now reduced the number of such pools in the region.

In the past, the stream course itself tended to be littered and sometimes partially dammed by woody debris. This debris not only created 'stream habitat structure' but was also a primary food source for various kinds of decomposer organisms that broke the debris structure down gradually over the years. Other organisms then fed on the decomposers, leading to complex food webs linking bacteria and algae to aquatic insect larvae and fishes (Tavares-Cromar and Williams 1996) in an ongoing cycle.

In cross-section, the old-growth stream course was relatively narrow and deep. It was narrow because the roots of shrubs and trees partially stabilized the banks, which were often undercut by currents. Frequent tree-falls and root wads presented obstacles that these streams had to flow around. It was deep because of deflection of the current downward by woody debris. In response to the relatively low flow, these streams had short meanders. High flows occurred in spring, but the frozen ground prevented carving of larger meanders. Potentially erosive power from high flows was reduced exponentially when the rivers overflowed their banks. Because of the large reservoirs of water in the aquifers and the slow rate of sub-surface flow, the streams were fed by water year-round, at a fairly steady rate. The spring floods were superimposed on these regular sub-surface flows.

Groundwaters below more than a few metres of soil remain at an approximately constant temperature year-round. Flows from springs are normally some 1° to 2°C warmer than the region's average annual air temperature. Thus, springs in Toronto region streams are about 10°C year-round, and the small locales where strong springs enter small streams remain at about that temperature throughout the year. Away from such spring locales, the surface water reaches freezing temperature in winter and about 18° to 20°C in summer. In larger streams of old-growth forests, as in the Toronto region, water temperatures tend to be stratified on a daily basis, with temperature varying least near the bottom.

In the middle to lower reaches of the larger streams, on the clay plains downstream from the moraine, the influence of groundwater is smaller. These streams were sufficiently wide and shallow that they were not as fully shaded by streamside vegetation. The temperature of these streams approached air temperature during warm periods. Sunlight striking the water and nutrients transported from upstream supported photosynthesis and the growth of aquatic algae.

Some surface flows entered streams in the middle to lower reaches following heavy spring rains, causing water levels to fluctuate more in middle to lower reaches than in the upper ones. In flat areas, the stream formed natural dikes along the banks, behind which floodwaters were trapped. Wetlands in these parts of the watersheds helped moderate flood flows through attenuation and slow discharge of groundwater. Wetlands also retained and cycled nutrients, before releasing them downstream or into sediments, thus becoming habitats highly diverse in plant and animal life.

In the past, both brook trout and Atlantic salmon were abundant in old-growth streams of the Toronto region.

With respect to acidity, the waters were almost neutral. The calcium carbonate in the stones from the bedrock neutralized any acids produced when plants decompose. The water was generally clear and free from suspended sediment. Exceptions to high water clarity occurred where strong plant decomposition released brown tannins to stain the water.

Streams such as those described above are close to being ideal ecological habitat for species of the salmonid family that spend their whole life in a stream, such as brook trout, or some migratory salmonids that use the stream for spawning and nursery habitat, such as Atlantic salmon. In the past, both of these species were abundant in old-growth streams of the Toronto region.

ECOLOGICAL DISTURBANCES WITHIN OLD-GROWTH ECOSYSTEMS

Old-growth ecosystems in the Toronto region were adapted to the natural disturbance regime in which they had evolved. For example, the occurrence of natural forest fires was expected ecologically, but precisely where and when a fire would break out was not predictable. Such a fire most likely did not create major widespread degradation. The valleys were too wet to burn well, and the ash from fires up-slope generally did not wash into streams to fertilize them since surface runoff was rare during the dry fire season.

Beaver constituted another major disturbance. Beaver cut down streamside trees and shrubs for food and structural material for dams. Thus they permitted short-lived, opportunistic plant species to thrive in the forest openings, and aquatic plants to grow in the well-lit beaver ponds. Beaver dams and ponds tended to be temporary, in part because the beavers eventually ran out of local food and departed. Sometimes spring spates washed out dams, contributing to woody debris tangles downstream. The beaver ponds also provided specialized local aquatic habitats (Sprules 1941) for both invertebrates and fishes (they were local habitats for members of the perch and sunfish families; see Chapter 12). The ponds also provided habitat for old-growth amphibians, turtles, waterfowl, and muskrats. These aquatic species were preyed on by old-growth terrestrial reptiles like snakes, birds such as heron and eagle, and mammals like weasel, mink, fox, and wolf. Drained beaver ponds provided wet meadows for rabbits and deer, which were preyed upon by wolves. Poplars and other hardwoods then invaded these meadows to create food and building material for beaver that returned to rebuild the dam and thus restarted the cycle.

Where streams strongly eroded the substrate, as in the ravines of the Lake Iroquois shoreline, the gravel was gradually moved downstream and created sequences of rapids and pools. These were preferred spawning locales for fishes such as walleye, white sucker, and lake sturgeon that thrived in the nearby bays of Lake Ontario. Turtles used gravel bars for egg laying.

Lower in the stream courses, where spring floods inundated coastal flats, northern pike and muskellunge spawned on the submerged dead vegetation; the rapidly developing young fishes escaped to larger, more permanently flowing waters as the flood waters receded gradually. Many amphibians as well as fishes served as food for reptiles, birds, and mammals.

As sketched above, the combination of annual and interannual fluctuations in lake levels, especially as intensified by flooding related to spring storm events, ensured that the coastal zone was permanently disturbed, from an old-growth perspective. Here a disturbance-dependent wetland ecosystem thrived, with many desirable opportunistic species to complement those of the old-growth parts of upstream valleys, thus swelling the region's biodiversity.

Along the more protected parts of lake bays, near rocky banks, and in beds of aquatic plants, the young and adults of sunfish, black bass, perch, walleye, pike, and muskellunge thrived. Mergansers, loons, ospreys, and eagles preyed upon these organisms. Reptiles and herons preyed on amphibians and small fish species in turn.

Offshore in the lake, lake whitefish, lake trout, adult Atlantic salmon, ciscoes, and burbot were abundant. These all preferred cold, deep water during summer. Occasionally, when strong northwesterly winds blew the warm surface waters southward across the lake in summer, the bottom cold water would reach the surface near the Toronto region shore. On such occasions the offshore species could be found in the upwelled cold water near the lee shore at the surface. At these times some of the nearshore warm water species such as spottail shiners would die, trapped by the cold water.

MAJOR EFFECTS OF HUMAN ENCROACHMENT

Prior to settlement by Europeans, the old-growth ecosystems of the Toronto region had accumulated and stored much valuable 'biological' material in several formats – as old individuals of native species, soil organic (nutrient) matter, surface biomass, and the gene pools of plants and animals. Such materials were important in maintaining the identity and integrity of the ecosystems in the face of natural disturbance regimes and the low-intensity use by Native peoples.

With permanent European settlement late in the eighteenth century, harvesting of natural resources became selective, with, at any given time, the removal of the most preferred individual organisms of the most preferred species that were readily available. These were used for domestic purposes or marketed commercially, and extraction proceeded more rapidly than such high quality individuals could be regenerated in the ecosystem. Consequently, the next round of exploitation took resources of lower quality, but again, the best individuals available at that time. Such selective sequential resource 'mining,' with each stage involving high-grading of whatever was still available, eventually led to ecological degradation and to economic losses. Eventually, there came a time when nothing of any value was left to be exploited.

Settlers also cleared and burned away most trees and understorey plants of the old-growth forest. Forest soils were gradually changed to agricultural soils, which increasingly shed some water across the surface and carried silt into streams. With time, some upland wetlands were drained and fields were under-tiled, allowing them to dry earlier in the seasons and thus improving growing conditions for agricultural crops. In some areas, streams were straightened to create fields of regular shapes, and new agricultural surface channels were dug to increase drainage efficiency. These 'improvements' exacerbated spring floods and delivered additional silt, nutrients, and agricultural chemicals to receiving streams. When natural water courses are replaced with drainage ditches, not only does the aquatic and riparian species composition of the area change, but community structure and function are altered; in particular, predators are lost (Williams and Hogg 1998). This process was exacerbated further when agricultural soils were altered to urban soils through erection of settlements, roads, and cities.

In natural watersheds, as we have noted, most of the water flowing in river and stream channels is derived from baseflow; overland flow is relatively rare, as the infiltration capacity of most soils is sufficient to absorb most rainfall. As such, the chemical nature of the water reflects that of the soil through which it has percolated. Further, passage through the soil tempers the effects of major rains by dampening increased discharge in the channel. In heavily urbanized regions, such as modern Toronto, however, where much of the soil surface has been covered with asphalt, concrete, and buildings, 'parking-lot hydrographs' result (Hynes 1975) (Figure 3.3). Consequently, when it rains, most of the water leaves as a short, violent surge of overland flow to the nearest channel, often carrying with it oils and other urban surface-accumulating chemicals as well as large objects. With less water percolating into the soil, the subsurface water table in many localities has dropped, and the amount of groundwater discharge has decreased. Thus, small wetlands and first-order streams often now dry out earlier in the year with, again, loss of aquatic habitat and biodiversity in the region.

Some of the larger streams in the region were cleared for transport of small craft. Medium-sized streams were dammed to produce waterpower for mills of various kinds. Settlement, especially forest-clearing, caused spates in streams to become more violent, leading to more frequent and intense flooding. Natural dikes were destroyed to speed the run-off from bordering wetlands

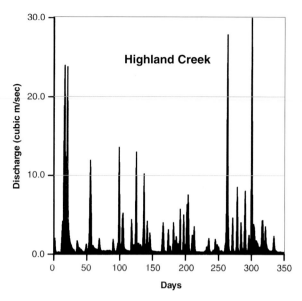

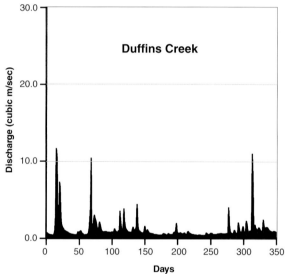

FIGURE 3.3 Runoff regimes of two watersheds subjected to similar weather events for the period 1 January (day 1) to 31 December (day 365) in 1995. Highland Creek, which is urbanized to a greater extent than Duffins Creek, illustrates a 'parking-lot hydrograph' of frequent short violent spates with periods of very low flow between them.

when they were flooded. Again, with less water percolating into the soil, the water table dropped, and the amount of groundwater discharge decreased. Thus, small wetlands and first-order streams dried out in summer.

With 'modern progress,' humans produced increasingly large amounts of wastes of more kinds and of more artificial and unnatural qualities. These were vented in the cheapest way – into the water, air, and soil – in the nearest convenient locale. Streams in particular became severely polluted in many reaches by human wastes, sawdust, woollen mill washings, farmyard run-off, and wastes from dairy production. Some of this material entered the groundwater; a recent survey of 1,300 farm wells in southern Ontario showed that approximately 40 percent were contaminated with farm chemicals and/or faecal coliform bacteria (Rudolph, Goss, and Rudy 1992).

During the urbanization of Toronto, soil from basement excavations was dumped into wetlands, thus obliterating them and intensifying cross-surface flow rates. Along the Toronto waterfront, excavated fill from the foundations of the large, downtown office towers was used to extend the shoreline many metres into the lake. Thus the Ashbridge's Bay coastal wetland complex of several hundred hectares was filled in almost completely. On the other hand, fill was used to create the Leslie Street Spit and Tommy Thompson Park, as well as four smaller landfill projections along the Toronto region coast: Colonel Samuel Smith Park, Ashbridge's Bay Park, Scarborough Heights Park, and Bluffers Park. These shoreline projections now provide new habitat for colonial water birds, small mammals, reptiles, amphibians, and for near-shore fish species. Hence, if planned correctly, some development practices can actually benefit fishes and wildlife.

Old-growth species have had time to come into evolutionary equilibrium with the natural disturbance regime of their habitat. They are specialist species that maintain an old-growth ecosystem's identity and integrity. Such species are *not* well adapted to the extreme disturbances created by humans – extractive harvesting, opportunistic clearing, haphazard construction, indiscriminate pollution, and wetland land-filling. On the other hand, rapidly reproducing small-bodied, generalist species *are* adapted to extreme disturbances. Hence, the overall effect of development was a transformation to a poorly organized, ragged ecosystem dominated by short-lived, low value, and often exotic species more tolerant of human abuse.

Stream banks, ravines, and wetlands were more difficult to exploit for forest products and urban development, so some semblance of the old-growth plant species persisted on the land part of this network. But stream clearing, dam building, water pollution, and wetland draining and in-filling greatly altered the water parts, except in some relatively inaccessible upper tributaries on the moraines, at the base of the escarpment, and in some large coastal wetlands. One of the consequences of the increased flooding, especially at times when the ground was not frozen in spring, was that the streambed changed radically. The meanders became larger laterally and longitudinally, thus leading to a re-carving of the watercourse. The streambed itself became wider and shallower. Undercut banks disappeared and the banks became eroding slopes. Massive amounts of silt and gravel were washed downstream into the coastal marshes at the river mouths. For example, the lower section of Highland Creek, which now has a mean annual depth of less than 0.5 m, was capable of navigation by 97 t schooners prior to 1840 (Williams and Williams 1997).

The stream flow rates fell to ever lower levels during dry summer periods. Eventually the flows were insufficient, and barrier bars, formed by along-shore currents in the lake, closed off the coastal marshes from the lake. Atlantic salmon, returning to spawn in the streams in autumn, were thus prevented from entering them.

Among fish species, Atlantic salmon and brook trout have been particularly sensitive to human interventions in the valley ecosystems. Atlantic salmon runs to greater Toronto region streams were already dwindling by the mid-nineteenth century because of overfishing and stream degradation. Atlantic salmon did not return to spawn in the Don River after 1853 (Huntsman 1944) and had disappeared from all Lake Ontario tributaries by the end of the nineteenth century. Brook trout also have disappeared from the more developed, downstream parts of all the watersheds. Indeed the brook trout has come to be accepted as an integrative indicator of the combined levels of all adverse human effects on the stream parts of old-growth ecosystems in the Toronto region (Hallam 1959, Wichert 1994).

> *A recent survey of 1,300 farm wells in southern Ontario showed that approximately 40 percent were contaminated with farm chemicals and/or bacteria.*

REACHING A LIMIT AND THE BEGINNING OF MORE SERIOUS REFORM

Starting in the mid-nineteenth century some informed leaders argued for and tried to initiate less harmful practices with respect to many human activities in the Toronto region. These conservation efforts gradually intensified over the decades as outbreaks of severe human disease were attributed to polluted water and as the harm done to fisheries from pollution and overfishing was recognized. The actions then undertaken – extending sewers downstream and into the lake, chlorinating drinking water, building fish hatcheries to mitigate loss of spawning habitat, and so on – did not correct the problems. They only alleviated some of the symptoms of degradation in a temporary fashion.

The corrective actions were all too little and far too late. New abuses were added to the only partially abated earlier problems. By about 1950 much of the Toronto region was an ecological slum. At the heart of this was an abuse-tolerant ecosystem that had self-organized into a network of highly degraded locales and that was actively expanding at the expense of such vestiges of the old-growth landscape that had managed to survive.

The reform process that started in Toronto in the 1940s implicitly recognized that the association of aquatic species integrated the state of the whole complex of ecosystems in the region. Conservation authorities were created with a major mandate focused on aquatic ecosystems. Another focus was on the sandy moraine areas that had been deforested and used unsuccessfully for farming before they were abandoned; reforestation of these aquifer-recharging areas was undertaken.

One of the first activities of the government of Metropolitan Toronto, founded in 1953, was to improve the collection and treatment of human sewage in the Metro area. Prior to 1953 more than 20 local sewage-treatment plants collected sewage and discharged treated effluent directly into the streams. In general the effectiveness of

The History of the Don River Watershed

Archaeological evidence suggests that the Don River watershed has been continuously inhabited since the glaciers retreated from the region about 10,000 years ago. After occupation by Palaeo-, Archaic, and early Woodland Indians for thousands of years, the region was inhabited by Iroquoian peoples, who became the first settled agriculturalists in the area. French fur traders were traversing the region by the second decade of the seventeenth century, although it was not until 1688 that the Don River first appeared (unnamed) on a map of the Great Lakes.

In 1787, the British made a provisional purchase of the region (including the watershed) from the Mississauga Indians. Within four years the region had become part of the newly created province of Upper Canada. As a precaution against American invasion, Governor John Graves Simcoe in 1793 moved the capital of the province from Niagara to a new settlement, christened York, that was arranged in a 10-block pattern on Taddle Creek, just west of the mouth of the Don River. York grew quickly: in 1834 it had a population of over 9,000 and was incorporated as Toronto.

The burgeoning settlement strongly altered the original ecosystem of the Don watershed. By 1852 nearly 40 flour, textile, and lumber mills had been established in the watershed, and agriculture was extensive. The last spawning run of Atlantic salmon in the Don was in 1853. At the end of the century, with industrialization and urbanization progressing apace, the lower stretch of the river was channelized and widened for shipping, and most of the land adjoining the lower Don was urbanized.

The growing population and industry brought with it increased pollution. Between 1912 and 1918 the largest lake-filling project in North America was undertaken to alleviate pollution of Ashbridge's Bay by creating new industrial land – indeed, all construction south of Front Street rests on former lake bed. After the Second World War suburbia rapidly began to stretch up the Don valley. By the mid-1960s, new transportation routes were deemed necessary to serve the suburban population, and the Don Valley Parkway was constructed, destroying much of the natural ecosystem remaining in the valley. As the city of Toronto continues to grow – to a population of 2.4 million in the late 1990s – over 80 per cent of the Don watershed has been developed. This figure may rise to more than 90 per cent in the next few years.

Although the watershed has been partly cleaned up – there are no longer paper mills and paint factories, and 30 of the 31 sewage treatment plants have been removed – it is still subject to two major pollution sources: overflow from combined sewers in the older urban areas, and stormwater runoff. The water quality currently does not meet acceptable standards in terms of total phosphorus, faecal coliform bacteria, and metals such as lead and copper. Despite the environmental damage due to overdevelopment and pollution, the watershed does support some wildlife: 176 bird and 25 amphibian species have been recorded, and 18 fish species are known.

sewage treatment was low. Some plants operated at high efficiency, removing up to 95 percent of the harmful substances, but treated only about 50 percent of the waste generated in the sewage-catchment area. Other plants treated all of the waste generated in the catchment area but removed only about 50 percent of the harmful substances. In 1953 Metro Toronto began to systematically decommission the small, local sewage-treatment plants, and build a trunk system to carry much of the sewage waste for treatment at large regional treatment facilities with improved treatment technology along the Lake Ontario shoreline. This partial strategy, implemented gradually over more than 40 years, resulted in some improvement of the water quality of urban streams and the lakeshore.

Meanwhile conservation authorities, created opportunistically by the province following passage of enabling legislation in 1947, provided opportunity for correcting or mitigating some of the most severe hydrological problems. Authorities also designated many parks along streams for both recreational purposes and conservation reasons. The Metropolitan Toronto and Region Conservation Authority and the Credit Valley Conservation Authority were created in 1957 to address many of the above concerns and have acted to improve the quality of the valley and wetland ecosystems.

Since 1950 the most degraded reaches of Toronto region streams and also the polluted Toronto Bay waters have been partially rehabilitated. But many areas that were farmed in 1950 have since been developed further into commercial and urban areas and into transportation corridors, with some further loss in the quality of stream and wetland ecosystems in those large areas of the watersheds.

Some actions taken ostensibly as a corrective were ill considered. Building concrete channels for streams that were subject to intensive flooding and erosion owing to urbanization actually intensified the flooding problem downstream and increased the danger to humans there.

Along the Toronto waterfront, excavated fill from the foundations of the large, downtown office towers was used to extend the shoreline many metres into the lake.

With better understanding and management of run-off, some of this concrete can be removed. Except in the moraine areas and some protected natural areas, many first- or second-order streams no longer flow in summer, and many that do flow have been buried in culverts underground. Any submerged, smooth concrete surface such as a channel wall or a bridge buttress provides ideal attachment sites for blackfly larvae. Infestations of blackflies can get worse if the channel is also receiving discharged sewage, which provides increased food for the larvae (Williams and Feltmate 1992).

Starting late in the 1980s, the degraded Toronto Bay and the degrading coastal zone of Lake Ontario were given more attention, with leadership by the Royal Commission on the Future of the Waterfront and the various municipal planning offices led by the Metropolitan Toronto Planning Department. It has come to be generally accepted that desirable ecosystems will self-organize where they are provided opportunities free from the more inane kinds of abuses by humans (Regier 1992). In contrast, undesirable ecosystem slums self-organize where abuses occur. All levels of government now recognize that the whole network of desirable aquatic ecosystems – aquifer recharge lands, rivers, streams, riparian corridors, wetlands, ponds, lakes, coastal zone, lake shore – need protection. However, for a variety of reasons, most high-level government agencies are attempting to place responsibility for the rehabilitation and preservation of ecosystems on lower levels of government and also on individual citizens and volunteer groups. Some of this protection can be achieved through zoning of the whole water-related network of cores and corridors with explicit connections to the land-related network of cores and corridors of relatively natural habitat (Kaufman et al. 1992). More protection can be achieved through better husbandry of human activities everywhere in the watersheds. Every user should be expected to take responsibility for that use and be fully accountable for its ecological consequences.

Part 2
From Wilderness to City

4

Native Settlement to 1847

Conrad E. Heidenreich and Robert W.C. Burgar

The Native occupation of the Toronto area began some 11,000 years ago, when caribou hunters entered a boreal forest environment south of the receding ice sheet. Over the centuries that followed, these resourceful people adapted their ways to changing environmental conditions. In order to survive they formed a close relationship with the natural environment in which they lived, a relationship built on an intimate knowledge of their surroundings that transformed itself into a spiritual bond of respect and reverence. Although relationships to the environment changed over the centuries from caribou hunting with spears to complex agricultural villages, the bond between Native peoples and the land they occupied was a deeply ingrained part of all their cultures.

This chapter is a brief treatment on human relations with the natural environment through settlement and subsistence. It is about adaptation.

PALAEO-INDIANS, c. 9000-7000 BC

It is certain that the first human beings entered what is now Ontario not later than 10,000 BC. At that time, the huge Laurentide ice sheet had melted back to just north of the French River-Lake Nipissing axis. South of the ice sheet, climate and vegetation conditions were similar to modern subarctic Canada. A spruce forest covered the north shore of Lake Admiralty (the precursor of Lake Ontario) and most of southwestern Ontario. Farther north, this forest opened up into a spruce-lichen woodland that graded into a tundra near the ice margins. Lake Admiralty stood at least 90 m below the present lake level, with a shoreline about 23 km east of Hamilton and 15 km south of Toronto. The mammalian fauna was composed of typical subarctic species, such as caribou and musk ox. Mastodon and mammoth were present in early Palaeo-Indian times, but there is no certain evidence that they were hunted.

The archaeological evidence suggests a small population of perhaps 200 people, split into extended families, following migratory caribou herds that ranged from the spruce forest in winter to the lichen-spruce woodland in the summer. From the tools left behind by these people (mainly fluted spear points, scrapers, and chopping and cutting tools), it is fairly certain that they were hunters with spears. Unfortunately, due to the present high level of Lake Ontario, much of the evidence for Palaeo-Indians in the Toronto area lies on the lakebed, up to 90 m under water. What evidence there is for Palaeo-Indians suggests that they migrated through the Toronto area between their winter and summer ranges. On these migrations they would stay at small camps such as the Esox site near Wilcox Lake on the Oak Ridges Moraine and the Westlake site near Bolton (Figure 4.1).

As glaciers continued to melt, the land rebounded, lake levels rose, the climate moderated, and forest conditions changed. By about 8500 BC spruce was more confined to wet, low-lying areas, and red and jack pine became dominant. As the climate continued to warm over the next 1,500 years, a white pine forest became established in southern Ontario, followed by deciduous trees such as oak, elm,

Mastodon and mammoth were present in the Toronto area in early Palaeo-Indian times.

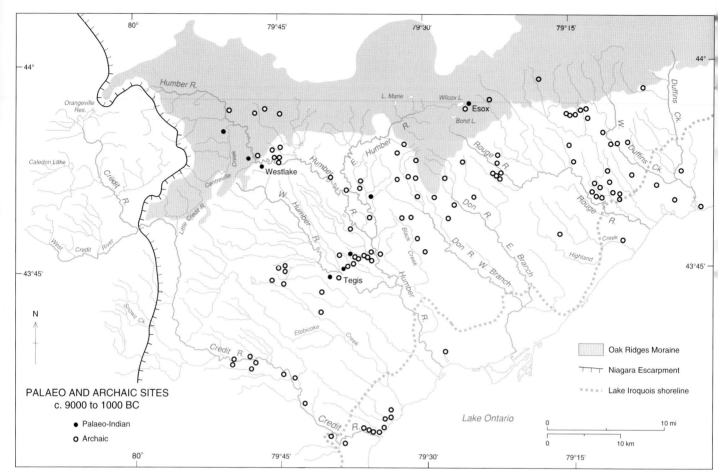

FIGURE 4.1 Distribution of Palaeo-Indian and Archaic sites according to survey data recorded by the Toronto and Region Conservation Authority and other agencies. The surveys on which these data are based are incomplete east of the Rouge and west of the Humber watersheds. Most archaeological sites in the built-up areas were destroyed before they could be recorded.

maple, beech, and birch, while jack pine, red pine, and birch grew on the Canadian Shield to the north. Lake Admiralty was now only some 60 m below the present lake level.

The Palaeo people adjusted to this changing environment. A greater variety of tools evolved, and some of the old forms, such as the fluted spear points, fell into disuse. These people were still spear hunters following caribou, but they probably made use of other animals, fish, and berries.

ARCHAIC, c. 8000–1000 BC

The Archaic is a long and complex period ending rather arbitrarily about 1000 BC, when pottery begins to appear at archaeological sites. It is marked by higher lake levels, warmer temperatures, and deciduous forest conditions. Human groups adapted themselves to these changes by creating greater diversity in their subsistence economies and technologies.

By 8500 BC the coniferous forest had become a mixed forest dominated by pine, birch, and oak. Hemlock largely replaced pine by 8000 BC, with elm, maple, and beech becoming increasingly important. By about 5000 BC the Toronto area was covered by a mainly deciduous forest of hemlock, oak, elm, maple, beech, and hickory. Hemlock continued to be present until about 3000 BC, when it abruptly disappeared for about 1,000 years. Throughout this period, Lake Admiralty rose steadily, reaching about 40 m below the present lake level by 7500 BC and near modern levels by about 1000 BC. By about 2000 BC the climate had continued to moderate to the point at which a purely deciduous forest became established along the north shore of the lake.

After about 7000 BC heavy woodworking tools such as polished stone adzes and gouges entered the tool kit. Netsinkers and an increasing orientation of habitation sites to lakeside and river-mouth locations demonstrate the seasonal importance of fishing, while the woodworking tools strongly suggest the manufacture of dugout canoes. By 4000 BC, marking the beginning of the Laurentian Archaic substage, an astonishing variety of copper tools make their appearance.

By the Terminal Archaic (c. 1200-1000 BC) there is a marked trend toward smaller stone points, possibly suggesting the use of the bow and arrow. A dog burial at the McIntyre site (c. 3500-2000 BC) near Rice Lake is the earliest evidence for dogs in Ontario, although their presence is suspected among the earlier hunting cultures. Habitation sites have a strong lakeshore orientation from spring to fall and produce bone tools, many of them used for fishing. Fish bones have been recovered, indicating that sturgeon, trout, pickerel, pike, bass, sucker, catfish, and drum were caught. In addition there is evidence that duck, turtle, and beaver were successfully hunted. Seeds from berries and nuts are also present. Habitation sites in these locales tend to be large, indicating not only larger human groups but also recurring occupation of productive places. The few house patterns that have been delineated suggest oval dwellings up to 15 by 10 m in size.

A dog burial dating from c. 3500-2000 BC near Rice Lake is the earliest evidence for dogs in Ontario.

EARLY AND MIDDLE WOODLAND, c. 1000 BC-AD 700

The division between late Archaic and early Woodland is based largely on the introduction of pottery making into Ontario from the south about 1000 BC. This was a time when the climate became slightly cooler, permitting the present Great Lakes-St. Lawrence mixed forest to become established along the north shore of Lake Ontario. White pine, yellow birch, sugar and red maple, red oak, basswood, beech, elm, butternut, and ash characterized this forest. The major large game species were white-tailed deer, elk, black bear, and, farther north, moose.

In the Toronto area the early Woodland groups are termed Meadowood (1000-400 BC) (Figure 4.2). Archaeological data suggest that bands of about 30 to 40 people intensively exploited spring spawning runs at river mouths with nets, weirs, and spears. Social interactions between neighbouring groups seem to have been frequent, and, culturally, the Meadowood people show strong relationships to similar groups in what are now southern Quebec and New York State. The introduction of cultivated plants to the middle Woodland peoples about AD 800 gradually transformed them into the agriculturalists of the late Woodland period.

LATE WOODLAND, c. AD 600-1650

The late Woodland period itself is usually divided into early, middle, and late, marked by the introduction of maize (AD 600-900), the wide acceptance of maize, bean, and squash (AD 900-1400), and finally the evolution of fully agricultural societies and their eventual destruction about AD 1650. It is during the late Woodland period that the various Ontario Iroquoian groups evolved from the middle Woodland societies. This was a period during which the climate became somewhat warmer and moister, no doubt aiding in the northward diffusion of southern domesticated plants.

By AD 900 the first clusters of villages with a heavier reliance on maize cultivation appeared. It would seem that maize cultivation was adopted very slowly and did not occur everywhere at the same time. The Toronto area, for example, continued to be occupied by hunting and fishing peoples until about AD 1100.

The introduction of cultivated plants and the creation of larger, more permanent settlements meant that, for the first time, localized changes were made in the forested landscape of Ontario. People had begun to change their subsistence organization from food collecting to food production. Trees were girdled by removing the bark around the base of their trunks to kill them, and all brush and weeds were burned to create fields. Lacking domesticated animals except the dog, none of the Native Ontario agricultural societies had fertilizer until Europeans introduced other animals. The wood-ash that

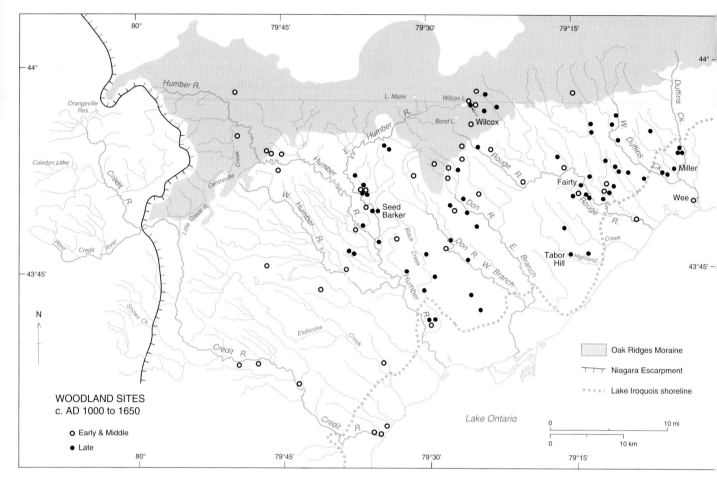

FIGURE 4.2 Distribution of Woodland sites recorded by the Toronto and Region Conservation Authority and other agencies. The data are incomplete east of the Rouge and west of the Humber watersheds. Most archaeological sites in the built-up areas were destroyed before they could be recorded.

was left on the fields after burning provided some nutrients, but eventually the soils gave out and new fields were cleared. When productive fields got too far from the village, or firewood was no longer available, the village had to be moved. Gradually the forest would reclaim the site of the former village and fields.

After about AD 1000, populations began to increase. This led to larger, more formally planned villages and houses, more centralized decision making, and larger cooperating social groups. Between about AD 1250 and 1330, clusters of small villages extended through the deciduous forest zone north of Lakes Erie and Ontario. By AD 1100, bean, squash, and sunflower had diffused to Ontario from the south and became crops along with maize and tobacco. The villages were occupied year-round, although a substantial part of the population would still be gone from spring to fall in outlying fishing and hunting camps.

Villages averaged about 1 ha in size, containing up to 300 people. During this period, the classic Iroquoian longhouse was developed. The development of the longhouse, occupied by related families, and the more regular layout of the interior of the villages into groups of parallel longhouses, probably representing related lineages, suggest increasing complexity in social organization and decision making. It is probable that the matrilineages and decision-making powers of a village council, so typical of the Iroquoian groups (as later recorded by Europeans), evolved at least by this time.

From 1400 to 1600, villages rapidly expanded in size and number. Villages of 2 ha in area became common, the largest reaching 6 ha, with populations over 2,000 and perhaps 1,800 ha of land under cultivation. These large villages seem to have been regional strongholds for clusters of smaller villages. These clusters can to some extent be identified with the forerunners of the historic Iroquoian tribes mentioned later by French explorers and missionaries. Three of the many village clusters fall into the greater Toronto area: a group near Pickering, including the well-known Draper site, that were occupied by people ancestral to the historic Huron; a group of more diverse sites in the upper Humber valley that were also occupied by ancestors of the Huron; and a cluster of villages in the Milton area, probably occupied by the Aondironnon, one of the Neutral tribes.

Judging from the rapid growth of villages and houses, as well as the formation of regional clusters of villages, the fifteenth and sixteenth centuries were a period of rapid cultural evolution toward increasing social and political complexity. During this period, villages joined into tribes with related decision-making structures. During the sixteenth century and perhaps earlier, some of these tribal units formed tribal confederacies, created for mutual defence against common enemies. In 1639 the Huron told French missionaries that the first two Huron tribes settled together about 200 years earlier (c. 1440), and were joined by two other tribes about 1590 and 1610.

Beginning in the first half of the sixteenth century, the ancestors of the Huron and Petun began a slow northward movement out of the Pickering area and the Humber valley. This movement was more or less complete late in the century. Within about a generation and a half, well over 10,000 people in at least ten villages had moved north, away from the lands between the Oak Ridges Moraine and the north shore of Lake Ontario. The stories told to Champlain by the Huron, that some of their ancestors had moved out of the Trent

Iroquoian, Algonquian, and Similar Names

The terms Algonquian and Iroquoian refer to broad language families analogous to Germanic, Romance, and Slavic in Europe. Algonquian can be subdivided into a number of languages such as Cree, Ojibwa, and Micmac, and Iroquoian into Huronian and Five Nations. Some European equivalents in the Romance language family are Italian, French, and Spanish. Languages within families are mutually unintelligible. Just like the European languages, those constituting Algonquian and Iroquoian had regional dialects such as Ottawa, Algonquin, and Eastern Ojibwa among the Ojibwa speakers; Huron, Petun, and Neutral in the Huronian language; and Mohawk, Seneca, Onondaga, and Susquehannock in the Five Nations language. Although this is an oversimplification, especially for the Iroquoian language family, it does point out that there was enormous diversity among Native languages.

What confuses the issue is that linguistic terms are often used as political, cultural, or territorial equivalents. The groups speaking Huronian, for example, comprised three territorial confederacies; the Huron, consisting of five largely politically autonomous tribes; the Petun, of two tribes; and the Neutral, of about eight tribes. Each of the component tribes had a name for themselves and consisted of a cluster of villages within a territory, inhabited by people who had a common past and who considered themselves to be largely independent of their neighbours. The early French explorers called these tribes 'Nation.' The Five Nations language comprised the five tribes of the Iroquois Confederacy (Iroquois League) as well as the Susquehannock, who were at war with the League.

These confederacies (leagues) were essentially defensive alliances between tribes against mutual enemies. Similarly, at one time the Ottawa comprised at least four territorial bands and the Eastern Ojibwa at least twelve, each with its own core territory and its own name, such as the Amikwa, Mississauga, and Nikikouet of the Eastern Ojibwa. The difference between tribes and bands was one of cultural complexity; tribes being larger in population and socially, politically, and economically more complex than bands.

Because the Iroquoian groups of the eastern Great Lakes lived close to each other, interacted with each other, and seem to have had a common origin, they shared most of their culture traits. By contrast, the Algonquian speakers were distributed from the Atlantic coast to the Rocky Mountains and were culturally extremely diverse. A further complication is that all Native peoples were subjected to enormous changes after European contact, resulting in the extinction of entire languages (Huronian, for example), the disappearance of many territorial groups (tribes and bands), and the amalgamation of many others who now carry a common name, often that of the largest component group.

valley into Huronia at the southern end of Georgian Bay in the sixteenth century for greater safety from the League Iroquois to the south, indicate increasing warfare. By the time the French arrived on the scene, the pattern of warfare between the Huron-Petun alliance and the Iroquois confederacy was well established.

With the abandonment of the Toronto area, the north shore of Lake Ontario east of the Neutral in the Milton area became a hunting territory for the Neutral, some Huron, and perhaps the Seneca, who claimed the northwestern shore as their ancient hunting territory when they occupied it in the late 1660s. During the late 1630s, smallpox and measles devastated all the Iroquoian groups and many of their Algonquian neighbours farther north. Bent on bolstering their decimated population and absorbing other Iroquoian groups, the League Iroquois, armed with Dutch muskets they had obtained through trade, attacked the French and the Ontario Iroquoian and Algonquian groups. The Neutral Aondironnon were dispersed in 1647; the five Huron tribes in 1648-9; the two Petun tribes in 1649-50; and the rest of the Neutral in 1651-2. All had been offered membership in the Iroquois League. At least two Huron villages accepted and were permitted to move to Iroquoia, south of Lake Ontario. Those who refused were attacked, and the survivors integrated into the families of the League villages. In the late 1650s and 1660s, both the Erie and the Susquehannock were dispersed and partially absorbed. When the Tuscarora joined in 1713, Iroquois aims had been achieved: all Iroquoian-speaking peoples, with the exception of two small Huron-Petun remnants at Detroit and Quebec, now belonged to one Iroquois Confederacy. A peace between all the League Iroquois (except the Mohawk) and the French, along with some of their Native allies, was concluded in 1654.

THE HISTORIC PERIOD, AD 1650-1800

The Iroquois, 1650-1701

About the mid-1660s, the Iroquois were running short on furs within easy travel of their main villages. Furs had by now become necessary to purchase goods from the Dutch and English. Now at peace with New France and its Native allies, the Iroquois pushed their hunting and trapping farther into southern Ontario and westward along the south shore of Lake Erie into areas whose populations they had defeated and to a large extent absorbed some fifteen years earlier.

Contemporary French maps show clearly that the village locations of the Iroquois du Nord, as these Seneca, Cayuga, and Oneida were called, were very carefully chosen. All were at strategic points that connected the north shore of Lake Ontario to the interior lakes and river system, which could be followed to the rich beaver grounds and the northern Algonquian fur producers (Figure 4.3).

The village of Teiaiagon ('it crosses the stream') was located on Baby Point on the east side of the Humber River (Figure 4.4). The western ends of the present Baby Point Road and Baby Point Crescent in Toronto roughly

The Direct Historical Approach

It is customary among archaeologists and prehistorians to begin their studies with the earliest known cultures in an area and gradually advance through time toward the more recent ones. Reapproximations of these cultures have to be made from the fragmentary evidence recovered from the excavations. In eastern Canada and other parts of North America, however, scholars are fortunate to have an abundance of documents and maps written by explorers, fur traders, and especially missionaries, in which they describe the material and non-material culture of many Native groups before they underwent radical culture change as a result of European contact. By matching the archaeological cultures of the early European contact period with those mentioned in the historical records, a much more complete study of the people can be put together than from either of these sources alone.

Once the historical culture is understood we can attempt to trace it back gradually through time into the period for which there is only archaeological material. Knowing some of the details of a culture at one point helps in making meaningful reapproximations for another time for which there is only fragmentary information. This research procedure, which involves moving cautiously, step by step, back in time from the known to the unknown, is called the direct historical approach.

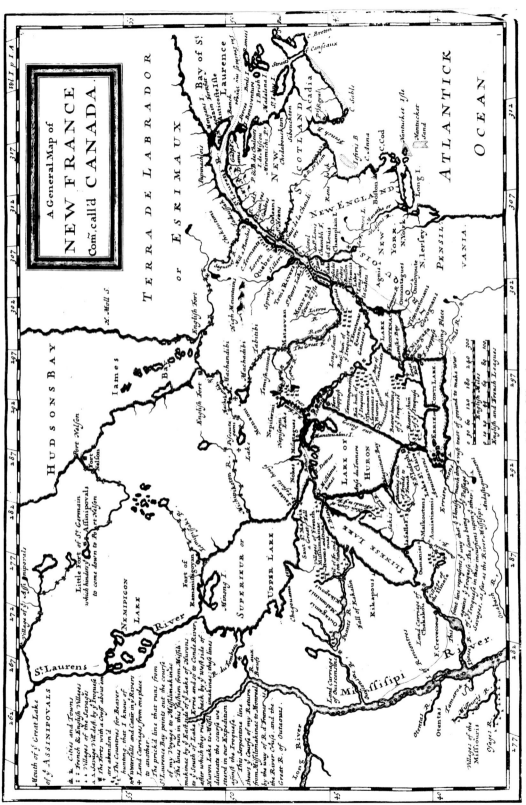

FIGURE 4.3 *A General Map of NEW FRANCE Com̃, call'd CANADA*, by Louis-Armand de Lom D'Arce de Lahontan, first appeared in his book *New Voyages to North America* in 1703. The map was based on observations made between 1683 and 1690 when the author was in Canada. Although cartographically inferior to most maps of the period, it is unique in showing the major beaver hunting grounds of the Iroquois and Natives allied to New France. (National Library of Canada C94713).

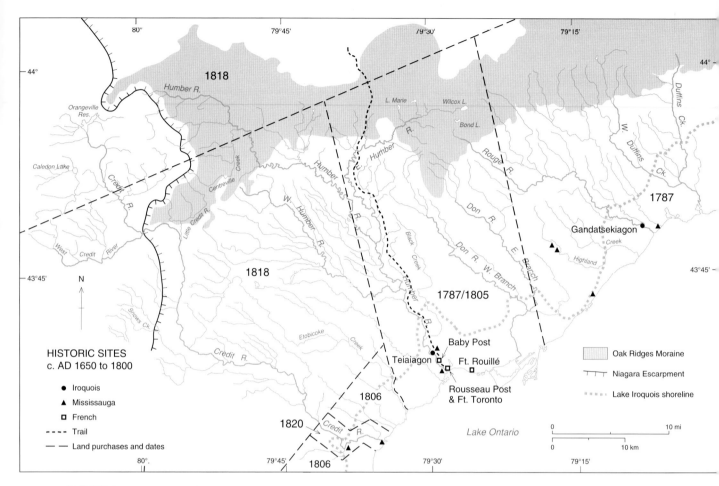

FIGURE 4.4 Distribution of historic Iroquois and Mississauga village sites, French forts, trading posts, and land purchases. The Mississauga sites are underrepresented and the location of the Humber trail is only approximate.

outline the former village. At this place, shallows in the Humber provided a ford for the main east-west trail along the north shore of Lake Ontario. The same shallows marked the end of canoe travel up the Humber River, necessitating a portage of about 45 km to the west branch of the Holland River before canoes could proceed to Lake Simcoe. The village itself stood on the high bluff of Baby Point, protected on three sides by the steep river bank, a classic location for an Iroquois village. The size of the village is not known. Unfortunately, modern-era houses were built on it before it could be properly investigated.

Although trading and beaver hunting were important, the main subsistence activities were agriculture (maize, beans, squash); fishing with nets, spears, and weirs; and hunting. Contemporary Frenchmen traded corn at Teiaiagon and noted that both the Humber and Rouge Rivers were excellent salmon streams. They also commented favourably about the abundance of other fish and fowl in the bay and among the islands of what is now Toronto Harbour. The major game animal was white-tailed deer, which was hunted in mass fall drives during the rutting season in the acorn-producing areas of the Oak Ridges Moraine. In 1615, Champlain witnessed a hunt near Rice Lake, where 120 deer were taken in 38 days, and in 1687 French and Native hunters combined to kill 200 deer in a few days in the Ganaraska River watershed. Meat from these hunts was not preserved but taken to the villages and consumed at feasts. Outside the rutting season, deer hunting was carried out by small groups of men, but was not nearly as productive.

The French built forts at opposite ends of Lake Ontario – Fort Niagara in 1668 and Fort Frontenac in 1673 – thus hoping to close both ends of the lake and prevent the Iroquois from carrying their furs to the English of Albany on the Hudson River. These forts were bitterly resented by the Iroquois because they were located on lands that, in the words of the great Onondaga orator Otreouti, 'the Great Spirit has dispos'd of in the favour of our ancestors.' Both sides became determined to control southern Ontario, and during the 1680s French-Iroquois relations rapidly deteriorated. In 1687 Governor Denonville launched an attack on the Seneca. Although the French and their Native allies captured some Iroquois at Gandatsekiagon ('among the birch trees') and Ganneious the majority of the Iroquois had abandoned their settlements on the north shore of Lake Ontario and had returned to their villages on the south shore. Yet the Iroquois continued to hunt in the area, where they came into increasingly bloody contact with southward-moving Algonquian people, especially the Ojibwa and Mississauga, intent on using southern Ontario for the same purpose.

By the late 1690s, continued harassment by French troops and France's Native allies had reduced the Iroquois warriors from their customary 2,600 men to about 1,200. It became clear to all concerned that a peace was desirable. Abandoned by the English at the Treaty of Ryswick (1697), the Iroquois were left to make a separate peace with the French and France's Native allies. After several years of sporadic fighting, the Iroquois began negotiations with a number of Algonquian groups, and on 4 August 1701 concluded a peace settlement with the French and 22 Native groups with whom they had been at war. Part of the settlement was shared hunting on the north side of Lake Ontario. In separate deals, the Ojibwa, Mississauga, and other Algonquian groups demanded access to the traders at Albany across Iroquois territory. The Iroquois were in no position to turn down these proposals. On 19 July, while peace negotiations with the French were going on, the Iroquois 'gave' southern Ontario to the English king. It was a desperate move by which the Iroquois hoped to gain English military support to regain or protect their northern hunting grounds. But it was to no avail. Unlike the French, the English were not about to commit troops to aid their Native allies. What the negotiations leading to the Treaty of 1701 achieved was peace between the Iroquois and the French as well as France's Native allies.

Ojibwa and Mississauga, 1701-1800

The peace of 1701 opened south-central Ontario to settlement by Ojibwa and Mississauga. In 1708 the Iroquois entered into an agreement with three Mississauga villages, permitting these people to settle on Seneca lands on the north shore of Lake Ontario. The hunting grounds of southern Ontario were now effectively controlled by the northern Algonquians, who rapidly occupied them, including the abandoned Seneca villages at Teiaiagon and Gandatsekiagon.

French documents show that by the 1720s there were three main Mississauga villages on the north shore of the lake – at Kenté, Toronto, and Niagara – with a fourth on the shores of Lake Simcoe. These were estimated to have about 150 men of fighting age, giving the four villages a probable population of about 750 people. These Mississauga controlled the subsistence and fur resources from the east end of Lake Erie to the Bay of Quinte, and from the north shore of Lake Ontario to the margins of the Canadian Shield. They increasingly traded with the English at Albany and, after 1725, at Oswego. In 1720 a post had been opened by the French for the Mississauga at Toronto, but fur returns were so low that the post was closed ten years later.

The Iroquois village of Teiaiagon was located on Baby Point on the east side of the Humber River.

With the closure of the Toronto post it became hopeless to rely on only Niagara and Frontenac to stem the flow of furs to the English. In 1749, fur returns at Oswego listed 25 Mississauga canoes bringing 175 packs of fur. Natives from the Great Lakes, who used the Lake Simcoe-Toronto portage, accounted for an additional 51 canoes and 357 packs. Since the standard pack was about 45 kg the lack of a post at Toronto was costing the French at least 22,600 kg of fur a year, 8,000 kg coming from the Mississauga. Between May and July 1750, the French built a new Fort Toronto, a small storehouse surrounded by a palisade, on the east bank of the Humber River near its mouth. During those two months, the French obtained some 79 packs, far exceeding their expectations. A decision was therefore made that a larger post was needed, and in September 1750 a fort was begun near the entrance of Toronto Harbour (the present Exhibition grounds). The new fort, some 50 by 55 m and built of oak timbers, was named Fort Rouillé after the

French minister of the marine. By 1757, Fort Rouillé averaged about 150 packs of fur per year. Its express purpose was to stop the bleeding of furs to Oswego by any means possible. Too often the simplest and cheapest method of keeping the Mississauga from going south with their furs was to supply them, and all other Native traders who happened to pass by, with abundant quantities of brandy. In 1757, one missionary described Fort Rouillé as 'un cabaret d'eau-de-vie.' Toronto was quite simply an outlet for booze.

In 1754 the Seven Years War broke out. With Fort Frontenac destroyed by the English in 1758 and Niagara under siege in July 1759, orders were given to burn Fort Rouillé to prevent it from falling into English hands. When the English arrived at the site a year later, they reported that a tract of 120 ha of cleared land surrounded the ruins of the fort, that deer were extremely plentiful, and that the timber near the fort consisted of large oaks, hickories, maples, and some poplars. The Mississauga in the area promised henceforth to be loyal to the English.

Brief references to the Mississauga of the Toronto area, and related people in southwestern Ontario, enable us to form a rough outline of their relationship to the natural environment. They were essentially fishers and hunters, with a vital part of their diet coming from collecting and from casual maize horticulture. The seasonal availability of foods, and the variable location of these, necessitated a great deal of flexibility in group size and settlement patterns. Like all people with this kind of life style, past and present, the Mississauga knew their surrounding environment intimately because they were totally dependent on it. This intimacy stretched beyond subsistence to a deep spiritual relationship that, in their eyes, made them as one with their surroundings. More than one European echoed the observation of Father Louis André in 1670 that 'no people can be more attached to their native soil than the Indians.'

By the 1780s the Mississauga had two main villages, one near the mouth of the Humber and the other at the mouth of the Credit River, as well as a couple of smaller ones scattered throughout the region. There were small seasonal encampments elsewhere, such as those around Toronto Harbour for fowling and fishing, and those on the Oak Ridges Moraine for fall hunting.

In 1757, one missionary described the French Fort Rouillé, on the present site of Toronto, as 'un cabaret d'eau-de-vie.' Toronto was quite simply an outlet for booze.

To the Mississauga, the fall was essentially a preparation for winter. From about September to the beginning of November, men and women would depart in small groups to the mouths of the creeks and rivers flowing into Lake Ontario for the fall fish runs and for fowling. The Credit, Humber, Don, and Rouge were well known as especially productive salmon streams. Once the fish runs abated, the men would go north to the Oak Ridges Moraine for late fall hunting. This hunt coincided with the rutting season for deer, which congregated in large groups in the acorn-producing areas to fatten for the winter. Deer were hunted in mass drives or chased into water where they could be speared from canoes. Some attempt was made to preserve meat for the winter by smoking it.

Seasonal activity dominated the lives of the Mississauga. Especially important was the manufacture of clothing for the winter from deerskins if cloth goods were not available from traders. Some winter fishing was carried out in Toronto Harbour. Elizabeth Simcoe noted ice tents, whose occupants were jigging and spearing pickerel and muskellunge. The first important spring activity that drew all the families out together was sugaring. It could begin in late February, but according to Elizabeth Simcoe it was a March activity in Toronto. This was also the season when geese, ducks, and wild pigeons returned. Toward the end of May, maize was planted on light sandy loams near the villages. During the summer months, when the environment was more productive, people were together in larger settlements. Subsistence activities covered all sectors – fishing (especially salmon), hunting, gathering of berries and, by the end of the summer, butternuts. Acorns were consumed only as a winter famine food. Maize was harvested in late August. It was not cultivated as intensively as it was among the Iroquoian groups and was never a staple among the Mississauga until they were confined to reserves in the nineteenth century. The shoreline in and around Toronto Harbour and the Island was an especially productive area for fishing, fowling, and collecting. Elizabeth Simcoe wrote that the Mississauga also considered the Island 'so healthy that they come there and stay here when they are ill.'

Following the American Revolution and the arrival of the United Empire Loyalists in the Kingston and

Niagara areas, Governor General Lord Dorchester found it expedient to secure the north shore of Lake Ontario by treaty from the Mississauga. The logical beginning for this project was to purchase the most desirable part first: Toronto Harbour and the portage to Lake Simcoe. On 23 September 1787, a preliminary treaty was signed with three Mississauga headmen who represented their band: Wabukanyne, Neance, and Pakquan. Although this treaty extinguished their claim to the Toronto area for £1,700, the exact tract of land was not specified. In August 1805, the Mississauga met with agents of the Crown at the mouth of the Credit River to get a more exact idea of what they had sold. The parties eventually agreed upon a tract of some 100,000 ha. At the Mississauga's insistence, the western boundary ran along the east bank of the Etobicoke Creek. Specified in the treaty was that they retained 'the fishery in the said River Etobicoke ... for their sole use.' In September 1806, they sold 28,657 ha, lying between Etobicoke Creek and Burlington Bay, for £1,000. They retained all fishing rights to Etobicoke, Twelve Mile, and Sixteen Mile Creeks, as well as the low grounds adjacent to these creeks that they had under cultivation – in all, 4,427 ha. They also reserved 'the sole right of the fishery of the River Credit with one mile on each side of the said River.' Similar reserves were laid out along the lower reaches of Twelve Mile and Sixteen Mile Creeks. The Mississauga, like other Native groups, probably thought that the treaty was an agreement to share their hunting range with the English, of whom there were very few at the time that the treaty was signed. The fact that they insisted on keeping exclusive fishing rights and planting rights in the core areas of their territories meant that they were not willing to share those important resources. What they apparently did not mind sharing or relinquishing was hunting in the vast forests north of Lake Ontario. Neither the Mississauga nor the English could have foreseen the rapidity with which the area was settled and transformed into a cultivated landscape.

Elizabeth Simcoe

Elizabeth Posthuma Simcoe, née Gwillim, was born in Whitchurch, England, near the border of Wales, probably in 1762 (baptismal certificate), although later in life she claimed it was 1766. She died at Wolford Lodge, Devon, in January 1850. Orphaned at birth, she was raised by her mother's sister and, like other ladies of the landed gentry, received an education in the languages (she spoke French and German in addition to her native tongue), music, drawing, and the social graces. By all accounts she was a lively young lady, sharp featured, about five feet tall, who loved riding, dancing, sketching, reading, and good conversation. In 1782 she married her aunt and uncle's godson, Colonel John Graves Simcoe, by whom she had eleven children.

In 1790, George III appointed her husband lieutenant governor of the new province of Upper Canada, and on 26 September 1791 they sailed for the port of Quebec with their two youngest children. In June 1792 they left for Niagara, where they lived until July 1793, when they took up residence at Toronto, renamed York a month after their arrival. Mrs. Simcoe stayed at York on three occasions: 29 July 1793 to 11 May 1794; 13 November 1795 to 29 April 1796, and 29 June 1796 to 21 July 1796, when they began their return journey to England.

During her sojourn in Canada, Mrs. Simcoe kept a fairly detailed diary. Her entries reveal an observant person interested in all aspects of nature, especially flowering plants, trees, fish, all forms of wildlife, and the local Mississauga people. In addition to her written descriptions, she made sketches of the natural landscape and gradually emerging settlements. On occasions she also made sketch maps of her husband's travels. Her descriptions and sketches are spontaneous rather than studied; she simply re-

Elizabeth Posthuma Simcoe, née Gwillim. (National Archives of Canada C-139975)

corded what she saw for her own entertainment and the education of her children in England. They are the earliest depictions of nature in the Toronto area.

Although the exact boundaries of the Toronto Purchase were in dispute until 1805, a decision had been made as early as 1788 to build a town there, and in August of that year a plot was surveyed by Alexander Aitkin. Building at the site commenced in July 1793, with the arrival of Lieutenant Governor John Graves Simcoe. He was very impressed by the potential of the site for a town: the excellent defendable harbour, the route up the Humber and through Lac aux Claies (which the lieutenant governor renamed Lake Simcoe in honour of his father) to Georgian Bay, and a good hinterland of timber and fertile soils. On 26 August 1793, following his policy to replace Native with English names, Simcoe formally changed the name of his month-old settlement from Toronto to York.

As English settlement proceeded outward from Toronto (renamed such in 1834), the Mississauga gradually retreated to their reserves and their lives began to deteriorate. Early in the nineteenth century their hunting north of the lake was gone because settlement had destroyed the game. The 'exclusive right to the fishery' was never enforced, with the result that Europeans robbed them of their fish. The Mississauga were dwindling rapidly through starvation, alcohol abuse, and disease. In 1820, at one of their lowest points, they were approached by agents of the Crown to sell their reserves. One parcel of 809 ha was sold outright for £50. Most of the remaining reserve, 3,537 ha, was sold for 20 shillings. Along with these sales, the Mississauga had to relinquish 'all woods and waters ... and especially all sole and exclusive rights of fisheries' on the Credit, Twelve Mile, and Sixteen Mile Creeks. All they in fact retained were 81 ha on the east bank of the Credit, where the government promised to build them houses and provide for their education and religious instruction. From later events, it is clear that either the Mississauga had not understood the implications of these transactions, or they had been told that the government was doing something very different. What they thought they had done was sell 809 ha and place the rest of their reserve in trust with the government to protect it from encroachment by white settlers.

In 1826, the Methodist minister Reverend Peter Jones, who was the son of a Mississauga woman and an English father, moved to the Credit Reserve to begin a school and religious instruction for his people. One of his first

Canada, Ontario, Toronto

Canada – This name was first mentioned by Jacques Cartier on his second voyage to the St. Lawrence River (1535-6). At first he thought that the word was the name for the area around the present city of Quebec, but later he learned from the local Iroquoian speakers that the word Canada meant *un village*.

Ontario – According to the eighteenth-century Jesuit priest and linguist, Father Pierre Potier, this Iroquoian word means 'big lake.' It is made up of *ontare*, which means 'lake,' and *io* meaning 'big or beautiful.' In view of the size of the lake, 'big lake' is more likely than 'beautiful lake.'

Toronto – First mentioned on maps of the 1670s and 1680s, the name was applied to Lake Simcoe as *Lac de Taronto*. The change in the second letter from *a* to *o* came much later and was probably a copying error. Father Potier rendered the word as follows: *T* is a prefix denoting locality and *gronto* means *une arbre dans l'eau qui sert de pont pour passer une rivière ... un pont* (a tree in the water that serves as a bridge for crossing a river ... a bridge). The linguist John Steckley contends that the word is Mohawk, from *tkaronto* meaning 'where there is a tree or trees in water.' Whether Huron or Mohawk, it is likely that the word refers to the fish weirs constructed by the Huron, and kept up by the Mississauga, across the Narrows between Lakes Couchiching and Lake Simcoe, as well as across the mouth of the Holland River. These were constructed of vertical posts and crosspieces, and were used as footbridges. The late-seventeenth-century French name for Lake Simcoe was Les Piquets (the stakes or posts). This name appears on some maps alongside Lac de Taronto. By the eighteenth century the French name for Lake Simcoe had changed to Lac aux Claies (Lake of the fish weirs). It is probable that the French names are rough translations of the Iroquoian original. The name Toronto probably became attached to the beginning of the Humber Trail at Lake Ontario because this was the route to Lac de Taronto.

acts was to goad the government into building the houses they had promised. He also tried to oversee the annual payments owed to the Mississauga from their earlier land sales, payments out of which they had been frequently cheated. Under Jones's direction and the active support of the band leaders, the community changed. By the late 1830s the village had grown to about 250 people in 40 houses. They had a sawmill, blacksmith shop, stores, school, chapel, and 332 ha under cultivation, with another 400 ha in pasture. Within about a dozen years the Mississauga had made the transition to become self-sufficient farmers, only to discover that they had no title to the land they had worked so hard to cultivate. Furthermore, the new lieutenant governor, Sir Francis Bond Head, was formulating a policy similar to the American Indian Removal Act (1835) whereby the Indians of southern Ontario were to be removed to the islands of Georgian Bay and farther west. Late in 1837, the Mississauga decided to send Peter Jones to England to plead their case. Although he spent a year petitioning various ministers and influential people, he was unsuccessful in gaining legal title for the Mississauga to the land and farms they occupied. Increasingly Mississauga land was being subdivided and sold off. At the same time, the salmon, which was still an important food and source of income, was declining rapidly with the building of mills on the Credit River. Increasingly desperate for a secure title and afraid of deportation, families began to leave for other reserves in Ontario, while others stopped working their land.

The Mississauga sold one parcel of 809 ha outright for £50. Most of the remaining reserve, 3,537 ha, was sold for 20 shillings.

Early in 1847, the Six Nations Iroquois on the Grand River Reserve heard of the plight of the Mississauga. The Iroquois remembered that a generation ago, when they had fled from the United States, the Mississauga had given them land. In a full council, the Iroquois now offered the Mississauga the southwestern corner of their reserve. The Mississauga gratefully accepted, and in April 1847 abandoned their farms on the banks of the Credit and journeyed west to build a 'New Credit' settlement on the Six Nations Reserve. With this exodus, the Native occupation of the greater Toronto area ended. In 1991, the Mississauga on the Grand River numbered 627 people.

5

Spatial Growth

Warren E. Kalbach

In 1991, the Toronto metropolitan area was the largest urban concentration of population in Canada, containing over 1.3 million dwelling units occupied by almost four million inhabitants. The core of this urban concentration, the City of Toronto, had taken about 200 years to increase from 241 people within an area of about a third of a square kilometre to an area of approximately 100 km² with a population of 635,400 people (Statistics Canada 1992, 1993).

LOOKING BACK

Prior to its settlement, most of what is now southern Ontario was heavily forested. The Carolinian forest — mainly of deciduous trees, with cedar and tamarack swamps — extended inland from the northern shores of both Lake Ontario and Lake Erie. There it blended with the Great Lakes-St. Lawrence forest region, consisting of a more northern mix of trees and including more conifers (Austen, Cadman, and James 1994). In addition to the extensive forests, other major impediments to rapid settlement in the area were the roughness of the morainic ridges, insufficient soil over the limestone, soil infertility, and generally poor drainage (Wood 1988).

A rather limited area along the northwest shore of Lake Ontario, and its tributary back country, showed consistent gains in human population, an indication of heavy settlement in the 'Golden Horseshoe' area in the years to come (ibid.). The general unattractiveness of the northwest shore region for settlement in the early years was reflected in the reactions of many on first arrival at the site of the old trading post at Toronto. The captain of the sloop *Mississauga,* who in 1793 surveyed the Toronto harbour for John Graves Simcoe, the first lieutenant governor of Upper Canada, described the site as 'rather untamed, standing in front of a backdrop of dense and trackless forests that lined the margin of the lake' (Berchem 1977). Just how dense and trackless the forests actually were was aptly demonstrated by Simcoe and his party, including Indian guides and surveyors, when they lost their way just short of their destination while attempting to return from an early scouting trip to the Holland River (Figure 5.1).

The lay of the land inward from the beach between the Humber and Don Rivers was described as a declivity that rose northward, dark with virgin forest (Mulvany 1884). The natural setting had little to offer beyond the security of its harbour and the excellent sanitary conditions resulting from its graded terrace. It was described as a place of perpetual gloom, damp and depressing, and as a dismal dump where the winters were ice hells and the summers so hot that the land was thought to be not far removed from the fires of damnation (ibid.). Elizabeth Simcoe, the wife of the lieutenant governor, was likewise unimpressed, as she reportedly found the place damp and depressing in its perpetual gloom, the mosquitoes bothersome, and the rattlesnakes repulsive. She believed that her husband's ill health was greatly aggravated by

> *Just how dense and trackless the forests around Toronto actually were was demonstrated by Simcoe and his party when they lost their way while attempting to return from an early scouting trip to the Holland River.*

FIGURE 5.1
A drawing based on the 14 October 1793 entry in Mrs. Simcoe's diary, showing Col. Simcoe with surveyors and Native guides, sighting Lake Ontario after becoming lost in the forest. (Berchem 1996. Courtesy F.R. Berchem and Natural Heritage Books)

the extreme variations in climate (Berchem 1977). Other than the possible attraction of free land, there seemed to be little to serve as an inducement to attract settlers into the area to help clear and develop the land.

LITTLE YORK RISES FROM THE MUDDY FLATS

With the arrival of the *Mississauga* at Toronto, Lieutenant Governor Simcoe and his Queen's Rangers set about the task of clearing the land in the vicinity of the present Canadian National Exhibition grounds, and constructing Fort York for their accommodation. Fort York, at its inception, was nothing more than the simple sum total of a trading post, shelters for several Mississauga Indian families, Simcoe's canvas house (previously acquired from the late Captain Cook), and a cluster of huts for sheltering the Queen's Rangers (Figure 5.2).

Simcoe's primary objective was to establish York at the site he had selected for the new capital of Upper Canada while extending Yonge Street north to Lake Simcoe as a trading and military road. In May 1794, the Queen's Rangers laid out 111 lots on both sides of the Yonge Street line and, starting from Lot 1 (which later became Eglinton Avenue), cleared about four miles of road to Lot 17, about a quarter mile north of what is now Sheppard Avenue. Further grants of land were given by Simcoe to improve Yonge Street between York and Lot 29 and from Lot 29 to the pine forest at Holland Landing (Figure 5.3). Work was slow and tedious, and the pace slowed even more when Simcoe was informed that he could no longer use the Queen's Rangers for land clearing and road construction. In the end, he had to resort to the few remaining soldiers at Fort York to help push his Yonge Street project through to Holland Landing. They did manage to reach their objective by mid-February 1796, but the resulting road was not much more than a very rough wagon track.

THE CITY BEGINS TO TAKE SHAPE

To connect the garrison at Fort York with the site of the city, which Simcoe had christened York in 1793, the main roads followed the lines set out by the deputy surveyor, Alexander Aitkin, in the original survey. The first streets were King and George, with most of the shoreline to the west of them set aside as a military reserve. The region north of Queen Street was divided

FIGURE 5.2
A drawing embellished from Mrs. Simcoe's 'Hutts of the Queen's Rangers, 1794' conveys a sense of the 'beginnings' of Toronto. (Berchem 1996. Courtesy F.R. Berchem and Natural Heritage Books)

into blocks one and a quarter miles square (Masters 1947). Although building had started in 1794, progress was slow. However, by June 1797 when the legislature met for the first time, it was able to do so in two recently completed brick buildings.

By 1803, York was still a small community of only 456 residents, but apparently was large enough and had sufficient economic activity for the new lieutenant governor, Peter Hunter, to initiate a weekly public open-air market on the site of the present St. Lawrence Market. By 1816, the population had expanded to 720 residents, and within two years King, Wellington, and Front Streets were being developed west of Yonge Street. The area between New (Jarvis) and Yonge was still mostly swamp but contained a number of houses and the town's first church, St. James.

It took at least another 15 years before a commercial section began to develop near the foot of Yonge Street. Houses were being built as far north as College along Bathurst and Spadina, and up to Bloor along Yonge Street. The main development was to the northwest,

In 1818, the area between New (Jarvis) and Yonge was still mostly swamp but contained a number of houses and the town's first church, St. James.

where Queen intersected Peter, John, and Graves Streets. To the east, Queen was broken by the swamps of Moss Park, but extended to the only bridge across the Don River.

The flow of immigration up the St. Lawrence River into Upper Canada, mainly from the British Isles, had pushed the population of York to over 9,000 and into incorporation as the City of Toronto in 1834 (Nader 1976; Mulvany 1884). By 1841, Toronto had grown to 14,200 inhabitants, and had established a position as Upper Canada's principal marketing centre. However, in spite of its progress there were still many who apparently would have agreed with the wife of a high official, who remarked that 'Toronto was still a little ill-built town, on low land, at the bottom of a frozen bay, with one very ugly church without tower or steeple, some Government offices, built of staring red brick in the most tasteless, vulgar style imaginable ... and the dark gloom of the pine forest bounding the prospect' (Masters 1947). Such criticism notwithstanding, the population again doubled before mid-century, making Toronto

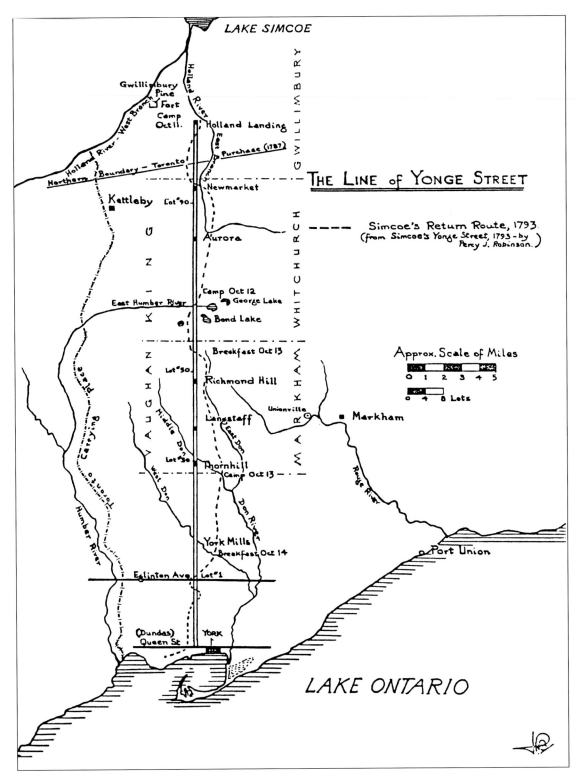

FIGURE 5.3 The location of York and 'The Line of Yonge Street' providing the setting for the subsequent growth and development of the Toronto area. (Berchem 1996. Courtesy F.R. Berchem and Natural Heritage Books)

the largest city in the western part of the province, and a major centre of trade and commerce.

TORONTO IN THE MID-NINETEENTH CENTURY

The mid-nineteenth-century census reported almost 31,000 people residing between the Don River and Garrison Creek, and north from Front Street by the lake to the dirt wagon track that was Bloor Street. Within these boundaries some 4,558 families and individuals resided in accommodation that ranged from elegant private residences for its more wealthy elite and successful merchants, to boarding houses, log cabins, and shanties for unskilled workers and recent immigrants.

Much of the heavier industrial activity required to serve the growing city and the surrounding area was located outside the City of Toronto in York Township and adjoining areas of York and Peel Counties. There were a few gristmills, foundries, tanneries, and distilleries within the city, but most of the 40 or so smaller factories and workshops scattered about the city employed only a few workers. These smaller shops produced a wide variety of basic consumer goods and products, ranging from carriages, guns, and rifles, to soap and candles, starch, oilcloth, and other necessities of the time (Canada 1855).

Housing in Nineteenth-Century Toronto

As today, variations in the quality and location of housing reflected, albeit roughly, Toronto's social class structure in the nineteenth century. The British elite and English gentry were awarded the most desirable land grants, and only the wealthy could afford the more substantial building materials of stone and brick. At mid-century, Toronto's housing inventory included 24 stone and 938 brick houses, 3,630 frame homes, 19 log houses, and 10 shanties. Additional business and public structures included 472 shops and stores, 110 inns and taverns, 22 schools, and 9 public buildings. There were almost two dozen churches: 5 Church of England, 6 Methodist, and an assortment of other denominations (Masters 1947; Board of Registration and Statistics 1853).

Continuing growth and increasing concentration of the population in the Toronto area, accompanied by competition between government, business, and private interests for advantageous locations, contributed to the emergence of distinctive urban areas. Early descriptions of Toronto make frequent references to its wholesale and warehouse district on Front Street, the financial institutions on Wellington, and business and retail establishments on King and Yonge Streets. The mansions of the upper ten families on Jarvis and of the nouveau riche on Spadina contrasted sharply with the dilapidated little cottages and shacks of immigrants and labourers that lined University Avenue and York Street, clustered in Cabbagetown, and filled the back lanes throughout the city (Mulvany 1884; Wynn 1987). While King Street was becoming a street of distinction in business, the major hustle and bustle still centred on the waterfront, and Toronto's wharves were accommodating heavy schooner traffic. Hotels and warehouses lined Front Street, daily stagecoaches left for Holland Landing, and a fish market added to the lakefront bustle.

By mid-century, the city's social structure comprised an older aristocracy and their children, a rising industrial/commercial and financial middle class of various Protestant persuasions, and blue-collar trades and labourers. The older aristocracy were those who had arrived with Simcoe, or soon afterward, and had been given the most desirable official appointments and positions in government and the best and largest land grants in the colony to provide support for a landed gentry and the Church of England. Besides holding key positions in government, many of those from the older established families were participating successfully in the various commercial and industrial activities of Toronto, while remaining active in the Church of England.

With respect to culture and cultural activities Toronto still seemed to be somewhat deficient. Too young to have any real indigenous culture, the city relied on competing British and American culture emanating from London and New York via the respective newspapers of these cities. Local newspapers such as the *Globe* were about the only means by which the population could be informed about political and cultural activities, but most of their space was devoted to American and European news, and advertisements from local merchants.

At mid-century, Toronto was still relatively small, young, and decidedly British. While serving as the capital of Upper Canada from 1793 until 1840, Toronto had established an economic and social infrastructure that allowed it to capitalize on its potential as a centre for trade and commerce, leading to continuing growth and development.

CONDITIONS FOR SUSTAINED GROWTH IN TORONTO

Beginning with the Yonge Street extension to Lake Simcoe, the colonial government developed a transportation network to facilitate settlement and shipment of supplies to settlers as well as the export of agricultural products, lumber, and exploitable natural resources from the Toronto region. By mid-century, Toronto was Upper Canada's largest wheat exporter, but its economic focus was changing from that of a commercial lake port to a centre of growing industrial activity and financial influence.

Between 1851 and 1881, the city's population almost tripled. Yet Toronto, as a supply and wholesale distribution centre, was hard pressed to keep pace with the increased economic activity resulting from the rapid expansion of railroads throughout the entire southern part of the province. During the same period, the population of Upper Canada, which became the province of Ontario at Confederation in 1867, doubled to almost two million, with about 86 percent settling in rural areas. Ontario's population density had also doubled, while concentrating along the northern shore of Lake Ontario, especially in the area of Toronto and the York and Peel Regional Municipalities.

Prior to 1890, manufacturing in Toronto was in its infancy, but as agriculture prospered in the region and the Prairies opened for settlement, the city developed into a major centre for the manufacture of farm equipment. The relocation by the Massey agricultural machinery company to Toronto in 1879 was only one of many moves by manufacturing firms to Toronto from smaller urban centres to take advantage of the city's growing and diversified labour force. By 1911, manufacturing had become the single most important industry in Toronto, employing 35 percent of its labour force and achieving parity with Montreal in terms of manufacturing employees (Nader 1976).

The discovery and exploitation of natural resources on the Canadian Shield and in British Columbia from 1890 on were the main reasons for Toronto's growing importance as a financial centre. The value of transactions on the Toronto Stock Exchange, founded in 1852 and incorporated in 1878, would ultimately challenge and exceed the combined total of the two Montreal stock exchanges by the 1930s.

GROWING PAINS

During the nineteenth century, most developing towns no doubt suffered from growing pains of one kind or another. Early Toronto appears to have been no exception. With the city slow to develop an adequate water supply, its fire-protection service was ill equipped to fight any of the large fires that periodically occurred. Moreover, its primitive privy-pit system of waste disposal and the presence of mosquito-infested swamps between the city and the Don River were not conducive to the maintenance of public health and prevention of infectious diseases. At the same time, some of the new public buildings were just beginning to be equipped with running water and toilets (Firth 1983).

By 1881, Toronto was spilling over into adjacent areas on either side of the city. Population in West and East York had reached 19,000 and 23,000, respectively, in an area of about 1,150 km^2, for an average density of 96 persons per 2.5 km^2. This overflow had already spawned a number of small service and shopping centres adjacent to the city in so-called dormitory settlements. As these places grew, they tended to become more dependent upon Toronto for the provision of water and sewer services. By 1861, the Village of Yorkville, just north of Bloor Street, had grown large enough to justify building the first street railway to carry commuters south to the business and commercial area of the city. By 1883, Yorkville had reached a population of 2,200 in an area of less than 2.5 km^2. The growing pollution of nearby streams was direct evidence that the village had already outgrown its own water and sewage disposal systems and was having difficulty finding a solution.

This problem was resolved when Toronto established a water storage facility on a hill north of Yorkville and the city council moved to annex an area including Yorkville in order to extend adequate water,

By 1861, the Village of Yorkville, just north of Bloor Street, had grown large enough to justify building the first street railway to carry commuters south to the business and commercial area of the city.

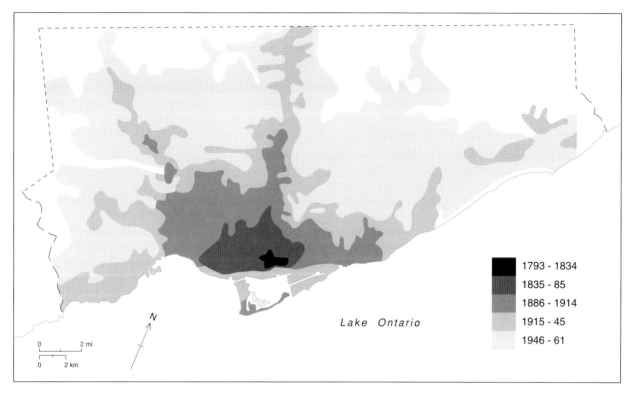

FIGURE 5.4 Growth and expansion of the Toronto metropolitan area population, 1793-1961.
(Modified from Kerr and Spelt, *The Changing Face of Toronto* [Ottawa: Queen's Printer, 1965], 97)

sewerage, and other services. This was the first annexation by the city since its incorporation, but certainly not the last. Between 1883 and 1912, through a series of annexations, the city's area increased to around 82 km^2, which, with continuing immigration, boosted Toronto's population to almost 400,000 (Figure 5.4). Such growth required a commensurate expansion of the city's services with respect to paving streets and sidewalks, as well as providing garbage collection, public street lighting, and police and fire protection.

In Ontario's more sparsely settled rural areas, the problems of obtaining water, disposing of garbage, and using privy-pits, while often having negative impacts on the environment, never seemed serious enough to cause public concern. But these same problems increased dramatically in the more densely populated and rapidly growing urban areas, and could not be so easily ignored. For example, in the early years, privy-pits were common in Toronto. They were generally connected by surface drains to the streets, or were emptied manually. The surface drains, in turn, emptied into nearby streams, unless connected to one of the larger sewers that had been constructed to flow east to the Don River or southeast into Lake Ontario.

The city had taken the initiative to construct sewers as well as to pave the streets and sidewalks, but as residential areas began to expand rapidly, sewers had to be extended or constructed on a voluntary and locally financed basis. It was difficult to keep up with the need. Despite sewer line construction throughout the 1870s and early 1880s, the city's engineers found that over half of the 11,000 homes they visited in 1885 were still using privies. Of these, 28 percent were full, and another 20 percent were classified as foul (Brace 1995).

Yet an increase in sewerage lines was not the entire solution. By 1891, an estimated 10 tonnes of untreated solid waste were being dumped into Lake Ontario each year. A by-law passed in 1908 stopped the practice of dumping raw sewage, and several attempts were made to pass legislation to improve the system by constructing sewer trunk lines, septic settling tanks, disposal plants, and a water filtration plant. By 1930, over 1,084 km of sanitary sewers and 104 km of storm and relief sewers, as well as additional private drains from new buildings

Urbanization and the Changing Built Environment

The City of Toronto and its surrounding suburbs have fanned out dramatically from the three and a half hectare site of the original settlement. By the end of the nineteenth century, urban encroachment was evident not only in the decline of Ontario's rural population, but more specifically in the reduction of rural farm acreage, as well as in the diminishing size of woodlands and farm woodlots in the areas adjacent to Toronto.

At the beginning of the twentieth century, only 24 percent of the population of Peel County was urban, and more than 90 percent of its land was taken up by farms. Although the population of York County was already more than ten times that of Peel in 1901, the pattern of decline in farm acreage in both counties was similar, dropping from about 90 to around 60 percent by 1981. Woodlands were equally endangered. Over the same period they had been reduced to just 3 percent of York's land area, and to just 2 percent of Peel's.

In Toronto, the built environment has essentially replaced the natural environment. On King Street, buildings that had reached seven floors in the 1880-90 period of redevelopment and expansion were dwarfed by 20-storey buildings by the eve of the First World War. Prior to the Great Depression, the Royal York Hotel, with 26 storeys, was the highest building in the British Empire. It became the elegant postcard symbol of 'Toronto the Good' for many years, even after taller structures had been built.

Transportation arteries form an important part of the built environment. Much of the urban space that has not been taken up by countless public and private structures has been paved over to provide sidewalks, streets, and expressways. The latter accommodate a staggering number of vehicles. Each day 750,000 vehicles enter the new City of Toronto (Metro Planning Commission 1995). The combination of traffic congestion and urban sprawl makes the average daily commuter time in Toronto 60 minutes, one of the longest for any Canadian city (Vincent 1998; Boothroyd 1998). Although public transit peaked in 1988 and has been declining since, Toronto residents used the transit system 372,000,000 times in 1996 (Boothroyd 1998).

to the street line had been constructed, including a new pumping station and filtration plant.

OVERFLOWING THE CITY

In 1931, the suburban population was less than a third of that within the city's boundaries. Within two decades, with the population of the city itself actually declining, the number of people living in the suburbs had surpassed that of the core. By 1981, the suburban population was more than four times as large as that in the city. This trend has continued and by 1996, when the suburban population within the Census Metropolitan Area (CMA) reached over three million, its population was five and a half times larger than that within the city proper.

The continuing but uneven growth and distribution of the population between city and suburbs, and the difficulty of the fast-growing suburbs in providing necessary services, ultimately led to corrective provincial legislation with the Municipality of Metropolitan Toronto Act of 1953. Under this act, Metropolitan Toronto became a federation of 13 municipalities, which were later consolidated into the City of Toronto and the boroughs of York, East York, Etobicoke, North York, and Scarborough.

Metropolitan Toronto assumed responsibility for the major regional services, which included debenture borrowing, public transportation, wholesale water supply, trunk sewers, regional parks, police protection, and expressways (other than the provincial highways and major arterial roads). Municipalities retained responsibility for neighbourhood parks, local water and sewer systems, garbage collection, fire protection, local streets, electricity distribution, and property taxation.

The major growth and urbanization of Ontario's population that have occurred in the Toronto area have resulted in the periodic adjustment of boundaries to maintain some degree of congruence between the relatively arbitrary political boundaries and those of the actual expanding and functionally interrelated urban community. The space occupied by the City of Toronto changed very little between 1951 and 1996, while the borders of the Toronto CMA were redefined at each subsequent census to ultimately include an area of almost 6,000 km^2. By the 1996 census, there were a little over 4,275,000 residents (Statistics Canada 1997).

The almost unavoidable consequence, arising from the problems of governing and servicing this continually expanding and complex urban population, occurred on 1 January 1998. The provincial government amalgamated the borough of East York, and the cities of Etobicoke, North York, Scarborough, and York with the City of Toronto to form Canada's largest city. According to the 1996 census, this newly created megacity would have a population of at least 2,390,000 in an area of about 650 km^2, creating an average population density of almost 10,000 persons per 2.5 km^2.

PRESERVING OPEN SPACES AND PARKS

Toronto's record of preserving or creating open spaces and parks has never kept pace with the city's growth and development. Visions of a planned city with green belts of undeveloped parklands and open spaces, along with wide tree-lined boulevards and lakeside esplanades, belonged to the early colonial governors and their planner architects.

Planning was tolerated only when rapid population growth threatened to create major public health problems because of inadequacies in safe water supply and waste disposal. Although interest in planning finally began to rise in the 1940s, it did not become very effective until the 1960s and 1970s, when public resistance developed in response to the seemingly concerted effort of developers and the provincial government to build more expressways in and around the city. Environmental groups gained sufficient support to slow down and even stop major projects by requiring environmental impact studies to be carried out before development could proceed (see also Chapter 16, 'From Acquisition to Restoration').

Toronto's expanding postwar economy and population exerted considerable pressure on the Metropolitan Planning Commission's land-use plans. The enactment and enforcement of building codes as well as land-use and zoning regulations since the Second World War have not achieved a utopian cityscape, but they have probably helped to minimize the worst excesses of rapid and unplanned urban development. Landfills have been converted to parks and golf courses. In an effort to keep people in the downtown areas and the streets user friendly after business hours, developers in the central business and commercial districts have been offered incentives to make provision for open space and green areas and to reserve a certain percentage of their building space for residential units.

Contrary to the situation that existed during Toronto's early years, when forests and undeveloped space were seen as obstacles to settlement, it now seems that there are not enough trees, parks, or green space to provide a healthy environment for city dwellers (Figure 5.5). Less emphasis is being placed on the maintenance of the traditional British formal parks, with manicured lawns and formal flower gardens, in favour of converting vacant lands and even existing parks into more natural forest and meadow landscapes. With the ageing of the population, there should be a continuing increase in the number of naturalists and walkers seeking out undeveloped areas for bird watching, nature study, and other relaxing activities. Yet, in the increasingly diverse urban population, there are many who still prefer the older style of formal gardens, such as Toronto's Edwards Gardens (Mandel 1998). With the changing fashions in photography, Edwards Gardens has become so popular as a setting for wedding photos that the city has had to charge a fee for appointments and set strict time limits on site use in order to limit congestion and to minimize public disorder.

Cloud Gardens

New parks, like Toronto's Cloud Gardens, are bringing new life to the downtown office canyons of stone, steel, glass, and concrete. Highly innovative and occupying only a small section of an existing downtown park, Cloud Gardens has been built on the roof of a parking garage, with a three-storey waterfall and a jungle of tropical trees within a glassed-in conservatory. Designed to convey the sense of a well-preserved urban ruin (Mandel 1998), Cloud Gardens is one of the more recent examples of a developer assuming an obligation to the city. In exchange, the developer was given the right to build an office tower that would have exceeded height limitations specified in the local zoning regulations.

With the renewed interest in the need for urban planning since the Second World War, urban planners and developers are beginning to work together in more creative ways to open up the city and to make the cityscape a more useful and user friendly environment.

FIGURE 5.5 Aerial photographs (1957, top, and 1989, bottom) provide a bird's eye view of the extent of Toronto's growth and development in just over 30 years. Non-built-up areas shown in green. (Northway-Photomap Inc. Coloured areas were identified by B.I. Roots and S.L. Pettigrew and are for general reference only).

PROTECTING THE QUALITY OF LIFE IN THE MEGACITY

As established landfill sites begin to overflow, solid and hazardous waste disposal has become an increasingly serious problem for the rapidly growing municipalities that constitute the new megacity of Toronto. Attempts to establish new regional landfill sites have been doomed to failure: no one outside the Metro Toronto area wants the city's mega-garbage in their backyard, nor do they have the capacity to absorb the sheer volume of solid waste being generated by the megacity's residents. Metro's system of seven transfer stations and two landfill areas were operating close to their maximum capacity in 1994, when they moved and dumped 1,800,000 tonnes of solid waste while diverting another 206,127 tonnes from their landfill sites through waste reduction and recycling programs. But, as the population increases, so does the volume of solid and hazardous waste needing disposal. There is a never-ending need for more efficient waste-reduction and recycling programs to ensure community health (Metro Planning Commission 1995).

THE SOCIAL AND ECONOMIC FABRIC OF THE MEGACITY LANDSCAPE

In Toronto's early years, its population was relatively small and largely British – the English, Irish, and Scottish predominated throughout most of the nineteenth century. The picture began to change during the early 1900s, when record numbers of immigrants began arriving from the European continent. By the Second World War, almost one-fifth of the city's population was not of British origin. One of the most dramatic demographic changes has been in Asian immigration. The 1851 Census reported no Asian immigrants living in Toronto. Even 100 years later, only 1 percent of the Toronto CMA population was of Asian origin. However, after 40 years of a more open immigration policy, the proportion had increased to 15 percent of Toronto's population by 1991. Such dramatic changes in immigration patterns greatly enriched the city's cultural milieu by increasing its ethnic diversity.

In 1993, an average of almost 119,000 persons a month sought and received general welfare assistance in Metro Toronto.

In addition, changes in immigration policies have affected the size and character of Toronto's labour force in response to the demand for the skills required to operate and maintain the city's evolving social, political, and economic systems and their supporting infrastructures. In 1991, there were one and a half million in the labour force. The trade and manufacturing sectors of industry employed over 200,000 workers each. Another 125,000 were employed in finance, insurance, and real estate, while health, social, and educational services accounted for another 182,000. Seventy-four thousand were working in construction industries; 72,000 in government service; 48,000 in communications and other utilities; 37,000 in transportation and storage industries; and not unexpectedly, only about 6,000 in primary industries. Even more revolutionary than the shift in employment from primary and secondary to tertiary industries has been the increasing participation in the labour force of women in general and married women in particular, even those with children still at home. The occupational characteristics of the female labour force still differ from those for men; it is still skewed in the direction of traditionally defined women's work, but to a much lesser extent than in the past. The megacity clearly needs to maintain a labour force of both men and women with the wide range of educational, technical, and professional skills necessary to support the increasingly complex urban community that Toronto has become.

THE MEGACITY: MORE OF EVERYTHING?

Socio-economic change and demographic trends will continue to challenge the megacity's ability to provide a suitable living and working environment for its population. With periodic economic instability and the restructuring of business and industry to compete in global markets, and increasing mobility of the labour force, more people than ever before are in need of assistance in finding employment, education, job retraining, affordable housing, health care, and respite care for the elderly and infirm in their families. Such needs are present in most Canadian cities: the uniqueness of Toronto's urban problems lies not so much in their nature as in their numbers. For example, in 1993, an average of almost 119,000 people a month sought and received general welfare assistance in Metro Toronto.

If Toronto continues to have such large numbers in need of welfare, jobs, or shelter, there is also a positive side to the slate. Toronto has the country's largest labour force. It has continued to be the favoured location for major private companies serving the national market and the primary destination of arriving immigrants. Its resources in general are reflected in the largest projected budget for any city in Canada – $5.9 billion for 1999. *Canadian Geographic* calls Toronto a megacity because it is bigger than any other Canadian city in every respect. Hogtown now embraces one of every 123 Canadians.

The new megatropolis of Toronto has more firefighters, more libraries, more everything. And, with almost 5,000 police officers and 470 doughnut shops within the city limits, there is almost one cruller-and-coffee shop for every 10 of the city's finest (Vincent 1998). From the present perspective it seems realistic to assume that the new megacity of Toronto will become increasingly culturally diverse, and that it will continue to expand into the twenty-first century as Canada's pre-eminent commercial, financial, and cultural centre.

Part 3
The Past and Present Natural Environment

6

Ecology, Ecosystems, and the Greater Toronto Region

Ann P. Zimmerman

This chapter is intended to provide some basic background on ecology generally, and on ecosystems in particular, to assist the reader to understand the significance of the later chapters that deal with the impacts of growth and development in the greater Toronto region on various groups of organisms, both plants and animals. This basic background would be applicable to almost any area of the planet that has experienced the heavy hand of *Homo sapiens* – and that means almost anywhere on Earth. However, we have attempted to winnow from the voluminous literature on ecology and ecosystems only those concepts and principles that seem to have the greatest relevance to the greater Toronto region and what has happened there over the span of time covered by this book.

Science is a way of looking at the natural world. There are many ways to look at the world and explain what we see, but scientists believe that natural events follow predictable patterns. And if we observe these patterns carefully and analyze them properly, we will be able to understand them, explain them, and predict what will happen to them in the future.

Ecologists ask a variety of questions. Why are there so many different kinds of organisms in the world? To what extent do predators control the numbers of their prey or numbers of prey determine how many predators there will be? What other factors impinge on organism densities? Why do male and female cardinals look so different while the sexes in blue jays are almost indistinguishable? Why are dandelions so common and lady's slipper so rare? Why do some female fish lay millions of eggs while female birds lay comparatively few?

Most ecologists are also concerned about degradation of the environment. Perhaps better than anyone else, ecologists recognize the short- and long-term implications of environmental change for the sustainability of organisms. We consider the special places of the greater Toronto region highlighted here to be special because they still have some degree of ecological integrity: most available habitats are occupied by native species rather than being overwhelmed by non-native, exotic species, and the feedback loops between the indigenous organisms and their environment are still operating to some extent. All these factors help to ensure the self-organization and repair fundamental to maintaining ecosystem processes.

We recognize that there are a number of criteria by which we could designate the ecosystems of the greater Toronto region that are still functioning as 'special.' Some of those criteria would be scientific; others might be aesthetic, economic, or spiritual. All these are integral to environmental protection.

ECOSYSTEMS AS THE FUNDAMENTAL UNIT OF LIFE

In the preceding section, we catalogued some questions that ecologists might ask. Ecologists also want to know how many fish we humans can catch before we interfere with the ability of the population to sustain itself. How many trees can be harvested in a forest before we interfere with the forest's ability to regenerate trees or impair other forest functions like water regulation?

The only difference between our first set of questions and the latter one is the explicit benefits humans are deriving from ecosystems. But ecosystems provide

Ecology and the Process of Science

It is easy to equate ecology with a concern for the environment. An environmental ideology is neither better nor worse than ecology, but it is different.

Science has specific protocols by which it tries to reach an understanding of the natural world. Those protocols compel scientists to take their observations about the world and recast them as hypotheses. Hypotheses are explanations for a particular observation, but they have to account for all the available information. Furthermore, they have to make predictions. It is the process of testing the accuracy of our predictions that is at the heart of the process of science.

numerous, less obvious benefits to humans. Ecosystems are responsible for the balance of oxygen and carbon dioxide in the atmosphere; they regulate the climate through their influence on temperature and precipitation; ecosystem processes produce soil and decompose dead organic material.

When we think about surviving, of what it means to be 'alive,' we tend to think in terms of individual organisms. But remaining alive depends on more than just the individual organism. The air we breathe, the water we drink, the food we eat is the net result of the interactions of many organisms in a variety of locations over long periods of time. Ecosystems are special because they are the fundamental unit that permits life on Earth to continue.

In an article published in the journal *Nature,* a group of ecologists and economists (Costanza et al. 1997) identified 17 ecosystem 'services,' whose total estimated annual value was $33 trillion, twice the gross national product of all the countries on Earth. The authors' point was that humans should account for the costs of losing an ecosystem service in advance of contemplating its destruction.

As an example, Costanza and his colleagues estimated that ocean sediments provide an annual carbon dioxide removal service that would cost carbon emitters over $1.2 trillion annually if they had to pay to remove similar amounts of carbon from their effluents. Yet we continue to treat the oceans carelessly, failing to recognize the value or 'natural' capital they represent. The in-filling of the Ashbridge's Bay wetlands in the last century begins to take on new significance when Costanza et al. point out that the decision costs us close to $15,000 (1994 US dollars) each year for every hectare of wetlands lost. Those expenditures are needed to repair storm damage, dredge sediments, remove nutrients, and re-stock fish – just some examples of the services provided by wetland ecosystems. Figures 6.1 and 6.2 show the dramatic amount of in-filling that the area has undergone.

ECOSYSTEMS: EARLY HISTORY

This book introduces you to a variety of ecosystems that on first glance appear to have little in common. Close your eyes and imagine walking through a forest ecosystem. For most of us that picture will be dominated by trees. But what images emerge if we think about the Lake Ontario ecosystem? Blue sparkling water may be our first thought but, if we think about lake organisms, we're most likely drawn to the fish, the organisms at the top of the food chain. Fish and trees represent the majority of the biomass in their respective systems, so is it just overwhelming biomass that grabs our attention? But why is biomass distributed so differently in these systems? If mass is a defining criterion, can we reconcile the dynamics of a natural ecosystem with the massive but inanimate elements of the built environment? We might begin to wonder if it's possible to make any generalizations about ecosystems!

Reconciling such a diversity of examples was exactly the situation faced by the early ecologists. The perspective that the earth was not immutable, that it had changed over time, was startlingly new in the middle of the nineteenth century. Most of the discoveries that now underpin modern biology lay in the future. Darwin had just published *The Origin of Species* (1859). Mendel's work in genetics was published, but would be ignored for another 40 years. Genes and chromosomes were unknown. The structure and function of macromolecules like chlorophyll,

> *A group of ecologists and economists identified 17 ecosystem 'services,' whose total estimated annual value was $33 trillion, twice the gross national product of all the countries on Earth.*

▲ FIGURE 6.1
Ashbridge's Bay, 1880s. (City of Toronto Archives SC244-248)

◀ FIGURE 6.2
Aerial view of the Ashbridge's Bay area today. (City of Toronto Archives SC244-1440)

proteins, and DNA would not be understood for 100 years, a time frame spanning two, three, even four generations of scientists.

Earlier ecologists did not do 'bad' science just because they were technologically limited. As scientists do today, they made careful observations, developed vocabulary, and tried to draw global generalizations from their particular (and often regionally peculiar) environments. Just as they could not have imagined how ecology would develop, it is equally hard for us to envision the tools and insights of the next hundred years and hence to know what coming centuries' versions of 'truth' will be.

Most of us do not have access to the sophisticated technologies that today's ecologists use, nor to the accumulated body of knowledge prerequisite to asking questions at the discipline's cutting edge. But we can observe the world just as those early ecologists did by simply looking around the greater Toronto region. We can try to develop our information bases by asking questions, reading literature, watching television, listening to radio, and surfing the Internet. We can rise to the request at the end of this chapter to reconnect with the natural world and help in the accumulation of the huge databases that ecologists often require. Perhaps easiest of all, each of us can take advantage of the ecologists and naturalists who authored chapters in this book and allow them to look over our shoulders as we walk through and read about the greater Toronto region's special places.

SOME IMPORTANT IDEAS IN ECOLOGY

Science is about accumulated knowledge. Ideas and hypotheses that turn out to be wrong are discarded; ideas that continue to work are expanded, refined, and further tested. In understanding and appreciating where ecology is today, it is instructive to look at where it came from – the ideas that are still working, still generating new hypotheses, still advancing our understanding. The accompanying list, like any other, is probably biased, but we have tried to highlight ecologists whose ideas help all of us think about the natural world in a variety of different ways. We believe you will see echoes of their science in the chapters that follow, both in the words that are used and in the ways the authors of those chapters look at their respective systems within the greater Toronto region.

Ecosystem Ecology

Of course, all ecology is about ecosystems, but ecosystem ecologists are those scientists concerned about the properties of the complete system: the organisms and their physical-chemical environment. Some of the scientists who have shaped our thinking about systems at the level of the complete system are Frederic Clements and his antagonist H.S. Gleason, as well as Charles Elton, Raymond Lindeman, and Eugene Odum.

Frederic Clements was a plant ecologist. He was born in Nebraska just as agriculture was altering the face of the Great Plains. In 1916 he published his *Analysis of the Development of Vegetation,* in which he suggested that ecosystems are discrete, highly integrated units. It was Clements's view that ecosystems functioned as 'supra-organisms,' responding to disturbance in predictable ways and acting as a unit rather than as a collection of isolated, individual species.

Most ecologists consider the process of ecosystem disturbance and recovery – 'ecosystem succession' – to be a serial replacement of species that is 'non-seasonal, continuous, and directed' (Begon, Harper, and Townsend 1996). The familiar pattern by which abandoned agricultural fields in southern Ontario gradually fill in with herbaceous plants, then various softwoods, followed by hardwoods, is an example of succession. The traditional, Clementian view of disturbed systems was that they would pass through an orderly series of stages, eventually reaching some climax or equilibrium state in which there might be some species turnover, but a relatively consistent set of species would persist until some new disturbance restarted the process.

An alternative view to Clements's discrete community was presented by his contemporary Henry S. Gleason, a British ecologist. Gleason's experience was in systems long exposed to human interventions, though at much smaller scales than the disturbances taking place on the American plains. In Gleason's view, species' co-occurrences were simply due to chance: species arriving at the site prospered if they had similar environmental requirements, perhaps light, temperature, or nutrient levels or resistance to predators. Even if plant species co-occurred at numerous sites, they were simply species with a tolerance to or even preference for similar site conditions.

The Gleasonian model is one of species existing along environmental gradients, with no particular equilibrium or 'climax' condition. In this model, biological change continues as long as environmental change continues, stopping when environmental change ceases. The collection of species we find in a particular location is predictable only from their respective physiologies and the degree of disturbance on-going in the environment. Under a regime of severe or frequent disturbance, the biological community would be made up of species with wide tolerances and/or good dispersal/colonization skills (often referred to as 'generalist' or opportunistic species). When disturbance is mild or rare, the biological community should favour organisms with skills more closely matched to the specific environment, perhaps species with the ability to compete effectively for light or nutrients or with clever mechanisms for protecting their seeds from herbivores (so-called 'specialist' species).

Rather than being right or wrong, Clements and Gleason provide ecologists with two models against which they can test hypotheses about the distribution of organisms and their response to disturbance. Plant ecologists find little evidence that plants are selected as a group of species or that there are discrete boundaries between one community and the next. Plant communities seem to change along continua as predicted by Gleason. Animal ecologists, whose experiences include numerous examples of animals quite rigidly linked to the presence of other plants or animals, may be more open to a Clementian view.

Some ecologists maintain that the only dichotomy between Clements and Gleason is one of scale. Gleason's view was a very fine-scaled perspective of the English countryside, while Clements was looking at almost half a continent. Viewed at a continental scale, we can see where Clements is right. Northeastern North America is clearly a zone of mixed deciduous forest and, in the absence of human intervention, the species making up that forest are reasonably predictable. Nevertheless, as the pollen histories from the Humber valley show us, Gleason is also right. The relative numbers and frequencies of species at specific locations within the mixed deciduous forest have constantly shifted, with little evidence of specific plant alliances.

A.S. Watt is credited with harmonizing these issues of scale in his now classic 1947 paper in the *Journal of Ecology*, 'Pattern and process in the plant community.' Watt's perspective was that communities are neither fixed assemblages of species nor mere random co-occurrents in mutually satisfactory environments. He saw the landscape as composed of patches in different stages of recovery from disturbance. So as fire skips over the prairie or spruce budworms devastate a conifer stand, a mosaic of patches result, constituting a dynamic landscape.

Are there 'patterns' present in the distribution of organisms within a system? Are there communities of species characteristic of particular habitats? Are there species

Succession at the Leslie Street Spit

Succession can be described in terms of primary and secondary processes. Ecologists tend to reserve the term primary succession for the colonization of 'new' surfaces without seed banks (e.g., a newly exposed sandbar in a river or a recently delivered load of construction fill at the Leslie Street Spit). Secondary succession refers to the recovery or return of vegetation to an area following a disturbance. In secondary succession, well-developed soils and a seed bank remain in spite of factors that have removed vegetation (e.g., fire, disease or insect infestation, logging or land clearing).

As you walk around the Spit from newer to older areas, you should be able to find first successional stages where wave action is breaking down rubble into sand, later-stage examples of the early grassy 'pioneer' species, and other areas that are in transition from grasses to small shrubs to trees, with concomitant reductions in the grasses. These different plant 'communities' are referred to as seral successional stages.

While it is a generalization, earlier seral stages often have lower diversity (fewer kinds of organisms) than later seral stages. Even if you can't identify all the organisms you see, try comparing the number of species present in similar sized areas in different seral stages. How do you think the Spit will change over the next 10, 20, or 50 years? If you can visit the Spit often, you might consider taking a photograph at the same spot at regular intervals to record the changes. You'll find additional information on the Spit at the Toronto Regional Conservation Authority's website: http://www.trca.on.ca/ttp.html

whose presence is a prerequisite for the presence or absence of others? If we remove the agents of disturbance, do ecosystems eventually return to some climatically dictated equilibrium? These are difficult questions to answer, but the search for those answers reveals an element of the Clementian/Gleasonian rift less easily reconciled than simple issues of scale.

Gleason was a reductionist. His approach to understanding the collection of species present in a particular place was to tease out their individual environmental requirements and map them onto the characteristics of a particular habitat. Reductionist science proceeds via controlled, replicable experiments.

Unfortunately, ecology only rarely provides situations that can be replicated and controlled. Attempts to create replicate ponds of differing nutrient status in order to evaluate their role in aquatic plant communities are almost always subverted as each supposed replicant develops a unique flora and fauna. In trying to evaluate the impacts of an industrial discharge on a stream, we often sample above and below the outfall. Yet what happens below the outfall depends on both the effluent and its interactions with the ever changing chemistry of the stream water itself. How can we 'know' what effects are due to effluent alone when our samples are not independent from each other? How precisely can we quantify environmental exposures (e.g., the actual amount of a contaminant an organism absorbs from heterogeneous food sources or the total body burden of contaminants acquired from a variety of natural and anthropogenic sources)? How do we predict the environmental risks of new technologies (e.g., the risk of 'escape' of genetically engineered genes into the wild gene pool)? Large-scale ecological processes like climate change are extremely difficult to replicate, while it is simply impossible to go back and test the ecological implications of the extinction of an organism, say the passenger pigeon.

An alternative view of ecosystems emerged in 1927, when British animal ecologist Charles Elton published his book *Animal Ecology*. One of its important observations was that, in any particular place, there are always larger numbers of plants than there are plant eaters. Elton suggested that this relationship could be visualized as a pyramid, with plants providing the base of the pyramid upon which successive layers of plant-eating herbivores and herbivore-eating carnivores could be piled.

From Elton and the ecologists who followed we understand the two main ways in which organisms derive energy. Autotrophs (plants and a few specialized photosynthetic bacteria) 'feed' themselves (from *auto* for 'self' and *trophic* for 'feeding'). With only a few exceptions, autotrophs are able to 'capture' electromagnetic radiation or 'solar' energy from the sun and convert it to chemical energy through photosynthesis. Heterotrophs or 'other' feeders obtain their chemical energy by eating plants or animals, whether living or dead.

Elton's ruminations on the food and the enemies of animals led him to think carefully about the roles that organisms play in their environments. It was Elton who gave us the term 'niche': that specific combination of physical, chemical, and biological conditions that is unique to each species.

Eltonian pyramids and niches were important conceptual advances in ecology. They provided explanatory hypotheses for the structure of ecosystems. Refinement of the niche concept led to decades of work on another ecological tenet, competition and its corollaries – competitive exclusion versus coexistence, a discussion we will return to in our consideration of population ecology.

Many of the most important generalizations in our understanding of ecological relationships did not emerge until the 1950s, when Eugene Odum (with his brother Howard) popularized the use of the common currency of energy to compare different kinds of ecosystems. Much of the credit for the idea of studying ecosystems as a function of energy flow belongs to Raymond Lindeman. As a young ecologist, Lindeman operationalized the Eltonian pyramid by mapping the flows of food (in the form of living and dead organisms) between adjacent layers of the pyramid, or 'trophic' levels, and expressing that flow in units of energy. Lindeman had just taken a position at Yale University after completing his PhD and was in the process of rewriting the ideas from his thesis for publication when he died from complications of a liver ailment. His classic 1942 paper, 'The trophic-dynamic aspect of ecology,' was published posthumously, and we can only wonder if we would view ecology differently had he lived.

It is not surprising that the Odums found the universality of an energetics approach appealing. All of a sudden it didn't matter that the plants in the forest were huge and woody, while the plants in the lake ecosystem were so small they couldn't be seen without a microscope. Both performed the same function: capturing solar energy through the process of photosynthesis, turning it into chemical energy, and, as they were eaten, moving that energy up the food chain.

Whether we accept the Odums' ideas in their entirety or not, production by autotrophs and the efficiency with which that production can be transferred up the food chain help us understand why the biomass of the forest ecosystem is dominated by trees while that of the lake is dominated by fish. Trees have enormous requirements for energy. They need energy to produce leaves or needles, roots, woody tissues, and reproductive structures. The microscopic algae of the lake ecosystem need none of these things. Yet the same amount of solar energy falls on both systems. The amount of energy that the algae need to keep for themselves is very small compared to the trees, so in aquatic systems there is simply a higher percentage of energy available to be transferred to the herbivores – so much more in fact that aquatic ecosystems can often support two or three more trophic levels than are commonly found in terrestrial ecosystems.

Entropy and Ecosystem Health

The Odums treated ecosystems as so-called 'black boxes,' within which the flows of nutrients or energy were the unit of description. Identifying particular species was not necessary. Ecosystems can also be treated as black boxes with respect to their information content, entropy, or ecological integrity. Entropy – or the thermodynamic state of 'disorder,' randomness, or decay – is a spontaneous process, counteracted by photosynthesis, which uses energy to create 'order.' Being alive can be thought of as a temporary hedge against the ultimate fate of individuals as well as that of the universe: maximum entropy or maximum disorder.

Since its formation, the entropy of our solar system has increased – that is, its original quota of high quality, low entropy energy is gradually dissipating to a lower quality, high entropy, less ordered state. It is the presence of life on our particular planet that has temporarily slowed the march to disorder through the structure of ecosystems. Entropy is a difficult variable to measure, but only autotrophs, with their unique ability to turn high entropy, disordered inorganic carbon into lower entropy, ordered organic carbon stand between us and entropic doom. As we restrict the number and diversity of autotrophs, we in turn restrict the number and diversity of the animals that depend on them. We increase the disorder of the planet and move it closer to its final, entropic fate.

POPULATION, BEHAVIOURAL, AND EVOLUTIONARY ECOLOGY

Why are red-tailed hawks relatively common and red-shouldered hawks relatively rare? Manitoba maples are everywhere. Black maples are not. Fireflies over a darkening field delight us for just a few short weeks in the summer. We can't seem to get rid of silverfish, and Chapter 11 tells us they were here long before the dinosaurs. Ecologists use the term 'population' to mean the numbers of organisms of a particular species within a defined area at some specific time. Asking why populations behave as they do forms the substance of population ecology and its companion disciplines, behavioural and evolutionary biology.

Charles Elton flirted with population biology in conducting the first numerical analyses of the Hudson's Bay Company trapping records for lynx and hare. However, Herbert Andrewartha and L.C. Birch really established the foundation for modern population biology with the publication of their 1954 treatise, *The Distribution and Abundance of Animals*. Andrewartha and Birch worked primarily with insect populations prone to becoming pest species.

Silverfish were here long before the dinosaurs.

From their work, it is clear that the density of a particular population can be accounted for by only four variables: births (or hatching or germination), deaths, immigration, and emigration. Focusing on the contribution each of these factors makes to the total population helps generate hypotheses about what determines the average number of individuals in a particular habitat.

In its simplest form, the determination of population size is a function of the 'carrying capacity' of the environment. Rivalry for resources increases as the numbers of individuals in the population increase. At some point the demand for resources (from the combined numbers of individuals) outstrips the supply of resources, and the population can no longer grow. It has reached the carrying capacity of the environment.

If we think about all the resources that could contribute to carrying capacity for any one trophic level – e.g., light, nutrients, water, or grazers for plants; or food, access to mates, refuge from predators, and so on, for animals – and then think about all the species of organisms in any environment, it is likely that our list of organisms will be longer than our list of resources. Hence it isn't just individuals within the same population that may be limited by resources (intra-specific competition), but different species of organisms may also be competing for access to the same resources (inter-specific competition).

In the 1930s, G.F. Gause, a Russian ecologist, published a now classic set of papers on competition between three species of the single-celled micro-organism *Paramecium*. Gause was able to show that each of the three species (let's call them 1, 2, and 3) grew well in the laboratory in water-filled test tubes containing a food supply of yeast and bacteria. In other words, the three species could individually prosper under similar conditions, or appeared to occupy the same niche. However, when Gause put the different species together, things got interesting.

When species 1 and 2 or 1 and 3 were grown together, species 1 was always the ultimate survivor. Species 2 or 3 simply disappeared. However, when species 2 and 3 were grown together they 'coexisted' indefinitely. When Gause examined his cultures more closely, he found that species 2 and 3 had divided up their test-tube habitat. Species 2 tended to feed on the bacteria, which were suspended in the test-tube, while species 3 tended to feed on the yeast, which had settled to the tube's bottom. By dividing up the habitat and feeding on different food sources, the two species no longer occupied the same niche.

Gause expressed his findings as the competitive exclusion hypothesis: no two populations requiring identical resources (or having the same niche) can coexist indefinitely. Competitive exclusion, however, presents a problem: being a superior competitor and excluding other species would seem a recipe for having very few species on Earth. At first glance, it seems reasonable to expect a few extremely competitive plants, one or two herbivores, a carnivore, and a decomposer in each different habitat. In fact, this is exactly what we see in our human-stressed urban landscapes.

Where, then, does the earth's biodiversity come from? How can there be millions of species on the earth? How can the thousands that exist in just the special places discussed in this book coexist? And can humans learn to work with natural, ecological processes rather than against them, in ways that promote diversity and ecological integrity?

Even a casual walk through a natural habitat in the greater Toronto region will reveal organisms interacting. We see different species of woodpeckers hammering away on the same trees in High Park; several species of tadpoles wriggling along in the shallows of a pond in the Don valley; a variety of pollinators all visiting the same flowers in our own backyards. If all this co-existence were a violation of the predictions of competitive exclusion, it would be a shaky hypothesis indeed. So how do organisms coexist, or, in a broader context, where does biodiversity come from?

MacArthur's Warblers and Competitive Exclusion

Ecologists have spent considerable time looking for exceptions to the competitive exclusion hypothesis. A bona fide exception to a hypothesis is extremely important in science. Science proceeds inductively — that is, from a series of observations to a generalization that must explain all of the observations. A replicable observation or experiment that contradicts the predictions of a hypothesis requires scientists to rethink the situation and develop a new, more inclusive hypothesis.

Almost from the time of Gause, ornithologists had wrestled with a competitive exclusion 'problem' posed by migratory spring warblers. Several species of warblers were often seen feeding on spruce budworm caterpillars on a single tree (i.e., apparently occupying the same niche). If this were indeed true, competitive exclusion predicts that inter-specific competition should have acted to prevent coexistence.

The late Robert MacArthur, a graduate of the University of Toronto and professor at Princeton University, resolved this apparent exception to the principle of competitive exclusion with a series of careful observations (1958). He divided the trees in which the warblers were feeding into sectors and showed that the five apparently co-existing species actually inhabited discrete locations within the tree (closer or farther from the trunk, higher or lower in the canopy). The warblers' feeding territories rarely overlapped, but where they did MacArthur showed that they were foraging in quite different manners. Some species flitted about picking off only obvious caterpillars; other species methodically searched the same areas, feeding off hidden caterpillars. MacArthur concluded that the birds were indeed occupying different niches.

The answer is not uncomplicated, and oversimplifying both the question and the answer can quickly produce tautologies: complete competitors cannot coexist, therefore there must be differences among coexisting species that permit coexistence.

Biologists generally believe that competition negatively affects populations by reducing elements like survivorship, growth, or reproduction. Individuals might 'escape' competition by behaving differently from their potential competitors. They might divide up the habitat – spatially, as MacArthur's warblers did, or temporally, with different species of tadpoles feeding at different times. Organisms might specialize on different sized food particles within the same environment (woodpeckers) or extract nectar from different parts of the flower (bumblebees, honey bees, and butterflies).

Organisms have also been shown to be flexible with respect to their niche, restricting their distribution to a narrower range of environmental conditions than they can actually tolerate. This is clearly shown in the interactions between Gause's *Paramecium* species. Species 2 and 3 share the same 'fundamental niche,' that is, the set of conditions under which each can exist. When food and space are limited, however, and hence competition might be presumed, they restrict their respective diets and locations in the test tube. The conditions under which each species exists when in the presence of the other is called its 'realized' niche. The realized niche of many species has been significantly reduced by 'competition' with human beings.

Organisms appear to have other options as well. Thomas Park's work tells us that the respective densities of competitors is as important as the frequency of environmental fluctuation. If favourable environmental conditions become unfavourable before a particular population can reach its carrying capacity, then its density may decrease while that of a competing species increases. Under a scenario of rapid environmental change, potentially competing populations may oscillate in numbers but never manage to exclude each other.

We might now broach a preliminary answer to where biodiversity comes from and how we might maintain it in our own habitat, the greater Toronto region. Those individuals that chance has favoured in ways that allow them to avoid competition would be expected to have higher survivorship, growth, or reproduction – they would be 'fitter' – than those individuals forced into energy-wasting competition. Natural selection should favour fitter individuals and, over time, the species with

Thomas Park's Beetles

Thomas Park spent most of his career at the University of Chicago experimenting with two species of flour beetles. One of his species liked cold, dry habitats while the other preferred a hot, moist environment. When forced to coexist, the beetle that was in its preferred habitat always outcompeted the other species.

Nothing unusual here: Park's experiments confirmed Gause's hypothesis of competitive exclusion. But Park was intensely curious about exactly how one species eliminated its competitor, so he began to vary the environmental conditions under which the beetles were forced to coexist. Two of Park's findings have particular relevance here. He observed that competitive outcomes were not consistently predictable under intermediate environmental conditions (e.g., cold but moist or hot but dry). The species that ultimately won depended upon the initial densities of the two species. Even the inherently inferior competitor could occasionally win if its starting densities were high enough. Ultimately, he found that the beetles could coexist if environmental conditions were variable.

behaviours permitting coexistence should come to dominate the system. Biodiversity then can be thought of as a by-product of evolution, a reflection of competition that occurred in the evolutionary past rather than the ecological present, a consequence of living in a complex niche within a complicated habitat.

Human activities in the greater Toronto region often act to reduce habitat complexity. We humans fragment landscapes and simplify habitat. As the diversity of habitat prerequisite to coexistence disappears, suites of species go with it. Some species' habitats disappear completely; others find the area of habitat remaining just too small to accommodate their needs and the needs of a competitor, so at least one species disappears. A habitat with a paucity of niches – our all too familiar expanses of mown grass interspersed with concrete – has been referred to elsewhere as an ecological slum. In reality it may be more like an ecological desert, a depauperate environment occupied by a similarly depauperate fauna, the widely dispersing, generalist species so familiar in our urban experience: crab grass, dandelions, pigeons, the Norway rat, the cockroach, the yellow jacket. These organisms exist in the same physical space, but they are hardly species interacting in ways that organize and stabilize the system.

PUTTING IT ALL TOGETHER

The Special Places, Landscape, and Conservation Ecology

Landscape ecology can mean a number of different things. In North America, the term is often applied to a new generation of architects or landscape architects – professionals whose understanding of natural environments and desire for ecological sustainability compel them to design and construct built environments in ways that protect and promote native vegetation and wildlife.

Ecologists also expropriate the term landscape ecology, but in this context as a scientific discipline emerging from ecosystem and population ecology. Landscape ecology focuses on large-scale 'spatiotemporal' patterns – that is, patterns existing across space, through time. Monica Turner has described landscape ecology as a reciprocity between spatial pattern and ecological process (Turner 1998). Landscape ecology helps us understand change in ecosystems and the implications of human activities for that change.

D.R. Foster's 1992 publication, 'Land-use history (1730-1990) and vegetation dynamics in central New England,' is often cited as an example of the multidisciplinary approach taken by landscape ecologists. Foster demonstrates that although a forest ecosystem has re-emerged in Massachusetts since the decline of agriculture, it is not the same as the original forest. Current New England forests have higher proportions of birch, red maple, and oak and fewer hemlock and sugar maple than the original old-growth forests, with concomitant implications for nutrient cycling and animal distribution.

Issues that landscape ecologists deal with – habitat fragmentation, landscape connectedness, and patch dynamics – are relevant to another emerging scientific discipline: conservation ecology. Gray Merriam, a professor at Carleton University until 1996, looked at the implications of habitat area and connectivity (a word he coined) for the population dynamics of forest-dwelling birds and small mammals in the Ottawa area. Merriam, his students, and colleagues demonstrated the role of fencerows (narrow strips of trees, shrubs, and grasses growing along the edges of fields) in facilitating animal movements among forested patches. Using a combination of field study and modelling, they suggested minimum sizes (4 km^2) and numbers of forested patches (five) required to sustain chipmunks.

Numerous studies since that of Merriam have demonstrated the reluctance of various species to cross open areas, perhaps due to increased risks of predation. We are also beginning to understand the role of system 'edge' for species richness. Edge habitats are those transition zones between two or more distinct habitats. Edge habitats are often extremely diverse, as they may contain species from both primary habitats, usually generalist species that can tolerate edge conditions. However there are also species that clearly cannot tolerate edge habitats. The forest edge, for example, will differ from the interior of the forest in having higher light regimes, drier/warmer microclimates, and higher winds. Awareness of edge effects is being incorporated into the design of parks and preserves, with the recognition that more compact shapes produce more interior habitat for the same area of preserve. Awareness of edge effects should also inform decisions about fragmentation of intact habitat patches by roads, transmission corridors, or pipelines.

Many forest species may not be able to disperse through edge habitats. Nevertheless, linking habitat patches with 'corridors' – really just edge extensions – is being advanced as a solution for habitat fragmentation and the associated species losses that accompany it. Will species that could not live permanently in edge habitat still utilize edge corridors to move between suitable habitat patches?

Holling and Ecosystems

Dr. C.S. Holling, a graduate of the University of Toronto and currently a professor at the University of Florida, Gainesville, is one of the leading ecologists of the modern era. He is co-ordinating a number of research projects around the world on the mechanisms by which ecosystems might recover from human and other impacts. He is testing the hypothesis that ecosystems do not lose their coherence gradually when subject to a harmful impact. Rather, they decline gradually to some 'thresholds' point, after which they collapse rapidly and calamitously if the stressors are not removed.

Research has shown that an isolated patch of habitat will have fewer species than an area of similar size in a contiguous landscape. In 1970, Richard Levins, a mathematical ecologist, created a model suggesting that a local population lacking sufficient habitat to remain viable could still maintain itself if it could link to individuals in other patches. Levins called these 'linked' sub-populations a 'meta-population.'

Despite Levins's convincing mathematics, ecologists are only beginning to collect the empirical data needed to test his hypothesis. If some species do exist as meta-populations, what patch sizes do they require? How far apart can patches be? Are physical corridors necessary to maintain connectivity? Hows wide do the corridors have to be? If ecologists know little about the complex biogeography prerequisite for species conservation, we know almost nothing about how these issues could apply in the greater Toronto region. What is the role of the special places or other patches of natural habitat in maintaining species in the greater Toronto region? What species currently here already exist as meta-populations? Could currently extirpated or threatened species be restored or rehabilitated using a meta-population model?

We do not yet have the data we need to answer the questions we've raised; yet we seem to persist in behaviours that continue to homogenize the environment. Our greatest challenge is to 'package' the burgeoning human population (in the greater Toronto region as well as on the planet) in ways that have the least impact on natural ecosystems. Maintaining special places is not just about aesthetics or recreation or ecosystem services. The ecological integrity displayed in the special places is the only way we have to bring self-organizing, self-regulating structure into the biosphere. We need to protect the areas of integrity that still remain and learn how to bring integrity back to areas where it has been lost. Most important, all of us need to understand why this is necessary. Peter Raven (1995) expressed it well when he wrote, 'At the deepest level, the most critical environmental problems, from which all others arise, are our own attitudes and values. We are totally out of touch with the world we live in, and until we reconnect and readjust, solutions to environmental problems will continue to be stopgap ones.'

We need to protect the areas of integrity that still remain and learn how to bring integrity back to areas where it has been lost.

What One Person Can Do

- Use this book as an opportunity to get in touch with wild areas of the city by visiting one or more of the special places.

- Enjoy the variety of living things with which we co-exist by learning to identify them. Bookstores and libraries carry a variety of field guides, or you could create your own personal field guide using the web-based resources of the Royal Ontario Museum: (http://www.rom.on.ca/ontario/fieldguides.html).

- Volunteer to assist censusing frogs (http://www.zoo.utoronto.ca/~natalie/frogpage.html), butterflies (http://www.naba.org), lady beetles (http://www.schoolnet.ca/VP-PV/labybug/) or the birds that visit your feeder (http://www.news.cornell.edu/Chronicles/2.12.98/ bird_count.html).

- Increase the wildlife value of your outdoor space by planting native vegetation. Visit the Federation of Ontario Naturalists' website for renaturalization ideas – http://www.ontarionature.org – or consider joining the National Wildlife Federation's Backyard Wildlife Habitat program (http://www.nwf.org/nwf/habitats/index.html). You can view a slide show on native plants at the Great Lakes Basin website (http://www.epa.gov/greenacres).

- Join one or more conservation-oriented organizations whose mandates mesh with yours. A list of non-governmental organizations can be found at http://www.web.apc.org/~oen/groups.html. This website lists NGOs for the Toronto area, for Ontario, and across the country. The Toronto Field Naturalists are one organization specifically oriented to the greater Toronto area.

7

Vascular Plants

James E. Eckenwalder

The climate of Toronto ensures that, in the absence of disturbance and special site conditions, forests will cover the landscape. What is left of Toronto's natural forests is confined largely to the city's major ravines and their tributaries, the Humber, Don, and Rouge River systems, and Etobicoke, Mimico, and Highland Creeks. Precious little is left of the forests of the uplands, where we now live, work, and play. Even less remains of the distinctive wetland plant communities of the uplands, ravines, and shoreline regions that have been largely eliminated ('reclaimed') and converted to urban uses by dredging, filling, and landfill operations. Nonetheless, the landscape of urban Toronto has numerous trees concentrated along roadsides and lot lines and in yards, parks, and cemeteries. This urban forest and the other urban plant communities of the city's uplands have numerous structural differences from the natural forests. The anthropogenic communities that are created by people may have more species in total than the natural forests, but they are structurally less complex and the trees typically don't form a continuous closed canopy.

Although the vegetation of Toronto is not as complex as that of some tropical regions, the remnant native forest communities here typically consist of as many as ten layers of plants, each adapted to different light and humidity levels, flowering and fruiting seasons, and pollination and seed dispersal methods. White pines form a scattered supercanopy emerging above the continuous canopy of about a dozen most frequent deciduous hardwoods and less abundant and diverse conifers. The canopy also contains the crowns of woody vines like wild grapes and Virginia creeper. It overshadows and protects, but also competes for resources with the successively shorter layers of subcanopy trees, tall and low shrubs, perennial herbaceous plants, and mosses and lichens.

From the tallest trees to creeping flowers at their roots, these forests, and most of the non-forest plant communities here as well, are populated primarily with vascular plants. All of our most familiar land plants, except for mosses and their relatives, are vascular plants – that is, plants that contain cellular plumbing (vascular tissue) for transporting water and water-soluble materials such as minerals and sugars. Vascular plants are the most varied of the land plants, embracing ferns, horsetails, club mosses, and spike mosses (neither of which is a true moss), conifers, and flowering plants, these last the most numerous and diverse subgroup. Vascular tissue, in the form of wood, is also the mechanical support that permits tree growth and is thus responsible for the stature of our forests.

The green land plants that lack vascular tissues – the mosses and liverworts – cannot build self-supporting stems more than a few centimetres tall, although they may grow so densely in favourable conditions, like sphagnum bogs, that they form mats that are metres thick, of which only the tips are alive, green, and functioning. Such habitats are quite rare in Toronto, so almost all of the biomass of green plants consists of vascular plants.

The city straddles the transition between two of the four major latitudinal bands of vegetation in Ontario: the southernmost deciduous forest region, or Carolinian Zone; and the mixed forest region, or Great Lakes-St. Lawrence Zone. These regions are climatically determined, primarily by the extremes of winter cold and by the length of the growing season. The city, of course, creates its own, slightly warmer than expected mesoclimate (see Chapter 2), so we can grow tenderer trees and shrubs here than we might otherwise. However, this effect has apparently not yet had a measurable impact on the native plant communities, though it surely will in future, especially in concert with global warming.

Most of the remnant forest stands in the ravines are fairly typical of the mixed forest, often dominated by sugar maple, beech, basswood, white cedar, hemlock, and white pine, but deciduous forest species like sassafras and black oak may be found in lesser abundance, particularly in drier soils, such as those that characterize High Park. A very few locations in the city, such as the boggy West Don Parkland south of Finch Avenue, support a few boreal forest species that are characteristic of northern Ontario, including balsam fir, tamarack, and black spruce.

Because of pollution, competition from invasive alien species, and both purposeful and inadvertent disturbance, the preservation of the native flora of the region is roughly directly proportional to the height of the species. Thus, there is a higher percentage of our tree species than of the other growth forms, and the lichens and mosses are, perhaps, least persistent. The latter were far more susceptible than were the vascular plants to the increasing air pollution that followed intensifying industrialization and heavier automobile traffic after the Second World War. Even a relatively innocuous and passive activity like walking in the woods tramples the forest understorey and suppresses the regeneration of many species. As a consequence, Toronto is not a great reservoir of rarities. Instead, the native plants represent a typical expression of the most robust and tolerant species that are widespread in the surrounding region. The city now supports only about one-third of the native vascular plant species of the larger local region listed by Riley (1989), and many species listed by Scott and Ivey in 1913 are no longer found here. For instance, sand cherry (*Prunus pumila*), poison sumac (*Toxicodendron vernix*), burning bush (*Euonymus atropurpureus*), and swamp blueberry (*Vaccinium corymbosum*) all seem to have vanished from Toronto with the disappearance of most swamplands from the lower Humber valley.

Since most of the landscape is dominated by human activities, the majority of the plant cover in the city consists of anthropogenic communities, in which alien species predominate. So the vascular flora of Toronto consists predominantly of two largely distinct components: the increasingly but unevenly depleted typical forest flora of the mixed forest zone; and the increasingly rich array of cultivated species and weeds from cool temperate regions around the world that decorate the decreasing land area of gardens, parks, and boulevards.

Within these two broad source categories, there are many specific origins. These derive, in part, because of the environmental heterogeneity of the city, leading to distinctive plant communities under different circumstances, each with individual histories. Considering the urban communities, for instance, many of our most common weeds of disturbed habitats, including the notorious ragweed (*Ambrosia artemisiifolia*), are probably remnants of the agricultural Toronto that was still prominent before the First World War. In contrast, the extensive use of salt on roads, resulting in a specialized assemblage of salt-tolerant species (halophytes) such as saltgrass (*Puccinellia distans*), spearscale (*Atriplex patula*), and salt aster (*Aster brachyactis*) along roadsides and pavements, is much more recent, largely since the Second World War. Even more recently, there has been an explosive increase in the number of specialized garden clubs, in the variety of exotic plants that are grown, and the types of gardens in which they are grown. This trend has also included a heightened appreciation of native plant gardening and a corresponding great increase in the numbers and kinds of native plants grown.

Compared to the wide ecological occurrences of introduced plants, the origins of plants in the natural communities are much more tied to the kinds of environments in which they grow. They sort out by sunlight, temperature, moisture gradients, and rates and types of natural disturbance into communities with sources in different parts of the mixed forest, deciduous forest, or even boreal forest zones. There are substantial differences in the composition of the communities of the ravines depending on whether they are situated on the edges, slopes, various heights of terraces, the bottomlands, or the river levees. In general, but not exclusively, cooler, moister sites favour mixed forest species, like hemlock, while warmer, drier sites support deciduous forest species. Boreal forest species are confined to

> *Even a relatively innocuous and passive activity like walking in the woods tramples the forest understorey and suppresses the regeneration of many species.*

exceptionally cold and typically waterlogged sites that develop as bogs.

The ridge running north from Twyn Rivers Drive between the Rouge River and Little Rouge Creek is an excellent place to observe contrasting plant communities strongly associated with the effects of steepness and direction of slope. The northeast-facing slope overlooking Little Rouge Creek has numerous hemlocks (*Tsuga canadensis*) and a very sparse understorey of wood ferns (*Dryopteris cristata*). The sunnier southwest-facing slope is entirely deciduous, with scattered bitternut hickories (*Carya cordiformis*) and a rich understorey of native flowers. The flat land between the two slopes has numerous red maples (*Acer rubrum*) with a well-developed shrub layer including the roundleaf dogwood (*Cornus rugosa*). All of these communities are on the same soils and receive the same amount of rainfall, so the differences among them are the result of differences in drainage due to slope and in sunshine due to aspect. While the slopes have relatively free drainage, the tableland is waterlogged during the spring, as you might guess from the composition of its forest. Despite their equivalent drainage, the hemlock forest experiences much less drought stress than the hardwood forest on the opposite slope because the latter receives far more hot afternoon sunlight. Most of the large-scale variations in the composition of natural plant communities in Toronto are due to similar types of underlying environmental factors.

FERNS AND FERN ALLIES

The ferns, horsetails, and club mosses are unrelated groups of vascular plants that are usually considered together because they all lack seeds, reproducing by means of spores and a free-living gametophyte generation. Most plants and fungi have two distinct phases – referred to as generations – in their life cycles, with different numbers of chromosome sets. The sporophyte has two sets of chromosomes (like us) and produces spores that grow into a gametophyte, with just one set of chromosomes. The gametophyte produces the sperm and eggs that unite to begin the next sporophyte generation. In seed plants, the gametophytes are inconspicuous, hidden within the seeds and pollen grains. In the ferns and similar plants, by contrast, the gametophyte is a small green mat that requires abundant moisture to survive and reproduce the next conspicuous and familiar sporophyte generation. Seed plant gametophytes are dependent on their parent sporophytes but independent of free-flowing water. Even though they need available water for reproduction, ferns are still the second most diverse group of vascular plants, exceeding the conifers and other gymnosperms in species numbers both in Toronto and worldwide. Nonetheless, the ferns have been greatly reduced here compared to their original abundance. Only a few species are now common in the city, including the bracken fern (*Pteridium aquilinum*), the sensitive fern (*Onoclea sensibilis*), and the crested and marginal wood ferns (*Dryopteris cristatus* and *D. marginalis*). The lycopod group that includes the club mosses, spike mosses, and aquatic quillworts, although fairly common in Ontario, is virtually absent from Toronto. The lack of limestone bedrock outcrops in the city largely accounts for this. Only one of the lycopods was common before the First World War (Ivey 1913), the shining club moss (*Huperzia lucidula*), which has subsequently disappeared from High Park. Club mosses have long, evergreen runners that were heavily exploited in the past for Christmas wreaths.

The third major group of non-seed vascular plants is the horsetails, with a single living genus, *Equisetum*, found worldwide. Among the free-sporing vascular plants, the horsetails have fared best during the changes in Toronto. Unlike the lycopods, they are plants of open, disturbed habitats, and they thrive where water lies just beneath the surface of the soil. Species of *Equisetum* are abundant on unstable, sunny slopes of ravines, where groundwater emerges in the form of perennial seeps. They also abound in the damp swales of the Toronto Islands and the landfill peninsulas like Tommy Thompson Park and Humber Bayfront Park. The gametophytes of *Equisetum*, unlike those of ferns, are subterranean and are thus less vulnerable to disturbance and desiccation. Furthermore, the horsetails have extremely vigorous vegetative propagation via underground stems (rhizomes) and are virtually immune from animal grazing because of their high silica content, which had given them the alternative common name of 'scouring rushes' in the days before Brillo pads.

THE NATIVE AND NATURALIZED TREES OF TORONTO

A Meeting Place for Native Trees

Because trees have the greatest ecological impact on other plant and animal species in natural communities of forested regions, it is worth taking a look at the occurrence of selected native trees across the landscape of Toronto. Each of the approximately 50 species of native trees here has a distinct overall range in eastern North America, but it is possible to loosely categorize the position of Toronto within their total range into three general conditions. About 20 species, those conventionally described as Carolinian, are here at or near the northern limits of their local distribution. Another 20 species – many of which reach their peak abundance in the mixed forest zone, but some of which have much wider distributions – are well within the main bodies of their ranges. The remaining species are predominantly boreal species that reach their southern boundary near Toronto.

There are two main groups of native tree species near their northern range limits here. The first of these are species of the naturally disturbed soils of the river levees and adjacent floodplains. These include black maple (*Acer nigrum*), Manitoba maple (*Acer negundo*), black walnut (*Juglans nigra*), the rare red mulberry (*Morus rubra*), eastern sycamore (*Platanus occidentalis*), eastern cottonwood (*Populus deltoides*), and slippery elm (*Ulmus rubra*). Black maple is scattered sparsely as single trees or small stands on the floodplains and lower terraces of all three main river systems and is essentially confined to these habitats. Manitoba maple, in contrast to the uncommon black maple, is one of the commonest 'weed' trees of Toronto. In addition to its abundance in all of the ravine systems, it joins the silver maple (*Acer saccharinum*) and the exotic tree-of-heaven (*Ailanthus altissima*) and Siberian (often misnamed Chinese) elm (*Ulmus pumila*) as one of the most characteristic trees of property boundaries, poorly maintained flower beds and foundation plantings, and waste lots around the city. All are pioneer species in the wild that appear early in ecological succession, produce an abundance of wind-dispersed, winged fruits, and are tolerant of a far broader range of soil conditions than is typical of their natural floodplain habitats.

Eastern cottonwood, another of the floodplain species that is an ecological pioneer with wind-dispersed seeds, may also be found in waste places about the city. Its natural habitat is not confined to floodplains, since it is one of the most characteristic species of the shoreline sand dunes of the Great Lakes. It is the principal tree of the Toronto Islands and will be equally dominant on the colonized landfill of Tommy Thompson Park (the Leslie Street Spit) as the trees mature. A last example, black walnut, is also a characteristic urban species, appearing seemingly at random in little patches of ground such as untended wedges between roads and subway tracks. In this case, we can credit our large population of squirrels, which often forget at least some of the places they've placed their caches of nuts, rather than the floodwaters that move walnuts about in their typical floodplain habitat. Squirrels are similarly responsible for the wide establishment of the introduced ornamental horse chestnut (*Aesculus hippocastanum*), with its poisonous (to us), beautifully patterned nuts, which are one of the signs of beginning the school year for generations of children. Thus, many of the Carolinian floodplain species are not confined to their primary native habitats but, as ecological pioneers, have become firmly established as part of the flora of urban Toronto.

The second major group of species near their northern limits are upland trees that are found in drier soils than those on which the regionally dominant sugar maple thrives. These species include bitternut and shagbark hickories (*C. ovata*), blue-beech (*Carpinus caroliniana*), black and white oaks (*Quercus velutina* and *Q. alba*), and mitten-tree (*Sassafras albidum*). The latter three are all prominent in upper, southwestern-facing slopes of High Park, where the oaks form the canopy of an open community interspersed with prairie grasses and other herbaceous perennials and known as an oak savannah. Mitten-tree forms big groves here, but these are clones, genetically uniform derivatives from an original tree that have spread by root sprouting, a characteristic mode of increase in aspen trees as well. The hickories and blue-beech are scattered on dry slopes of the ravine systems and are uncommon in the city. In contrast to the floodplain species, none of this group is wind dispersed and none has become widely established in the urban setting, although shagbark hickory and the two oaks are planted to a certain extent in parks and as street trees. They also constitute an important part of the residual trees in many city neighbourhoods, such as the Beaches and Forest Hill.

Two of the northerly trees near the southern edges of their ranges are trees of the boreal forest that appear in southern Ontario in bogs, places with waterlogged soils that also accumulate cold air. These places, uncommon in the vicinity of Toronto, are quite stressful to most plants because of their accompanying acidity and stagnation, with reduction of nutrient and oxygen availability to the roots. Most of the plants here, including several members of the heather and rhododendron family (Ericaceae), are specialists in this kind of environment, and a few, like pitcher plants (*Sarracenia purpurea*) and sundews (*Drosera rotundifolia* and *D. intermedia*), get extra nitrogen by catching and digesting insects.

These highly specialized plants are not found in Toronto today, but the boreal trees that frequently accompany them, black spruce (*Picea mariana*) and tamarack (*Larix laricina*), are present in poorly drained portions of the valley of the West Don River. Two other boreal tree species that reach their southern limits near here are pioneers and have much broader ecological ranges in southern Ontario. While balsam poplar (*Populus balsamifera*) is widespread in wet soils throughout the city, white birch (*Betula papyrifera*), famous for its use in canoe construction, is also common and widely scattered in dry soils. Two more species with local southern limits in or near Toronto actually reach their greatest abundance in the mixed forest zone, although they extend into the southern boreal forest as well. Red pine (*Pinus resinosa*) is rare here on dry, fire-prone soils, while white cedar (*Thuja occidentalis*) can be found in many places with moist soils and a cool aspect. These differences in abundance and degree of environmental specialization among the trees that reach their southern range limits here underline the general principle that each species behaves independently in response to its own ecological tolerances, even if it seems to be characteristic of a predictable community of species.

In contrast to the varying abundance of the trees that reach range limits in or near Toronto, almost all of the species that spread well beyond the city in every direction are common here. They are the trees that best define the character of our remnant natural forests. Most of them reach their greatest abundance in the mixed forest zone that extends northward from the city. These include most of its major dominants, such as sugar maple (*Acer saccharum*), beech (*Fagus grandifolia*), white pine (*Pinus strobus*), and red oak (*Quercus rubra*), all of which dominate forests on dry to moist soils in Toronto as well. The dominants of wetter soils in the mixed forest, silver maple and white elm (*Ulmus americana*), while sharing prominence with the southerly floodplain species discussed above, are still very prevalent and important species. Other species, like red maple and white ash (*Fraxinus americana*), may be found in both upland and lowland forests, although typically more common in one or the other. A few other mixed forest species have fairly specific ecological requirements in Toronto. Thus black ash (*Fraxinus nigra*) is restricted to swamps, and hemlock generally to cool slopes, while large old trees of bur oak (*Quercus macrocarpa*) are a prominent feature of the lowermost ravine slopes at the edges of the floodplains of the large rivers. With few exceptions, the most common native trees of Toronto are those for which the city lies well within their main distribution rather than representing a range limit.

Naturalized Exotic Trees

Only a few of the numerous non-native trees that have been planted in Toronto have become established in natural forests here and could become self-sustaining. They represent a variety of different seed-dispersal modes and original homes. Norway maple (*Acer platanoides*) and Siberian elm are wind dispersed; white mulberry (*Morus alba*) and common buckthorn (*Rhamnus cathartica*) have bird-dispersed berries; horse chestnut seeds are buried by squirrels; and black locust (*Robinia pseudoacacia*) spreads by a variety of means, including clonal suckering. The maple and buckthorn were originally from Europe, the horse chestnut from southwestern Asia, and the elm and mulberry from eastern Asia. Black locust is native to eastern North America and has been recovered from the Don beds of the last interglacial period in Toronto, but it only got as far north as central Ohio and Pennsylvania before human intervention brought it back here during the present interglacial period. Vegetative propagation is also found in one of the most common tree willows of Toronto riverbanks, the hybrid crack willow (*Salix* X *rubens*), a hybrid between two European species that appears to spread when branches broken off by storms take root downstream from the parent tree. All of these established alien species are typically most prominent where disturbance is greatest, so they reach their greatest abundance in the downtown portions of the Don valley, like the ravines that punctuate Rosedale. All of them may be found throughout the city, however, in varying frequencies that reflect their individual ecological tolerances, just like the native species.

Many additional exotic tree species, as well as the species just discussed, may seed themselves into urban habitats. Inspection of vacant lots, hedges, property lines, foundations, and unkempt commercial plantings can reveal a wide variety of spontaneous seedlings (Figure 7.1). Tartarian maple (*Acer ginnala*), tree-of-heaven,

> *A few of the plants that live in Ontario bogs, like pitcher plants and sundews, get extra nitrogen by catching and digesting insects.*

FIGURE 7.1
Seedlings of introduced trees becoming established in the relative safety of a privet hedge. These include, from left to right, mulberry (*Morus alba*), wych elm (*Ulmus glabra*), and tree-of-heaven (*Ailanthus altissima*). (James E. Eckenwalder)

bean-tree (*Catalpa speciosa*), Russian-olive (*Elaeagnus angustifolia*), crabapples (*Malus* spp.), goat willow (*Salix caprea*), rowan-tree (*Sorbus aucuparia*), little-leaf linden (*Tilia cordata*), and wych elm (*Ulmus glabra*) may be most evident, but such unusual species as maidenhair tree (*Ginkgo biloba*), saucer magnolia (*Magnolia* X *soulangeana*), and corkscrew willow (*Salix matsudana* 'Tortuosa') can also be found. Many features of the essentially open communities found in these urban habitats permit the establishment of a wide variety of introduced species. Most important, perhaps, is the lack of the competition found in established closed plant communities, a reduced population of herbivores (and weeders!), and a relative abundance of water and nutrients as runoff from adjacent tended sites. Taken together, these factors make such neglected spots in the city a rewarding place to investigate the processes that lead to the naturalization of alien species.

In addition to direct seeding into disturbed and relatively natural habitats, some exotic tree species are capable of hybridizing with native species. Many of these hybrids are probably undiscovered, but some are quite conspicuous. The introduced white poplar (*Populus alba*), like all other poplars, is dioecious, that is, it consists of individuals that have only male or female flowers. It is most commonly found in Toronto as female trees that are more frequently pollinated by one of the two native aspen species, bigtooth aspen (*Populus grandidentata*) or quaking aspen (*P. tremuloides*), than by the few available males of the same species. The resulting hybrids are conspicuous in many places around the city, particularly in the ravines of the Don valley system and in the lakeshore parks, including the Toronto Islands. They stand out from their white poplar mother because they inherit brilliant white hairs on the underside of their leaves, and these contrast strongly with the bright green upperside, the two surfaces flashing in turn as the leaves flutter in the wind because of their flattened leaf stalks, so characteristic of the aspens.

Hybrids between Siberian elm and the native slippery elm are also fairly common in the floodplains of the rivers in Toronto and are conspicuous because of their intermediate leaf size between the tiny leaves of Siberian elm and the large leaves, up to 20 cm long, of slippery elm. Recognition of these hybrids is somewhat complicated because they could also involve two other introduced European elm species, English elm (*U. procera*) and wych elm. We can also anticipate hybridization between introduced and native species of willow (*Salix* spp.) and linden (*Tilia* spp.) as well as additional poplar species. As the number and prominence of introduced tree species in Toronto rise and the overall impact of human activities intensifies we can anticipate increasing numbers of hybrids at the interface between natural and urban habitats. Hybrid vigour may make such crosses particularly successful in coping with the stresses of an increasingly urbanized environment.

THE FOREST UNDERSTOREY: SHRUBS AND HERBACEOUS PLANTS

The patterns of geographical distribution, of habitat preference, and of rarity and abundance seen among the native trees of Toronto are also observed in the more numerous plants of the understorey and of the non-forest plant communities. As with the trees, the shrubs and herbaceous plants of the ravine slopes are better preserved than those of either the uplands or the valley bottoms, although there are more relatively intact floodplain communities than there are upland ones. The most common and widespread species throughout the city are usually those that have range limits extending far beyond Toronto, and these are typically plants of the mixed forest zone or those that are even more widely distributed in eastern North America. Species that reach distributional limits near Toronto – whether boreal species near their southerly boundary or southerly species near their northern boundary – are not only usually less common than species in the heart of their ranges here but are also more likely to be restricted to specialized habitats, such as the oak savannah of High Park or the boggy lowlands of the upper West Don River.

These specialized habitats, containing some of the rarer species in the city, are also ecologically marginal. They occur on sites at the wet and dry extremes of the moisture gradient, far from the mesic optimum in which moisture availability is relatively even throughout the growing season, without either pronounced droughts or periods of waterlogging. The dominant mesic forest species are simply more competitive on the more common and favourable mesic sites than the boreal or southerly species that are near the limits of their climatic temperature tolerances. Not surprisingly, the losses of species from the Toronto area since the First World War are largest among these locally rare species that had scant footholds here in the first place. The resurgence of wild lupines (*Lupinus perennis*) and other prairie plants with the managed burning of parts of the oak savannah in High Park, after decades of fire suppression and invasion by more mesic species, however, suggests that at least a few seemingly lost species might still be present in the form of a dormant seed bank within the soil. These long-lived seeds might renew such species if suitable conditions for germination and establishment returned to environments that have not been too altered for rehabilitation. Most of the lost species, perhaps as many as a quarter of the native herbaceous plants, are casualties of permanent habitat destruction and cannot be expected to return spontaneously.

THE IMPACT OF TORONTO ON THE VASCULAR PLANTS OF THE REGION

The Diversity of Plant Species

The three historical strata of vascular plants in Toronto – native, agricultural, and urban – have each left a mark on the overall diversity of vascular plant species here. The native flora of the old York County (Toronto together with the Regional Municipality of York north of Steeles Avenue) consists of about 1,000 species, of which perhaps three-quarters may be found in the city today. This figure includes 50 native tree species and about two and a half times as many native shrubs, so there are far more native herbaceous plants – about 600 – than woody ones. As noted above, far more herbaceous species than woody ones have disappeared with the development of Toronto, in part due to greater degradation of the understorey of our forest remnants compared to the canopy. The disappearances also have not been uniform with respect to original natural distributions. Boreal species have seen the greatest proportionate losses and mixed forest species the least, with deciduous forest elements in between.

The agricultural past of Toronto is reflected primarily in the array of weeds and other naturalized alien plants that are now a self-sustaining part of the flora. Almost 500 species of foreign plants have become established to varying degrees here. Just as with the native species, some are rare and some are so abundant that many people (and some seed companies offering 'wildflower' mixes) assume that they are native. Many of the common summer flowers of open ground throughout the city, like oxeye daisy (*Leucanthemum vulgare*), Queen Anne's lace (*Daucus carota*), butter and eggs (*Linaria vulgaris*), and red clover (*Trifolium pratense*), belong to this group of successful immigrants. Fully half of the naturalized weed species are members of just six plant families: 71 species of Compositae (daisies, dandelions, and ragweed, as well as many additional native species, including goldenrods and asters), 57 Gramineae (the grasses, also with many additional native species), 36 Cruciferae (mustards and rockets, with fewer than half as many native species), 27 Leguminosae (peas, beans, vetches, and clovers, also with fewer than half as many native species), 24 Rosaceae (cinquefoils and avens, and more than twice as many native species, including the confusing hawthorns and blackberries), and 23 Caryophyllaceae (pinks, campions, and carnations, with only two or three extra native species). Among these families, only Rosaceae includes native trees and shrubs here, although there are a number of introduced woody species of Leguminosae in Toronto as well.

The third and most recent floristic element is the hardest to quantify because it is experiencing the most rapid changes. This element consists of species that are currently cultivated in parks and gardens. It has grown explosively in the last couple of decades as gardeners have changed their approaches to gardening and suppliers have greatly expanded the range of plants offered. The corner grocery store may still sell the same few annual bedding plants that dominated many gardens in decades past – petunias, marigolds, geraniums, pansies, and impatiens especially – but the city's many garden centres display a wide array of annuals, perennials, shrubs, and trees. A few specialist perennial growers offer hundreds of species, and the diversity of some categories, like grasses and ferns, increases every year. Nor are commercial nurseries the only sources of plants today. Many gardening clubs have annual seed exchanges and plant sales for their members. The Ontario Rock Garden Society, for instance, most of whose members live and garden in and near Toronto, listed seeds of over 900 hardy species grown by members in its 1998 exchange. Others, like the Canadian Wildflower Society, with its headquarters in Toronto, promote the use of native species and may be responsible for the return of many species to the urban uplands through their seed exchanges and plant sales.

The total number of species in the cultivated flora is difficult to assess, but it is likely to be at least twice as large as the native flora, especially if we include the established aliens that are most common in urban habitats. Trees on the one hand, and shrubs on the other, provide conflicting estimates of the size of the cultivated flora in comparison to the native species. The plantings at Mount Pleasant Cemetery, which include most of the trees and shrubs that are currently grown in Toronto, though far from the total number of potentially hardy species that might be grown here, have over three times as many exotic trees as our roster of native species. On the other hand, there are slightly fewer alien shrubs than the total number of native species historically found in Toronto, though perhaps as many as those still extant in the city. There has been less emphasis on shrubs than on trees in the cemetery's planting program and so the trees of Mount Pleasant may better estimate the overall range of species grown in Toronto than do the shrubs. In combination with the offerings of garden centres and seed exchanges, then, the overall estimate of some 1,500 to 2,000 alien species on top of the native flora is probably realistic. As with each of the other elements of our present flora, these species vary in abundance and distribution, from single individuals of some esoteric connoisseur plants to billions of individuals of lawn grasses. The total flora is thus far greater now than during the agricultural phase of Toronto, when the number of species was only slightly more than that of the native flora, alien additions being partially offset by the elimination of native species with forest clearance and drainage of wetlands.

The Ontario Rock Garden Society, most of whose members live and garden in and near Toronto, listed seeds of over 900 hardy species grown by members in its 1998 exchange.

The Distribution of Vegetation in Toronto

Although the number of species in the alien flora of Toronto is perhaps double that of the native vascular plant flora, the physical structure of most of the urban plant communities is far less complex. While the native forests have multiple overlapping layers of continuous cover, most of the horticultural communities consist predominantly of just four discrete growth forms, mostly in separate patches. First are the trees, planted as isolated individuals or scattered so that their crowns rarely form a continuous canopy when along streets or in yards and parks. Although Mount Pleasant Cemetery and other collections boast 150 or more species of trees, about 35 are so common in the city as to be comparable in number of individuals to our native species, and this is a diversity approximately equivalent to that of the common native species. The second growth form consists of hedges and foundation plantings of shrubs, usually of uniform height and in the form of monocultures or plantings of two or three species together, although these are also good places to look for uninvited seedlings of many woody species. The third contains flower beds, containers, and borders, dominated by annual or perennial herbaceous plants of varying stature, but usually uniform or spatially graded, rather than layered and overtopping each other. There is relatively little vegetational difference between annual and perennial plantings in horticultural communities, because both can attain similar statures and densities during the growing season and both lack permanent above-ground biomass. However, perennials usually establish a photosynthetic presence earlier in the growing season and longer into the fall, providing more food for herbivores, while annuals typically have longer flowering periods, providing more consistent resources for pollinators and more colour for the gardener. Combined plantings can provide a nice balance. The last common growth form is that of lawns, exhaustively maintained at a uniform height whether they consist exclusively of lawn grasses or also contain an admixture of other low plants (lawn weeds), like white clover (*Trifolium repens*), dandelions (*Taraxacum vulgare*), plantains (*Plantago major*), and many others.

Most often, each of the above four growth forms is found in a solid block, either alone or adjacent to blocks of one of the other growth forms. The only pair of these growth forms that are frequently deliberately combined as layers of a single community are trees and lawns, the cohabitants of many front yards, parks, cemeteries, and golf courses. Mixed borders of shrubs among herbaceous plants are increasing in popularity but far from the norm in Toronto. Multiple growth forms also occur intermixed in vacant lots, where scattered trees, shrubs, and herbaceous perennial weeds (like burdock, *Arctium minus*) become established among annual weeds and clumps of grasses. Tall grasses and perennial pasture plants (like red clover) dominate in those areas of the city, such as hydro right-of-ways, that are mown only once or twice a year. It is difficult for trees and shrubs to become established in these artificial and largely ungrazed meadows.

ECOSYSTEM PROCESSES IN TORONTO

Although there are striking differences in the total number of vascular plant species inhabiting Toronto today compared to pre-colonial times, and also in the complexity of vertical stratification in the vegetation, the effects on ecosystem functioning are quite varied. Obviously, one of the major ecological impacts of green plants is through the photosynthetic conversion of inorganic carbon dioxide to sugars and other biologically significant organic molecules. The rate of this process at a given latitude, defined as net primary production (the difference between a plant's rate of photosynthesis and its use of the fixed carbon in respiration), is roughly comparable for different continuous vegetation types. For Toronto, this means that an original forest and a well-tended lawn both have a net primary production of about 500-1,500 g per m^2 per year, with the lawn, perhaps, toward the lower end and the forest toward

Some Cultivated Equivalents of Wild Growth Forms

Although trees, hedges and shrubs, flower gardens, and lawns are the four main growth forms in urban Toronto, there are also a few other less common community types. Vines and climbers on fences, trellises, and arbours provide an echo of the lianous Virginia creeper and grapevines of the natural forests, but they are usually much more restrained, both in height and in spread.

Increasingly popular ponds, pools, and other water features in urban gardens and parks provide homes for many aquatic and semi-aquatic plants and associated animals, but they tend to emphasize conspicuous emergents and floating leaved forms such as water lilies (*Nymphaea* spp.) instead of the wide variety of other aquatic growth forms. They also represent just a tiny fraction of the urban landscape — much less than the wetlands of the original vegetation — but more than a generation ago.

Permanent plantings of spring-flowering bulbs, like tulips, daffodils, and hyacinths, are somewhat comparable to the native spring ephemerals, like the trilliums and trout lilies, that flower and fruit before the deciduous trees above them expand their leaves and reduce the light reaching the forest floor. True bulbs are very rare among our native plants, however, being more characteristic of seasonally dry environments than frozen ones. A more anomalous and rather barren horticultural community is the rose garden, unlike anything in the original vegetation of Toronto, with its sparse thorny twigs emerging out of bare ground.

the upper end. The big difference between primeval and present Toronto for carbon assimilation is in the proportion of the landscape covered by plants. For the city as a whole, with the proportion of land covered by buildings and roads and other pavements, we can estimate a modest reduction in photosynthetic food production of only about 10 to 20 percent compared to the original landscape or the agricultural landscape that intervened.

Another ecosystem function of primary concern to people today is the removal of carbon dioxide as a greenhouse gas from the atmosphere. Here, the transformation of Toronto from forest to field and then city has had a larger impact than the modest reduction in primary carbon assimilation. This is where the reduction in vegetational stratification and the predominant conversion from trees to herbaceous plants has an obvious impact. While lawns may have roughly the same net primary production as forests, most of the resulting biomass is not accumulated, but is removed by mowing and quickly returned to the atmosphere through composting and other decay processes. Trees and shrubs, in contrast, add much of their fixed carbon in long-term storage to their own bodies in the form of wood. A young aspen tree 10 m tall, for instance, can add an annual volume of wood equivalent to a cube 45 cm on each side, or about 10 percent of its volume. The wood accumulates year after year as the tree grows, and the carbon dioxide is only returned to the atmosphere when the tree dies and begins to decay. However, a lawn of equivalent biomass, because it is mown to constant height and density, basically stores no additional carbon in a year. The density of trees in Toronto is now perhaps only 10 percent or less of what it was before the development of the city, so the ability of this area to remove the carbon dioxide it produces in heating buildings and powering motors in automobiles, lawn mowers, and other machines has been drastically reduced, but is still probably greater than it would be if we were still predominantly agricultural. Our production of carbon dioxide also removes much of the oxygen that is released during photosynthesis, and even more is consumed by respiration in living animals and plants and by decay of dead ones. Thus, the reduced storage of carbon in a Toronto of lawns compared to a forested Toronto is also mirrored in a lower net production of oxygen.

A third important ecosystem function affected by the existence of Toronto, nutrient cycling, is presently undergoing considerable change. In the original forests, nutrient cycling involved the *in situ* decay of each season's leaf fall and winterkilled herbaceous above-ground parts by micro-organisms and soil invertebrates. In the process, a great deal of organic matter was incorporated into the soil in the form of humus. Agricultural Toronto altered this relationship through export of some nutrients as human and domesticated animal food and quicker release of others, without organic matter accumulation, via burning of crop residues. Until very recently, ordinary nutrient cycling was progressively restricted by urbanization. Lawn clippings, yard waste, and fallen leaves were all carted off to landfill sites and soil fertility was replenished with chemical fertilizers and imported topsoil, manure, and peat moss. All that has changed since the 1980s, with a massive shift to individual and municipal composting and the increasing popularity of mulching mowers. Although the initial motivation for this shift in practice was diversion of bulk from overburdened landfill sites, it has also begun to restore some semblance of normal cycling to our urban landscapes. This emerging system, with its centralized composting at the scale of both the municipality and the individual lot, is far from a natural cycle. However, it does substantially reduce the need for totally external inputs in maintaining the long-term fertility of city soils and the urban plant and animal communities that they support.

The same increasing sensitivity to environmental issues that made backyard composting an easy sell for many Toronto residents has also encouraged a reduction in pesticide and water use. As with composting, municipalities were attracted to alternative forms of pest and water management, in part to save money. The popularity of organic approaches and relatively benign pest controls reverses a trend that saw ever increasing uses of herbicides and insecticides in the decades after the Second World War, especially on lawns. These toxins had a profound effect on wildlife in the city, indiscriminately eliminating beneficial insects along with pest species and thereby reducing food for insectivorous birds and mammals.

With the decline in pesticide use, many residents are noticing an increase in the diversity of butterflies, bees, and other garden visitors. The insects supported by the horticultural plant communities of urban Toronto are rather different from those of the original forests, however. This is due in large part to the differences in species composition and sources between the cultivated and native plants, mediated by their secondary defensive compounds. While every plant species has a unique combination of defensive chemicals in its leaves, with related species generally having similar types, herbivorous insects have followed two major strategies to overcome their repellent or toxic effects. Generalists can feed on a wide array of plants, though not particularly thriving on any because of an inability to fully detoxify each species. Specialists overcome the chemical defences of a particular plant, and even rely on its defensive compounds for their own protection (as with the monarch butterfly, *Danaus plexippus,* feeding on milkweeds, *Asclepias* spp.), but they are incapable of using any other plants as food. Every plant species supports both generalists and specialists in its native habitat. However, the huge diversity and abundance of exotics in Toronto, which have generally left most of their specialists behind, means that the increasing range of herbivores in our gardens and parks with the reduction in pesticide use are mostly generalists, rather than the predominant specialists of the original forests.

The shift from specialist to generalist herbivores in contemporary Toronto is an example of the chemical structuring of animal communities by plant communities. In this context, it is useful to point out that at least some of the defensive chemicals produced by plants are pesticides, but they are internal and affect only those herbivores that actually eat the plant, unlike the indiscriminate action of our external sprays, which kill any insect with which they come in contact, whether they are herbivores, predators of herbivores, or beneficial pollinators.

The plant cover of Toronto also provides a physical structure for feeding, roosting, and nesting sites. The simpler vertical structure of urban plant communities compared to the original forests suggests that dependent animal communities will be less complex than they were before the emergence of Toronto, but richer than during the agricultural phase, which was structurally even less complex than at present. One important structural element of Toronto's natural forests that is virtually absent from most of the urban communities is dead trees. Snags provide sites for cavity-nesting birds and mammals, as well as an abundant supply of wood-destroying insects upon which woodpeckers and many other birds thrive. No matter what the ecological sensitivities of urban residents, standing dead trees are never likely to reappear as a major component of Toronto's upland plant communities because of the safety hazard they pose.

THE FUTURE OF TORONTO'S VASCULAR PLANTS

There are conflicting pressures in modern-day Toronto that make predictions about the future uncertain. Increasing population, commercialization, and development must reduce the space in the city available for occupation by vascular plants and hence the overall photosynthetic capacity of the city. This is likely to affect trees, the biggest plants, disproportionately compared to smaller shrubs and herbaceous perennials. Since trees support the greatest number of animal species, we can expect further reductions in animal biodiversity with reductions in the number of trees. Patches of land can become too small to mow effectively, so the total cover of lawns is also bound to shrink in the future.

With the decline in pesticide use, many residents are noticing an increase in the diversity of butterflies, bees, and other garden visitors.

By contrast, increasing environmental concerns will make people demand that their municipalities become more responsible stewards of public spaces, and this approach is likely to save maintenance costs as well. There will be increasing pressure for public landscaping to emphasize native plants and naturalistic arrangements of them. Private gardens should continue to see increasing use of native plants, but also an ever richer array of exotic trees, shrubs, and herbaceous plants from all over the temperate world. If the increasing acceptance of wildlife in our urban spaces continues, the horticultural communities of the city may well become structurally more complex, with naturalistic gardens following ecological principles. This trend, should it materialize, will also have important effects on ecosystem functions, including water management, nutrient cycling, and production of greenhouse gases. The future plantscapes of Toronto will reflect some as yet unknown balance between our increasing numbers and our increasing desire to lessen our negative impact on the plants and animals with which we share the city.

8

Mosses, Liverworts, Hornworts, and Lichens

John C. Krug

The mosses, liverworts, and hornworts are three groups of small photosynthetic plants, and are collectively known as bryophytes. These three are only distantly related but are grouped together because they all have the same type of life cycle. In all three cases, a gamete-producing generation, the gametophyte, is followed by a short-lived spore-producing generation, the sporophyte, which alternates with the gametophyte generation. All three groups are non-vascular plants. That is, they have virtually no specialized water- and food-conducting tissue, unlike flowering plants, and no roots. Thus, they also have no stiffening tissue and no firm attachment to the substrate. These characteristics limit their size, so most bryophytes lie prostrate or flattened on the ground; those that grow upright are usually only a few centimetres high. They form the lowest vegetation layer on the ground, or provide a thin cover on rocks and logs or on tree trunks. In exposed sites, they are often intermingled with lichens.

The 'plant' that you see when you observe a bryophyte is the gametophyte, which is a long-living vegetative structure that, in most bryophytes, produces distinct male (antheridia) and female (archegonia) sexual organs. (The hornworts are an exception to this general rule, since they do not have distinct archegonia.) The antheridium produces numerous motile sperm cells, while the archegonium is usually a flask-shaped organ containing a single egg cell. A film of water is required to enable the sperm to swim to the archegonium and fertilize the egg cell. This method of fertilization is another habitat-limiting factor for these plants. The sporophyte, which produces the spores, develops from the fertilized egg cell and remains attached to the gametophyte, and in some instances is completely dependent on it. In other cases, the sporophyte may have limited photosynthetic activity, but it still remains attached to the gametophyte. The life cycle (Figure 8.1) is completed when the spores are carried by air currents and, on germination, develop into a new gametophyte. The non-flowering vascular plants, such as ferns, have a similar life cycle, but the emphasis is completely different. In these plants, the sporophyte is the 'plant' you see, and it becomes completely independent of the gametophyte.

THE NATURE OF THE PLANTS

The Hornworts

The hornworts, represented by *Anthoceros* (Figure 8.2), are a very small group of inconspicuous thalloid (flattened) plants, with sporophytes consisting primarily of a horn-shaped capsule. The capsule is photosynthetic and grows and releases spores over an extended period. Typically these plants inhabit bare

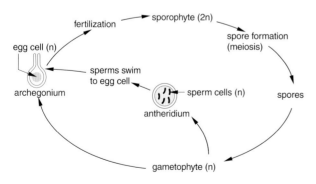

FIGURE 8.1 Typical bryophyte life cycle; n = haploid chromosome number and 2n = diploid chromosome number.

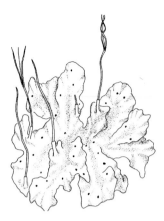

FIGURE 8.2 Hornwort, *Anthoceros laevis* (× 1.75). (Gilda Bellolio-Trucco illustration published courtesy Canadian Museum of Nature, Ottawa)

soil under seasonal cultivation; only four species are known in Ontario (Ley and Crowe 1999).

The Liverworts

Some liverworts are thalloid (Figures 8.3, 8.4) and may form extensive mats on bare, inorganic soil. The more evolved ones – such as *Marchantia* – may possess a system of internal air chambers for gas exchange. However, 75 percent of the liverworts are leafy (Figure 8.5). They are bilaterally symmetrical and form mats on rotting logs, rocks, tree trunks, or organic soil on the forest floor. The cells of these leafy liverworts are compara- tively large, which makes these plants appear more translucent than mosses. In most cases, these liverworts produce a simple, short-lived sporophyte with a capsule that splits into four valves, releasing spores. Within the capsule, hydroscopic cells called elaters alternately absorb water and dry out, and thereby eject the spores into air currents for dispersal. The more complex thalloid liverworts have antheridia and archegonia borne on stalks raised above the vegetative plant. The sporophytes develop on the archegonia-bearing stalks (archegoniophores). Liverworts occur in a variety of habitats but are unlikely to be found in very dry places, except for *Frullania* spp., which are found quite high up on living tree trunks. Two thalloid species in the greater Toronto region are aquatic. The total number of liverworts in Ontario is 171, of which 32 are thalloid.

The Mosses

All mosses are leafy; their cells are generally smaller than those of liverworts, so they tend to be stiffer, larger, and shiny, as well as less translucent. Their leaves may have primitive midribs. A few species are aquatic. Some mosses are prostrate (Figure 8.6) and form mats, whereas others are erect (Figure 8.7) and form carpets or turfs. The capsules of the sporophyte are long lived, sometimes photosynthetic, and have complicated mechanisms to shed spores. The exception is *Sphagnum* (Figure 8.8), of which there are six species listed for the greater Toronto region. They have simple capsules similar to

---

How Plants Get Their Names

How do plants get their names? In 1753 the Swedish botanist, Carl Linné, or Carolus Linneus as he is better known today, published a book entitled *Species Plantarum,* which means simply species of plants. In this book Linneus used a binomial system for naming plants. This system essentially divided plants into larger categories, or genera, based on sexual criteria. Within each genus were included plants with more trivial similarities that were called species.

Usually the names of the plants were in Latin and were derived from Greek or Latin roots. Over the years more colloquial or common names have developed for many plants. Thus, *Betula papyrifera* is called paper birch and *Impatiens biflora,* touch-me-not or jewelweed.

Originally Linneus used a single genus, *Fungus,* for all fungi. Subsequently this was divided into a number of separate genera based on gross features. Unlike the flowering plants, very few common names have developed for the fungi. The exception is for the larger fungi, especially the mushrooms, to which was attached a certain mysticism. Thus, *Amanita muscaria* was known as the fly Agaric, from the mushroom genus *Agaricus,* and the Ascomycete *Xylaria polymorpha* was called dead man's fingers.

The same is true of most bryophytes and lichens, very few of which have generally accepted common names. Some exceptions are British Soldiers for the lichen *Cladonia cristatella* and hair cap moss for *Polytrichum commune.*

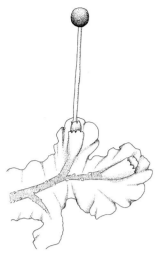

FIGURE 8.3
Simple thalloid liverwort, *Pellea epiphylla* (× 2). (Gilda Bellolio-Trucco illustration published courtesy Canadian Museum of Nature, Ottawa)

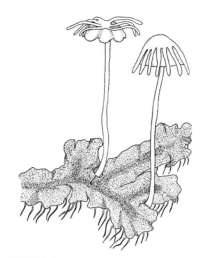

FIGURE 8.4
Complex thalloid liverwort, *Marchantia polymorpha* (× 1.8). (Gilda Bellolio-Trucco illustration published courtesy Canadian Museum of Nature, Ottawa)

FIGURE 8.5
Leafy liverwort, *Jamesoniella autumnalis* (× 9). (Gilda Bellolio-Trucco illustration published courtesy Canadian Museum of Nature, Ottawa)

liverworts and are highly adapted to living in wet areas such as bogs and swamps. Consequently, they are most vulnerable to changes in water table and drainage patterns. On the other hand, some mosses, such as *Polytrichum* spp. and *Grimmia* spp., are adapted to growing in extremely dry places, such as bare rock. For further information on morphology, life history, and classification of bryophytes see Schofield (1985) and Watson (1971).

The Lichens

Lichens are usually now considered to be symbiotic fungi. They are composed of a fungal partner (mycobiont), and a partner of either green algae or cyanobacteria, formerly considered to be blue-green algae. These two partners generally co-inhabit a vegetative structure, the thallus. In this symbiotic relationship the fungus derives the more obvious benefit. The chlorophyll-bearing partners (photobionts) carry out photosynthesis and provide the products to the fungus. The fungus, which is generally the dominant partner, carries out metabolism and biosynthesis of an assortment of unique chemicals or lichen acids.

Most lichens belong to a major division of the fungi called Ascomycetes, or sac fungi. Sexual spores in these fungi are formed by meiosis and are produced within an elongated specialized cell, called the ascus. A few lichens are Basidiomycetes, or club fungi, another major division of fungi. In Basidiomycetes the spores are produced externally on a basidium. All these structures are microscopic. Lichen reproduction, or lichen sex, usually takes place within cup-like structures (apothecia) attached to the surface of the thallus (as we see in *Xanthoparmelia*, illustrated on p. 121) or in flask-like structures (perithecia) embedded in the thallus. In Basidiomycetous lichens reproduction occurs in basidia located within the thallus. See Alexopoulos, Mims, and Blackwell (1996) for a more detailed account of fungal sex.

Lichens also reproduce asexually, that is, without meiosis, by means of special structures – soredia and isidia – borne directly on the surface of the thallus. These structures are air dispersed and contain cells of both the fungal and photosynthetic partners. This ensures that

How Lichens Survive

Probably one of the most significant characteristics of lichens is their ability to survive drying out for very long periods. This trait enables them to exploit substances, such as bare rock, not available to most photosynthetic plants, although a few species of mosses are well adapted to similar conditions. Surface temperatures of rock in exposed situations may vary by more than 100 degrees in the course of a year. Flowering plants in general are incapable of tolerating such wide fluctuations.

FIGURE 8.6
Prostrate moss, *Brachythecium salebrosum* (× 5). (Grout 1903)

FIGURE 8.7
Upright moss, *Polytrichum commune*, with capsule (hair-like covering sac) removed (× 1). (Grout 1903)

FIGURE 8.8
Aquatic moss, *Sphagnum acutifolium* (× 1). (Braithwaite 1880)

on dispersal cells of both partners are present and growth of a new colony is enhanced. This type of reproduction in lichens is more efficient, and hence more prevalent, than sexual reproduction in which a spore of the fungal partner must land in a place already occupied by a photosynthetic partner. Nevertheless, sexual reproduction is required from time to time in all lichens to maintain the genetic vitality of the lichen.

Asexual reproduction sometimes also occurs by simple fragmentation of the thallus, especially in fruticose

FIGURE 8.9
Section through apothecium of *Cladonia* sp., showing: A – spore-bearing layer; and B – underlying stratified layers occupied by the fungal partner; and C – photosynthetic partner (× 350). (John C. Krug)

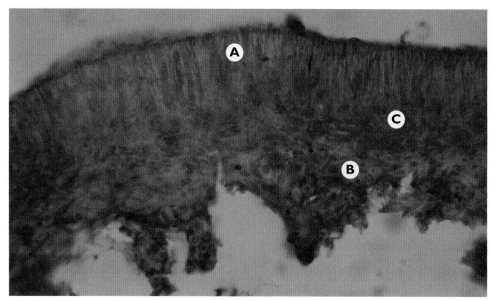

FIGURE 8.10 Foliose thallus of *Xanthoparmelia cumberlandia* with apothecia. The arrow shows the apothecia. (John C. Krug)

FIGURE 8.11 Umbilicate thallus of *Umbilicaria vellea*, a kind of rock tripe. (John C. Krug)

FIGURE 8.12 Fruticose thallus of *Cladina stellaris*, one of the reindeer lichens. (John C. Krug)

FIGURE 8.13 Crustose thallus of *Aspicilia cinerea*. (John C. Krug)

lichens such as the reindeer lichens (*Cladina* spp.), which grow on the ground and fragment readily in dry weather or when you step on them. Büdel and Scheidegger (1996) and Hale (1983) give further information on the morphology and reproduction of lichens.

Typically the thallus of many lichens is stratified, with the cells of the fungal and algal components in distinct layers (Figure 8.9). In some lichens the cells of the photosynthetic partner are randomly distributed. The thallus may be leaf-like, or foliose (Figure 8.10). In some cases of this type, such as in rock tripes, the thallus has a central attachment to the substrate (Figure 8.11). Lichens with a fruticose thallus can be shrubby, as in many reindeer lichens (Figure 8.12), or hair-like, as in old man's beard (*Usnea* spp.). In crustose lichens (Figure 8.13), the thallus is closely attached to the substrate. Some crustose lichens, such as *Lepraria*, are powdery.

HABITATS AND COMMUNITIES

Historically, southern Ontario was dominated by pine and hardwood forests. The present greater Toronto region, comprising most of Halton, Peel, and York regions and a portion of Durham, was no exception. A number of varying habitats would have been represented within this larger ecosystem, such as lakes, riverbanks, bogs, wooded valleys, hills, savannah, and cliffs. External factors like soil texture and acidity, underlying geology, and climate would have contributed to variations within the broad ecosystem. Many specific niches or microhabitats would have been apparent within the various habitats. Each of these microhabitats would have been dominated by an assortment of organisms competing with each other for limited food resources and dependent on each other in various ways.

These microhabitats would have been strongly associated with substrate. For example, there are distinctly different species on rough-barked as opposed to smooth-barked trees. Another factor is bark acidity. Acidity is usually higher on conifers than on hardwoods, but some hardwoods such as oaks have very acidic bark. The communities on such hardwoods show distinct differences from those on more neutral bark. The Europeans have developed a classification for the different communities of bark lichens (see Barkman 1958; James, Hawksworth, and Rose 1977; Hale 1983). Within some of these communities are associated bryophytes, although many of these bryophytes tend to be found more on the bases of trees and in bark fissures. Sometimes lichens and bryophytes compete for available substrate. This competition would certainly have strongly influenced the communities present in the original forest, and successional stages would have been apparent.

The diversity of bryophytes may influence the surrounding vegetation on the forest floor. Water dripping or running off during rains from overhanging vegetation provides nutrients that are absorbed by the bryophytes and certain lichens and are recycled in the ecosystem. Both bryophytes and lichens absorb moisture, slowing run-off water and keeping the ground moist. Bryophyte diversity tends to be lower in forests with extensive moss carpets – coniferous as opposed to hardwood forests (Schofield 1985). As trees fall and decay, the rotting logs add a moister component to the forest and are colonized by a different bryophyte community.

Rock cliffs and outcroppings, which occur along the lakes and rivers, harbour a different assemblage of bryophytes and lichens, depending on rock acidity, mineralization, porosity, and exposure. A similar classification to that mentioned for bark lichens has been developed for lichen communities on rock (James, Hawksworth, and Rose 1977). Shade, moisture, and texture as well as chemistry determine the component bryophytes. Crevices provide an ideal niche for bryophytes. Bryophytes also form distinct aquatic communities, with different species occurring in flowing water, at lake margins and in ponds.

Rock lichens are very important in plant succession. They colonize bare rock by producing holdfasts and thus, by interaction with the rock, gradually break down the substrate. The elementary soil so produced is suitable for colonization by other species, such as mosses, and eventually creates conditions in which seedlings of vascular plants can take root.

Recreation and Lichen Biodiversity

Opening forests for recreational use has had a drastic effect on the lichen and bryophyte flora, as the increased air circulation alters the microclimate beyond the limits of tolerance for many species. This tendency is very noticeable in the sparse flora on tree trunks and branches in 'managed' forests, where the trees have been thinned.

The delicate tissues of most bryophytes and lichens cannot withstand trampling, so all but a few species disappear from paths, thus removing the protection from erosion that they had provided. Trails become excessively muddy under wet conditions when the surface vegetation has disappeared, leading to more erosion. Extreme muddiness is especially characteristic of horse trails and of areas where mountain bikes are in use. The biodiversity of such areas decreases rapidly and many species become extinct.

THE HUMAN ELEMENT

With human colonization and the clearing of the forests, loss of habitat was initiated. The development of agriculture, industrialization, and urbanization further eroded original habitat. Much of the original flora was eliminated as a consequence, especially those elements poorly adapted to face competition from more aggressive species.

Both bryophytes and lichens were important components of these original microhabitats. The bryophytes tended to occupy the shadier and moister habitats (moist forests, especially at the base of trees, and shaded crevices in rock cliffs), while the lichens were in the more exposed and drier sites (trees, pebbles, and soil in open sites; and exposed rocky cliffs and outcrops). However, there are exceptions for both groups of organisms – *Peltigera elisabethae* and related species are typical of moist low-lying forests, while such mosses as *Hedwigia ciliata* and *Rhacomitrium canescens* frequent dry, exposed cliffs. Loss of habitat would have been followed by a loss of the bryophytes and lichens characteristic of specific microhabitats. For example, with the cutting of cedar from swamps for fence posts, the characteristic moss and lichen flora of cedar bark would have been very much reduced. Grazing, draining wetland bogs and marshes, and the spread of plant diseases, such as Dutch elm disease and certain coniferous rusts, also resulted in loss of habitat.

In order to realize the impact of this habitat loss fully, you have only to examine some of the classic collecting localities within the greater Toronto region from about 50 years ago. Some, such as the woods in Long Branch park, no longer exist, while in others, such as Sunnybrook Park, the community has been reduced to a few weedy or aggressive plants. Since microhabitat is especially important to bryophytes and lichens, the loss from such habitat erosion is substantial.

As European colonization took place some new habitats were created, and these provided a specific habitat for some organisms. With the cutting of pine forests and clearing of land, pine timber and stumps were used to create rail-and-stump fences around fields. These were populated by characteristic species of lichens, such as *Candelaria concolor*. Today such fences have largely disappeared and with them much of the characteristic lichen and bryophyte flora, although some of the more adaptable elements are found on similar substrates. The unpainted cladding placed on barns provided another substrate that was colonized by lichens, some of which were also found on fences. Cedar from swamps was cut and used for shingles on barns and houses. Again certain lichens and some bryophytes dominated this habitat. As building styles altered, such habitats changed, and with them the component flora. The introduction of granular shingles and gravel rooftops provided a habitat suitable for colonization by mosses and certain lichens.

As well, the clearings resulting from timber cutting provided a new natural habitat for colonization by bryophytes and lichens. The openings became sites for the development of *Cladonia* communities, as seen today in our northern boreal forest and in a few isolated sites in southern Ontario. Such clearings, as well as those on agricultural land, will provide suitable lichen habitat only when old and stable. Most of these *Cladonia* communities have now disappeared from the region. Mosses – such as *Tortula ruralis* and *Phascum cuspidatum* – also were able to colonize these openings and fields created for agriculture.

The introduction of such substrates as bricks, concrete, tombstones, and stone walls provided an opportunity for colonization by lichens typical of rock outcrops and cliffs. Species characteristic of both limestone and metamorphic rocks – such as *Acarospora furcata* and *Lecanora muralis* – have been able to invade such substrates (Hawksworth and Rose 1976). Mosses have also been able to colonize such habitats, especially in shaded sites, and may form extensive colonies between paving stones. *Tortula muralis* is found especially on such artificial substrates (Schofield 1985).

The introduction of such substrates as bricks, concrete, tombstones, and stone walls provided an opportunity for colonization by lichens typical of rock outcrops and cliffs.

With the destruction of the original forest, various kinds of managed forests have emerged, including conservation areas, plantations, and parks. These forests, even those planted many years ago, lack the varied microclimates of the original forest and consequently contain a much-reduced lichen and bryophyte flora. Unfortunately we have very little original forest remaining to serve as a basis for comparison, which gives added incentive to preserve what remains of the old-growth forest, such as the old-growth pine forests in Temagami. Rose (1976, 1992) demonstrated for Europe that certain species are characteristic of ancient forests, and he developed indices of ecological continuity. Species listed that would be typical of ancient forests throughout Ontario include the lichens *Cetrelia olivetorum, Lobaria pulmonaria,* and *Nephroma parile;* and the mosses *Brachythecium populeum* and *Dicranum montanum.*

The Effects of Pollution

Many cities are covered with smog and have polluted air in which one of the main components is sulphur dioxide, a gas that can adversely affect various metabolic processes in lichens (Richardson 1992). As a result, there are few lichens in heavily industrial and polluted areas, and those present are frequently reduced to a mass of asexual soredia. As you travel away from the source of pollution, the number of both lichen and moss species increases. This phenomenon has been documented for several European cities, and by LeBlanc and De Sloover (1970) for some lichens and mosses in the Montreal region. It is also true in the greater Toronto region. Data from such studies can be used to construct distribution maps of lichens around urban areas. These maps show a large reduction of lichens – a 'lichen desert' – toward the urban centre. Although there is little published information, definite lichen communities have developed on suburban trees as well as artificial substrates. Communities on trees include lichens species such as *Buellia punctata, Candelaria concolor, Flavoparmelia caperata, Hypogymnia physodes, Lepraria incana,* and *Parmelia sulcata,* while those on manufactured substrates include *Lecanora muralis, Phaeophyscia orbicularis,* and *Xanthoparmelia conspersa.* These species are moderately tolerant of pollutants and consequently have been able to develop suburban communities.

Besides direct atmospheric contamination, sulphur dioxide can be released into the upper atmosphere and transported from distant sources. (Sulphur dioxide from the nickel smelter at Sudbury, for example, reaches Toronto.) It combines with water in the atmosphere and is released as sulphuric acid in rain. The resulting acid rain alters the acidity (pH) not only of soil and bark but also of lakes. Although the most direct effect of acid rain is on lichens, there is also a noticeable decrease in certain mosses, such as the feather moss *Hylocomnium splendens,* and especially ones in water like *Eurhynchium riparioides.* Furthermore, some mosses typical of soil, such as *Polytrichum longisetum,* tend to grow submerged in acidified sites (Glime 1992). The most drastic effect is on lichens with a cyanobacterial photosynthetic partner, such as the dog lichen, *Peltigera canina,* as opposed to lichens with a green algal partner. Nevertheless, those lichens with green algal photosynthetic partners are also disappearing; even 30 years ago the lung lichen, *Lobaria pulmonaria,* for example, was relatively common in open woodlands and more distant sites in the greater Toronto region but now it is quite rare.

Because lichens often grow on drier sites, it is advantageous for them to become physiologically active when favourable conditions are present. So, in rainy weather lichens quickly absorb water, enabling them to live and reproduce. With adaptation for rapid water uptake, it follows that atmospheric pollutants will also be rapidly absorbed. Since the lichen thallus lives a long time, these pollutants accumulate, become toxic, and often kill the lichen. Likewise mosses are adversely affected by the uptake of pollutants and frequently become sterile. In the Montreal area there is a distinct reduction in the numbers of both mosses and lichens in polluted sites, which become dominated by weedy species such as the purple moss, *Ceratodon purpureus* (LeBlanc and De Sloover 1970). The same trend applies to the greater Toronto region.

Mosses are adversely affected by the uptake of pollutants and frequently become sterile.

Hawksworth and Rose (1976) used the presence of lichens in urban areas to construct a scale indicating sulphur dioxide levels. Although the key indicator lichen, *Lecanora conizaeoides,* is not present in the Toronto area, we find that *Lepraria incana* and *Parmelia sulcata* indicate the more polluted zones to some extent, while *Flavoparmelia caperata* [= *Parmelia caperata*] is found in zones farther from the pollution centre. Since all three lichens are also found in non-polluted areas, their sensitivity to pollution is only relative. We find that such factors as the roughness and acidity of the bark on which a lichen grows will have some influence on its distribution because these factors relate to absorption and buffering of sulphur dioxide. Lichens on acidic barks with low buffering capacity, for example, are more rapidly affected by sulphur dioxide than are those on alkaline barks with higher buffering capacities.

As well as sulphur dioxide several other pollutants, such as fluorides, have negative effects on lichens. Fluorides are associated with a number of industrial processes, including brick production, which was conducted for many years in the Don Valley. The effect of fluorides tends to be more localized on the lichen thalli than that of sulphur dioxide (Richardson 1992). There is also some evidence that nitrogen oxides and ozone may be detrimental to bryophytes and lichens. Other harmful pollutants are aromatic hydrocarbons, such as dioxins, and polychlorinated biphenyls (PCBs). The latter are commonly used in the electrical industry in transformers, and along with emissions from high-level transmission lines have an effect on lichens, although not as evident as with sulphur dioxide and fluorides. For a more detailed account of the effect of pollutants on lichens see Hawksworth and Rose (1976) and Richardson (1992); on lichens and mosses see Farmer, Bates, and Bell (1992).

Industrial processes often produce metals as insoluble particulates (Richardson 1992), and lichens frequently accumulate metals. Lichens can therefore be useful in measuring levels of metal contamination around industrial and urban centres. In some instances the metals, especially copper and zinc, can cause damage or destruction to the lichen thallus. Lichens may be eliminated, for example, on roofs with copper flashing (Hawksworth and Rose 1976) or near copper or zinc-coated wiring (e.g., telephone wires or zinc-coated wiring on trees) (Richardson 1992). In bryophytes, metal accumulates primarily in aquatic systems and has a direct effect on the component flora. *Scapania undulata*, a liverwort, will dominate in streams with heavy zinc content, whereas *Fontinalis antipyretica* cannot tolerate high zinc levels (see Glime 1992).

Industrialization has frequently had a significant effect on waterways. Acidification and organic effluents have caused water stagnation. These trends cause changes not only to the bryophyte flora but to the entire aquatic ecosystem, including the animals feeding on the bryophyte community (Glime 1978). Higher nutrient levels in effluents can cause an increase of the liverwort *Riccia fluitans* in non-acidified habitats. Enhanced nutrient levels may result in increased bacterial respiration, which may cause increases in certain mosses, such as *Hygrohypnum ochraceum* (Glime 1992). However, increased nutrient levels can also cause greater turbidity, resulting in decline of mosses with high light requirements, such as *Amblystegium riparium* (Glime 1992). There is some evidence that acidification

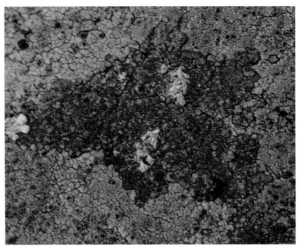

FIGURE 8.14 *Xanthoria elegans,* a characteristic lichen of 'bird rocks.' (John C. Krug)

leads to an increase in certain aquatic mosses – *Drepanocladus fluitans* and *Sphagnum* spp. – although in Ontario the increase in *Sphagnum* biomass is considerably less than elsewhere.

The Effects of Agricultural Practices

Another factor causing changes in the component lichen and bryophyte flora is agriculture. The use of animal manure, which is rich in nitrogen, has resulted in a distinct community forming on trees in open situations, such as fields. This community is also enhanced by dust laden with excreta. In such an environment, the lichen *Xanthoria fallax* is very characteristic. Another community, although not related to agriculture, develops on rocks where birds defecate. Such an environment, rich in nitrogen from the droppings, is dominated by *Xanthoria elegans* (Figure 8.14). On acidic rocks the corresponding lichen is *Lecanora muralis*. James, Hawksworth, and Rose (1977) include *Xanthoparmelia conspersa* as part of this acid rock community. Artificial fertilizers are also known to influence both lichens and bryophytes.

The use of herbicides, fungicides, and insecticides can inhibit the growth of lichens and bryophytes. There is some evidence that the effect is more severe on bryophytes. The current practice of using herbicides for weed control in parks damages

Lichens, which frequently accumulate metals, can be useful in measuring levels of metal contamination around industrial and urban centres.

bryophytes frequently. Some of these sprays also drift and wind up damaging bryophytes in semi-natural situations, such as wooded ravines and plantations. The use of copper-based sprays against plant diseases has certainly resulted in the decline of lichens in treated areas, such as on orchard trees. There is some evidence that insecticides may cause a reduction in lichens on rocky substrates (Brown 1992).

PRESENT-DAY BRYOPHYTE AND LICHEN FLORA

Thus, various environmental and industrial factors have contributed to reduction in the numbers of bryophytes and lichens present today in the greater Toronto region in comparison to the flora that would have been present in the original forests. Unfortunately we have no data on the numbers of lichens and bryophytes that would have been present in the past. However, much of the reduction would be due to loss of habitat, especially in cases where there is narrow tolerance in microhabitat requirements. Nevertheless, as pointed out for southern Ontario by Wong and Brodo (1992), some of the lichens are inconspicuous, and others are either disjunct or at the limit of their range, and thus were never abundant, which may explain in some cases the absence of collections in recent years. Similar arguments could be made for some of the bryophytes. Ireland and Ley (1992) have indicated that a number of the mosses reported from the region are southern in distribution.

Faull (1913) listed 101 species of lichens from the greater Toronto region, although the area covered was somewhat broader than the current region. Certainly some of these, *Teloschistes chrysophthalmus* and probably *Solorina saccata*, are no longer present. Wong and Brodo (1992) list 482 lichens for southern Ontario, of which 131 are recorded for the greater Toronto region. Undoubtedly some of the species common throughout southern Ontario, such as *Cladonia squamosa*, have not been recorded in the region but should be present here. In attempting to put some perspective on the changes in the lichen flora, Wong and Brodo compared the records for collections made prior to 1930 with those for more recent collections. Their data suggest a reduction of about 9 percent for southern Ontario between 1930 and 1960 and a further 2 percent reduction after 1960. The same trend, or perhaps an even greater one, would be expected in the greater Toronto region, as habitat loss and the effects of pollutants are probably more substantial. Data presented for Britain by Hawksworth and Rose (1976), based on studies of relic forests, suggest a reduction of over 50 percent of the original lichen flora.

In his treatment of mosses and liverworts, Graham (1913) recorded 14 liverworts and 49 mosses, supplemented with 8 additional mosses from the interglacial beds of Scarborough and the Don valley. However, he states that this list probably contains little more than one-third of the mosses and liverworts within easy collecting distance of Toronto. Some of these are no longer present, e.g., *Catharinea undulata* [= *Atrichum undulatum*] and *Climacium americanum*, if indeed the identifications are correct. Others characteristic of the bog flora that previously existed west of High Park, such as *Aphanorrhegma patens* [= *Physcomitrella patens*], are no longer present in this location. As the result of an extensive survey, Moxley (1939, 1940) lists 200 species and varieties of mosses from the Toronto region. Even so, several of those mosses recorded by Graham are not included in Moxley's list. If they were still present, it is unlikely that Moxley would have overlooked them. Ireland and Cain (1975) list 464 mosses for Ontario, including 208 for the greater Toronto region. Although this number appears to vary slightly from Moxley's, it should be remembered that the Ireland and Cain list is based on all known records, while Moxley's list is based on fieldwork undertaken in 1938-9. In a later publication Ireland and Ley (1992) list the total number of mosses for Ontario as 490, the increase being due to additional records.

Among the liverworts included by Graham (1913), it is doubtful that *Anthoceros* sp. (a hornwort) is still present, although more recent records are available for the others. Ley and Crowe (1999) record 171 liverworts for Ontario. Many of these are boreal or subarctic in

Bryophytes and lichens prevent erosion, recycle nutrients, provide a carbon reservoir, and are important in the development of natural succession of vegetation.

distribution, but 79 from this list have been recorded for the greater Toronto region.

There are no published data available for bryophytes comparable to Wong and Brodo's documenting the decline of lichens (1992), but like lichens, bryophytes have certainly been severely affected by loss of habitat. If we constructed data similar to Wong and Brodo's for bryophytes, there would also be a noticeable decrease in the flora.

Although we have no baseline data for the extent of the bryophyte and lichen flora in the original forests, the reduction in both elements has been substantial. Currently only a few isolated localities in the greater Toronto region contain any sort of reasonable bryophyte and lichen flora. These include Cold Creek, possibly parts of the Rouge valley, and the property operated by the University of Toronto at Joker's Hill. In order to study any sort of extensive bryophyte and lichen flora, you must travel farther afield, to the forested areas of Haliburton and Muskoka or the Bruce Peninsula, for example. In all likelihood, habitat erosion will continue to some degree as a result of pollution. It is essential that strong measures be taken to preserve the few remnants that are left of our bryophyte and lichen flora. Many people do not realize that bryophytes and lichens are an essential component of a naturally occurring, balanced ecosystem. They prevent erosion, recycle nutrients, provide a carbon reservoir, and are important in the development of natural succession of vegetation. Their small size usually causes them to be overlooked, but their elimination causes a serious reduction in biodiversity and changes in microhabitat, resulting in deprivation for the remaining organisms.

Fungi

David Malloch and John C. Krug

Fungi are a very diverse group of organisms and it is possible to discuss only a few of them in a single chapter. Two of the more conspicuous groups of fungi are treated here in detail: the mushrooms, and fungi that are responsible for plant diseases. The chapter is divided into two separately authored sections. The first outlines in a general fashion what characterizes fungi and describes their role in the ecosystem, then gives an account of some of the mushrooms seen on urban lawns. The second considers a number of diverse fungi, all of which cause disease on plants. First the classification and reproduction of the groups of disease-causing fungi is briefly discussed, information that applies as well to fungi in general. This is followed by consideration of a few of the important diseases and disease-forming fungi in the greater Toronto region.

Fungi in General and Mushrooms in Particular

David Malloch

David Hawksworth (1991) estimated the total number of fungal species on earth in 1992 would be about 1.5 million. He arrived at this number using several logical arguments, including known ratios of fungi to flowering plants in various parts of the world, rates of description of new species, local 'complete' surveys, and several other measures. Most mycologists accept this number as a conservative estimate. The interesting thing about Hawksworth's estimate is that it does not agree very well with the number of *named* species. In fact, comparing his estimate with the numbers that have been described leads to the conclusion that barely 5 percent of fungi have yet been discovered and named. Although this low number may be difficult to accept, it is similar to estimates for other small organisms such as nematodes and mites. Compare this with mammals, which are thought to be more than 90 percent named, and vascular plants, more than 80 percent named. It appears that biologists have rushed to name those organisms they can easily see and have ignored those they cannot.

If only 5 percent of the fungi have been named, we might suspect mycologists of all working on a few organisms and ignoring the rest. Surely then this 5 percent should be well known. Life histories, ecology, geographical distributions, as well as more esoteric information should be abundantly available for the favoured

5 percent. Unfortunately this is not so; even the most familiar of fungi have yet to be studied thoroughly.

With this background of neglect, fungi are difficult to fit into a context presenting the natural histories of larger organisms. For example, ornithologists are greatly interested when a bird is found a few hundred kilometres beyond its expected range and may almost stampede to the place where this has occurred. A mushroom found equally far beyond its known range surprises no one, not because mushrooms have wider ranges than birds but because mushroom ranges have never been properly documented.

A summary of the fungi of the Toronto region, then, is difficult. It can be presented in an attractive but misleading way or in an honest but wishy-washy way. Another approach is to present, in the manner of Hawksworth, an approximation of what is there, assembled from the tips of fungal icebergs we see protruding here and there around the city. This last approach is taken here, as the topic is first narrowed and then extrapolated from what is known for southern Ontario in general to predict what will occur in Toronto.

WHAT ARE FUNGI?

Fungi are considered by most biologists to form a kingdom of their own, parallel to those of animals, plants, bacteria, and the like. The kingdom of fungi includes a great diversity of organisms occupying a great variety of habitats. Because of their diversity it is difficult to state briefly exactly what characterizes fungi, but we might include these facts:

- Fungi are heterotrophic – that is, they obtain their energy by the oxidation of organic compounds. They are fundamentally different from plants, which are autotrophic and obtain their energy from light.
- Nutrition is by absorption. Nutrients are absorbed directly into the cell from the external environment. Most animals obtain nutrients by ingestion, taking food into their bodies, where it is digested and absorbed. Plants, being autotrophs, produce their food internally.
- The fundamental structure of fungi is a filament (hypha) in most species. A few fungi, such as yeasts and chytrids, may be single-celled, but these represent a very small percentage of the estimated 1.5 million species.
- Fungal cell walls contain chitin, the same substance forming the exoskeletons (shells) of lobsters, crabs, and insects. Plant cell walls contain cellulose, not chitin. Animals lack cell walls altogether.

Although fungi may seem to be as unlike ourselves as any organism can be, there are many things we have in common. Like all living things, fungi must obtain energy and must obtain foods to be used as the building blocks for their own bodies. They must reproduce and they must be able to travel through space and time.

Like humans, fungi utilize organic sources of carbon for the energy to power their life functions. Most of the sources of energy used by humans are also used by fungi. Substances such as sugar, starch, proteins, and fats are good sources of energy for both humans and fungi. When a mould appears on a slice of bread, it is because it has found it to be an acceptable source of energy and nutrition. We and that mould are equally able to digest bread, and, chemically speaking, go about it in a very similar way.

The total number of fungal species on earth is estimated to be about 1.5 million. Barely 5 percent of fungi have yet been discovered and named.

All living things are composed of substantial numbers of proteins. Our muscles, bones, hair, and brains, as well as our other organs, contain proteins. We could not continue to live and grow without proteins. Fungi also require proteins and often obtain these the same way we do. For example, when you eat some meat or beans you are ingesting a good source of proteins. These are cut by enzymes in the digestive tract into their individual components, amino acids. The resulting amino acids can then be reassembled into the proteins we need to form our tissues and organs. If you feed a fungus with some beans it will do exactly the same thing. Compare this with plants. If you give a piece of meat to a plant as its only source of protein building blocks, it will starve.

On the other hand, fungi, like plants, can also build their proteins from very simple sources. In fact, they can assimilate simple ammonium compounds from their environment and use these as the basis for synthesizing all their amino acids and proteins. This, combined with their ability to consume a great variety of carbon sources, gives fungi the ability to live under conditions leading to certain starvation for us. For example, given cellulose and a few minerals, many fungi are able to thrive and reproduce perfectly well. The ability to satisfy their nutritional needs with both the most complex and the simplest substances sets fungi apart from plants and animals and partially explains their great flexibility in dealing with environmental uncertainties.

Given these generalizations about fungal structure and nutrition, we are still left with the question of what they are actually *doing* in their environment. Miller (1995) outlined some of the roles likely to be played by fungi in ecosystems and discussed the difficulties involved in assessing these activities. Although still in the realm of generalization, Miller's outline does suggest a widespread and fundamental role of fungi in most ecosystems. More specifically, fungi can function as saprotrophs, feeding on dead organic material, have more active interactions with other organisms, and function as parasites or mutualists. As saprotrophs, fungi are involved in the decomposition of most dead material. They are abundant in dead wood, fallen leaves, soil, dead insects, dung. As parasites they cause the majority of plant diseases and are important, if not the most prevalent, agents of insect disease. They are also fierce predators of nematodes and other fungi. As mutualists they occur everywhere in partnership with algae to form lichens and with 95 percent of all plants to form mycorrhizae. Even the stomachs of cows and sheep depend upon the digestive activities of specialized fungi.

FUNGI IN SOUTHERN ONTARIO

In spite of a long tradition of studying fungi in southern Ontario, we are still not really certain what species occur here or what governs their distribution. As with most non-photosynthetic organisms, fungi are strongly influenced by vegetation. Fungi specialized for the decomposition of palms are as unlikely to occur here as birds associated with these trees. Southern Ontario is characterized by two great vegetational types: the Carolinian forest in the warmest parts and the Great Lakes-St. Lawrence forest elsewhere. These forests are dominated by deciduous hardwood trees such as oaks, beech, maple, and birch as well as several conifers. Conifers become more abundant to the north. Saprotrophic fungi are geared toward strongly seasonal events such as leaf drop and insect abundance. Long dry winters demand sturdy mechanisms for dormancy as well as sensitivity to signs of spring. 'How does it overwinter?' is a question frequently asked by mycologists. Most plants and insects have a variety of fungal parasites, and these parasites must be able to respond to seasonal changes of their hosts. Parasitic fungi must have life histories that mesh well with those of other organisms.

Fungal mutualisms in southern Ontario are well represented in our lichens and mycorrhizae. Our forests are characterized by a variety of trees having *ectomycorrhizae,* a type of mutualism involving the roots of plants and fungi reproducing by means of conspicuous mushrooms. Ectomycorrhizae are especially well developed in southern Ontario, resulting in a rich flora of large, conspicuous mushrooms. In contrast, tropical rainforests are not strongly ectomycorrhizal and produce far fewer large mushrooms than occur here. In this group at least, temperate forests, not the tropics, are the site of great biodiversity.

The ability to satisfy their nutritional needs with both the most complex and the simplest substances sets fungi apart from plants and animals and partially explains their great flexibility in dealing with environmental uncertainties.

FUNGI IN THE TORONTO REGION

We know too little about fungi in southern Ontario to be able to say how the Toronto region differs from the area as a whole. We are on the border between the Carolinian and Great Lakes-St. Lawrence zones, and some fungi characteristic of both areas will be found here. The summary of species that follows focuses attention only on fungi characteristic of the urban environment.

The Urban Environment

Although unlike its original state, the urban environment is nevertheless a lively biological region. It has its own peculiar soils, vegetation, and fauna. Wild plants and animals abound and interact with one another in complex ways. Most houses have gardens maintained in precarious equilibrium with invading 'wild' organisms. Parks may be numerous and approach a condition not unlike that found in areas outside the city. Fungi occur abundantly in these environments and assume the same roles they would in more 'natural' areas. It is no harder to find mushrooms in Queen's Park than in Algonquin Park. Fungi may even become established indoors and associate closely with other occupants, both plant and animal (including humans). Again we come up against our ignorance of fungi. Undescribed species can easily be found almost anywhere within Toronto. 'Rare' species occur throughout the city; a few years ago a large conspicuous fungus that had not been found in Canada before was discovered in High Park.

Fungi in Lawns

To deal with our lack of knowledge, even of urban fungi, and yet maintain some level of credibility this section narrows its focus even further and deals only with mushrooms, and only those known to occur on urban lawns in our region. Mushrooms on Toronto lawns are numerous, represent at least 100 known species, are complex in their environmental associations, and, most important, represent a true biotic element.

Lawns are a peculiar habitat, resembling overgrazed grasslands. Everyone in Toronto knows what constitutes a good lawn. This striving toward the perfect lawn by both private owners and public institutions leads to varying degrees of success and hence fairly predictable and classifiable results. There are perfect lawns, and there are weedy, dry, wet, sandy, ant- or grub-infested, and diseased lawns. We all know them well. Fungi also know them and have their preferences, not always in agreement with ours. Poor lawns, from our point of view, may yield wonderful mushrooms. As with other organisms in our region, some lawn mushrooms are highly seasonal and occur over a period of only one or two weeks, while others appear throughout the growing season. Some are common and can be found throughout the city, while others are known only from one small lawn. New records are waiting to be discovered and new species to be described.

Because a variety of lawns will produce a greater variety of mushrooms than a single lawn, it is not necessary to outline good localities in our region. Any lawn will do, and each will have its own characteristic mushrooms. In general, cemeteries are very good places to start. They have well-tended lawns shaded by large trees. Large city parks are also suitable. Because they are often not as perfect as the cemeteries, they have good individual characteristics. One can also search around public buildings of all types and around smaller residential lawns.

Although this book is intended for naturalists who are not primarily interested in the exploitation of nature, some comments should be included about eating mushrooms. Perfect lawns are often attained through the use of herbicides and insecticides. Fungi are often able to concentrate chemicals, and therefore eating mushrooms from lawns with unknown histories is not recommended. If you are certain that no such chemicals have been used and are confident you have correctly identified the mushroom, it is probably safe to eat lawn mushrooms known to be edible. On the other hand, mushrooms available from the supermarket are probably a safer alternative.

A few years ago a large conspicuous fungus that had not been found in Canada before was discovered in High Park.

THE MUSHROOMS

It is not appropriate to publish here a complete list of fungi for Toronto lawns. Instead a summary of those species most likely to be found with some discussion of their natural history is provided. Lawn mushrooms can be subdivided into groups in several ways, such as by taxonomic group, by their ecological function, by colour, edibility, and the like. For this purpose, they are arranged according to their season of appearance. Although some species can always be found during the growing season, most will be found mainly during a relatively short period.

Spring

Agrocybe dura

Agrocybe dura is truly a spring mushroom, appearing on Toronto lawns during May and early June. Only very rarely is it found at other times of the year. It can be a fairly large and conspicuous mushroom, with a cap measuring up to several centimetres in diameter. It is characterized by the fairly light cream to tan colours of its cap and stalk (Figure 9.1). There is a small annulus or ring on the stalk that will be obvious in most collections, although it may rub off if handled carelessly. Many specimens of *A. dura* will have the surface of the cap

FIGURE 9.1 *Agrocybe dura.* (David Malloch)

cracked like dry mud. The gills on the underside of the cap are yellowish tan to nearly white at first, but soon become brown as the spores mature.

Agrocybe pediades

Occurring throughout the season, *Agrocybe pediades* cannot really be considered a spring mushroom. The whole mushroom is yellow to tan. The gills are also yellow at first but soon become brown as the spores mature. The top of the cap is a little slimy or slippery when it is wet. There is a small annulus or ring on the stem. It could be confused with *A. dura* but is smaller and has a cap that remains nearly spherical for a long time, although eventually it will expand out to nearly flat.

Coprinus atramentarius

One of several inky caps, *Coprinus atramentarius* arises in clusters from buried and exposed wood during summer and fall. It resembles *C. micaceus* in its bell-shaped caps but is much heavier. The caps are usually covered with flat brown scales but can be grey and almost without scales. The flesh sometimes has the odour of freshly caught fish. The gills digest themselves, as they do in other inky caps. *Coprinus atramentarius* has earned some notoriety as a poisonous mushroom. It is in fact perfectly edible as long as you don't drink any alcoholic beverages along with it. If you do you will experience flushing, a rapid pulse, a 'tight' feeling, and other unpleasant symptoms.

Coprinus micaceus

This species is one of the earliest mushrooms to appear in the spring but it will continue to appear throughout the growing season after each rainfall. *Coprinus micaceus* is a wood-rotting mushroom that does especially well on the dead roots of streetside trees. When a tree growing on a lawn is cut down, the mycelium (fungal threads) of *C. micaceus* will invade the roots and send up great crowded masses of mushrooms whenever it rains. *Coprinus micaceus* is a typical inky cap mushroom. Inky caps differ from other mushrooms in dissolving their own gills at maturity. The spores are black and get caught up in the wet dissolving gill mass, yielding an ink-like fluid. *Coprinus* ink is black enough that it can actually be used for writing. Besides its habit of producing large clusters of mushrooms from buried roots, *C. micaceus* can also be identified by its light brown cap covered with a minute glistening powder.

Panaeolus foenisecii

Panaeolus foenisecii is without question the most common mushroom on Toronto (or Canadian) lawns. It appears in the spring and continues fruiting into the first frosts of autumn. When fresh, the whole mushroom, including the young gills, is dark chocolate brown, but it has a strong tendency to fade out to tan. It is small and delicate, with a cap ranging in size from a dime to a dollar. *Panaeolus foenisecii* is responsible for more calls to poison control centres from hospitals than any other mushroom in our area. Although it is not poisonous, it is very common – small children are frequently caught putting it in their mouths, causing parents and hospital staff to become frightened and to call for immediate help. In fact, this mushroom contains small amounts of psilocybin, a hallucinogenic substance, but these are so low that I have never heard of anyone having a neurological effect from it. There are one or two other mushrooms in our area that occur in the fall and are nearly indistinguishable from *P. foenisecii,* but these are much less common.

Psilocybe physaloides

This is a small brown mushroom (mushrooms like this are disdainfully called Little Brown Mushrooms, LBMs, by people seeking mushrooms for the table) growing attached to the lower portions of lawn grasses. The stems are not as brittle as those of *P. foenisecii* and the caps are a little thicker. It is typically a spring mushroom that may be found during the heat of summer. Some species of *Psilocybe* are famous (or infamous) as 'magic mushrooms.' Many are hallucinogenic and are used 'recreationally' for this purpose. It is now illegal to have hallucinogenic mushrooms in your possession in Canada.

Fortunately Toronto lawns are free of 'magic mushrooms,' and there is no fear of being apprehended by the Mounties for collecting *P. physaloides, P. foenisecii,* or mushrooms resembling them.

Summer

Agaricus campestris

Agaricus campestris is the well-known field or meadow mushroom. People all over the world collect and eat it. It is mostly white to very pale tan, except for the gills, which are pink at first and then chocolate brown. The stem has a fragile ring near its apex. There are several species of *Agaricus* occurring on Toronto lawns, but most are not as fragile as *A. campestris*. The most similar is *A. pampeanus,* originally described from the Pampa in Argentina but commonly occurring here as well. *Agaricus pampeanus* is slightly sturdier than *A. campestris* and can be distinguished with a microscope.

Calocybe carnea

Several species of lawn mushrooms are common in that habitat and rarely anywhere else. *Calocybe carnea* is one of these, and because of this limited habitat is rarely mentioned in popular mushroom guides. It occurs attached to the basal portions of lawn grass plants, often in clusters of two to three, and is characterized by its white stem and gills and bright pink cap. This is a small inconspicuous mushroom not much bigger than a button and can give you a lot of trouble when you try to identify it for the first time. Aside from the pink cap, it is quite nondescript.

Clitocybe dealbata

This species is one of the enemies of mushroom eaters. It contains significant quantities of the toxin muscarine and can make a person quite sick. It is especially dangerous because it resembles *Marasmius oreades,* the edible fairy ring mushroom, and may be collected by mistake. The caps are white, ranging in size from the diameter of a golf ball to that of a soft drink can (Figure 9.2). The stem and gill are also white. Like *M. oreades* it tends to grow in circular rings and arcs called fairy rings. It differs from *M. oreades* most clearly in the appearance of its gills. Those of *M. oreades* are rather distant from one another and turn up toward the cap when they meet the stalk; in fact they hardly connect to the stalk at all. The gills of *C. dealbata* are quite crowded and turn toward the ground where they meet the stem.

FIGURE 9.2 *Clitocybe dealbata,* or sweat-producing Clitocybe. (David Malloch)

Conocybe lactea

Conocybe lactea is our most characteristic summer mushroom. It becomes abundant in July, when the temperature and humidity are high, and will never be found in spring and fall. The white to slightly yellowish-white caps are usually the size of a quarter and occur at the tops of long, delicate stems. The gills are rusty brown as the spores mature. This species is particularly common on very well kept lawns. *Conocybe crispa* is a very similar mushroom, differing from *C. lactea* in having wavy or 'crisped' gills.

Conocybe tenera

Another characteristic summer *Conocybe, C. tenera* is smaller than *C. lactea* and is dull orange to tan throughout (Figure 9.3). The two species cannot be confused.

FIGURE 9.3 *Conocybe tenera.* Conocybes are often called 'dunce caps' or 'cone heads.' (David Malloch)

FIGURE 9.4 *Coprinus plicatilis*, the pleated inky cap. (David Malloch)

FIGURE 9.5 *Entoloma sericeum*. (David Malloch)

There are other orange *Conocybe* species on Toronto lawns, but these tend to be rather uncommon and to appear in the fall.

Coprinus plicatilis

Among our most beautiful lawn mushrooms, *Coprinus plicatilis* and its relatives look like tiny Japanese parasols in the lawn (Figure 9.4). These are exceedingly delicate and must be sought before the day begins to heat up. The caps are able to open widely for exactly the same reason that parasols can: the gills actually split from the outside of the cap to their edges, leaving a nearly flat surface like in the inside of a parasol. The caps are brown at first, but become grey when they open. They are small, ranging in size between a dime and a one-dollar coin.

Entoloma sericeum

Species of *Entoloma* are often more interesting under a microscope than they are in gross appearance. The spores are pink in mass and faceted like a cut diamond under the microscope. *Entoloma sericeum* is a dark steely-grey brown mushroom with a slightly silky cap (Figure 9.5). It is not usually very large, about the size of a two-dollar coin. The genus contains some poisonous species, and we have experienced some cases of *Entoloma* poisonings in the Toronto area.

Gastrocybe lateritia

Similar in appearance and season to *Conocybe lactea*, *Gastrocybe lateritia* is nevertheless a very distinct mushroom. Its principal claim to fame is that it is a gasteromycete – it does not forcibly discharge its spores away from its cap. Instead, the mushroom expands up from the ground and then falls back to decay in the lawn. It is not known why it does not discharge its spores like other mushrooms do, but its behaviour is so constant there is no doubt it is doing something useful. This mushroom has been known for only about 20 years. It was first discovered in Michigan and seems to have a range extending along the lower Great Lakes. *Gastrocybe lateritia* is very common in Toronto lawns and can be seen every July. To see it you must search in the morning, preferably before 9 o'clock during a warm wet spell of weather.

Marasmius graminum

This may be the smallest lawn mushroom in the Toronto area. The caps are brick red and no more than 3 to 8 mm in diameter. They are attached to dark wiry stalks and arise from the bases of individual grass plants. Although small and inconspicuous, it is a common mushroom and easily located if you lie in the grass and search among the individual plants. The spores of *Marasmius graminum* are colourless and the gills thus remain white at maturity. Unlike most lawn mushrooms, *Marasmius* species have the ability to dry up and then revive when moistened by rain.

Marasmius oreades

Often called the fairy ring mushroom, *Marasmius oreades* grows in large circular colonies, killing the grass at its advancing margin. Behind the margin, the grass comes back in a dark green healthy state. Ancient folklore has it that the *Marasmius* rings mark the site of fairy dances. At one site near Ottawa, a large *Marasmius* fairy ring won international attention as a UFO landing site. Caps of the fairy ring mushroom are cream to pale tan and

FIGURE 9.6 *Psathyrella candolleana.* (David Malloch)

are attached to tough whitish stems. The flesh sometimes has a garlic odour. Like most *Marasmius* species, *M. oreades* has the ability to revive from a dried state when moistened.

Psathyrella candolleana

This common mushroom is similar to *Coprinus micaceus* but differs in having gills that never digest themselves. The caps are grey brown, delicate, and attached to fragile white stems (Figure 9.6). The mature gills are dark brown to black. The mushrooms arise, like *C. micaceus,* from buried wood. This species illustrates a good point about mushroom identification: taxonomically difficult groups may occasionally be easier to resolve if you concentrate on one type of habitat, like lawns. The genus *Psathyrella* is a very difficult group in most habitats, from the boreal forest to the humid tropics, yet the species on lawns appear to be quite straightforward and easy to identify.

Fall

Agaricus arvensis

Often called the horse mushroom, *Agaricus arvensis* is the largest member of its genus on Toronto lawns. The large white fruiting bodies have an odour of anise and turn yellow with age or when cooked. The gills are dull pinkish at first but soon darken as the spores mature. This massive *Agaricus* can sometimes be abundant and can be seen from some distance. The stem has a ring that is less delicate than that of *A. campestris* but nevertheless can be easily rubbed off. *Agaricus arvensis* is a tasty mushroom yet is mildly poisonous to some people, causing stomach upset and diarrhoea. If you collect it on a pesticide-free lawn and wish to give it a try, just eat a little the first time to see if you can tolerate it.

Agaricus bitorquis

This is another large *Agaricus* species. It has very compact tissue, something like a white hockey puck. The cap and stem are white. The gills are dull pink at first and then brown. There is a well-developed annulus or ring on the stem that is clearly two layered. With its large size and compact flesh *A. bitorquis* is able to push its way through very hard earth and even asphalt. People who collect it for the table learn to feel the earth with their feet for telltale humps. It is a close relative to the cultivated mushroom.

Agaricus semotus

This is the region's smallest *Agaricus* species, with a cap seldom larger than a two-dollar coin. It is basically white with brown gills, although the cap is typically overlain with reddish to yellowish scales or fibres. Aside from its small size *A. semotus* can be recognized by its odour of almonds and by its tendency to turn yellow when dried or old. It is seldom as common as *A. campestris,* but can be locally abundant. There have been some fairly large fruitings on the lawns of Erindale College.

Agrocybe erebia

Agrocybe erebia is not much like its spring-appearing relatives. It occurs in clusters and has dark brown caps on white stems. The gills are light brown at first but darken as the spores mature. It is less common than other species of *Agrocybe* but occasionally attracts attention on lawns.

Coprinus comatus

Coprinus comatus – the shaggy mane – is among the best known of all mushrooms. The caps are egg-shaped to almost cylindrical and are brown to white; most are shaggy with brown scales (Figure 9.7). As with all the inky caps, the gills digest themselves and become reduced to a black inky mass. The appearance of *C. comatus* is a sure sign of fall in Toronto. The fruiting bodies often do not appear until after the first frost in late September. This species is sought for the table throughout the area where it occurs.

Hebeloma crustuliniforme

Hebeloma species are ectomycorrhizal: that is, they form mutualistic relationships with the roots of trees. They are therefore not strictly lawn mushrooms and could be eliminated from this discussion. On the other hand, when ectomycorrhizal trees such as pines, spruces, birches,

FIGURE 9.7 *Coprinus comatus,* or shaggy mane, is a type of inky cap. (Sherry Pettigrew)

FIGURE 9.9 *Inocybe decipientoides.* (David Malloch)

Inocybe decipientoides

Inocybe decipientoides is a poisonous mushroom, containing, like *Clitocybe dealbata,* significant quantities of muscarine. Like species of *Hebeloma, Inocybe* species are ectomycorrhizal and will only appear on lawns if the appropriate tree is present. *Inocybe decipientoides* usually associates with willows and poplars, but may occasionally also be found with oaks or even pines. The typical *Inocybe* mushroom has a brown silky to scaly conical cap and a white to brown stem (Figure 9.9). They are not strikingly attractive and are often avoided by people otherwise interested in mushrooms.

willows, poplars, and oaks are planted on lawns, certain, but not many, ectomycorrhizal mushrooms begin to appear. *Hebeloma crustuliniforme* commonly appears on lawns in the fall under spruces and to a lesser extent other trees. They are medium-sized mushrooms, 2.5 to 5 cm across, and occur in small groups near their tree (Figure 9.8). They are generally reddish brown to yellowish brown throughout and may have rounded sticky to slippery smooth caps. Several species of *Hebeloma* are known to be mildly poisonous.

Lepista nuda

Species of *Lepista* are similar to those of *Clitocybe.* In fact, some authorities prefer to discard *Lepista* altogether and deal with all of them as *Clitocybe* species. However, they have a recognizable appearance and many mycologists prefer to refer to them as a separate genus. *Lepista nuda,* or blewit as it is sometimes called, is a fairly large bluish to purplish fleshy mushroom. The spores are pinkish to nearly white, so the gills always keep their original purple colour. Blewits are common in pine plantations in southern Ontario, where they are eagerly sought as edible mushrooms. Again, because of the pesticide danger, eating these or other species collected on lawns is not recommended.

Lepista sordida

This species is blue to purple, like the related *Lepista nuda,* but is smaller and more delicate. Typically the caps are no larger than 5 cm and are quite thin. Their thinness allows them to dry out easily and lose their bluish colours; hence, they are often a dull pale brown during dry weather.

FIGURE 9.8 *Hebeloma crustuliniforme,* called 'poison pie.' (David Malloch)

FIGURE 9.10 *Lepiota cristata*, called 'brown-eyed parasol.' (David Malloch)

FIGURE 9.11 *Psathyrella epimyces*. (David Malloch)

Lepiota cristata

This little mushroom can become quite common in the fall. It has a white conical cap covered with brown scales (Figure 9.10). The stem and gills are white as well. There is a distinct ring near the top of the stem. *Lepiota cristata* commonly grows in dry sandy lawns in the High Park area but is common elsewhere as well. Like most *Lepiota* species, *L. cristata* has a strong penetrating odour that is hard to describe but not easily forgotten. If a microscope is handy, this species can be distinguished by its small conical spores, each with a little 'heel' at the bottom. It is said to be poisonous, but, in fact, eating *any* small *Lepiota* is a risky affair.

Lepiota rubrotincta

Lepiota rubrotincta is a very poisonous mushroom, perhaps the most poisonous of all lawn mushrooms. It may contain substantial amounts of amatoxins, the most dangerous of mushroom toxins. Although these mushrooms are small and not usually tempting to mushroom eaters, they may occur in numbers large enough to be collected. There was one case of poisoning by *Lepiota rubrotincta* near Windsor, Ontario, in the early 1990s, in which the people involved survived but suffered massive liver damage. The mushrooms are small, with caps the size of a one- or a two-dollar coin. The conical white caps are covered with pink to reddish brown scales. The stem and gills are also white.

Leucoagaricus naucinus

A typical fall mushroom, *Leucoagaricus naucinus* is fairly large and white throughout. The stem is bulbous and bears a well-defined ring near the top. It is commonly

FIGURE 9.12 *Mycena alcalina*, or alkaline mycena. (David Malloch)

collected and eaten around Thanksgiving but presents some danger because it is very similar to *Amanita virosa*, the deadly poisonous destroying angel. Unless you are very sure of yourself, it is best not to eat *L. naucinus*. *Amanita virosa* has been reported from lawns, but it is mycorrhizal (mutualistic) with the roots of trees and is not usually found on city lawns. The most conspicuous difference between *A. virosa* and *L. naucinus* is found at the base of the stem. *Leucoagaricus naucinus* lacks a volva, a cup-like structure enclosing the bottom of the stem of many species of *Amanita*. However, you must be careful when looking for the volva; if you pull the mushroom up rather than digging it out, the volva may be left behind in the soil.

Chlorophyllum molybdites

This species looks like a very large species of *Lepiota*. It has a conical cap covered with brown scales but is otherwise white. It is larger than a typical *Lepiota* species, having caps up to the size of a dinner plate. The most striking feature of *C. molybdites* is the green colour of the mature gills. In fact, the spores themselves are blue-green and cause the gills to become the same colour. *Chlorophyllum molybdites* is quite rare in southern Ontario but is occasionally found in the Toronto area. It is common in the subtropics and tropics and may appear in indoor tropical gardens. It is a poisonous species, and puts many people in the hospital around Washington, DC, and areas farther south.

Psathyrella epimyces

This mushroom is among the most striking of lawn mushrooms. It is a parasite of *Coprinus comatus*, the shaggy mane, causing the usually cylindrical cap of its host to become shortened and funnel-shaped (Figure 9.11). The fruiting bodies of *P. epimyces* then grow out of the centre of the 'funnel.' Most people would not even recognize the funnel-shaped structure as the cap of a shaggy mane except that healthy individuals usually occur nearby. When an infected shaggy mane is found, it is usually possible to note the location and come back the next year to see it again. This suggests that the shaggy mane lives as a mycelium (thread) underground throughout the year and simply carries the parasite with it.

Mycena alcalina

Mycena alcalina is a small yellowish brown mushroom with a delicate conical cap about the size of a dollar coin (Figure 9.12). The white stem is delicate but rigid, allowing the mushroom to bend over and then spring back up. The gills are white to grey at maturity. This is one of several *Mycena* species appearing in the autumn. Mount Pleasant cemetery is a good source of *Mycena* species in October.

Plant Disease Fungi

John C. Krug

Plant disease fungi are those that cause infections on plants. Along with bacteria and viruses, which also cause plant diseases, they are called plant pathogens – hence pathogenic fungi.

Fungi can attack a range of plant material and are termed either saprotrophs or parasites. Obligate parasites, or biotrophs, cause devastating diseases and are incapable of surviving on dead or decomposing material. Facultative, or weak, parasites, by contrast, can live either as saprotrophs – which can survive on dead material – or as parasites. They often do not cause extensive damage to the host plant. Plant disease fungi can be either obligate or facultative parasites. They are found in all major groups of fungi, although some families lack pathogens.

CLASSIFICATION AND REPRODUCTION

The various fungi are broadly grouped according to their mode of reproduction. This applies not only to plant disease fungi but to all fungi, without regard to their habitat or role in the ecosystem. There are three broad groups of fungi, the Basidiomycetes, the Ascomycetes, and the Zygomycetes. The Basidiomycetes, or club fungi, possess a club-like structure, the basidium (Figure 9.13A),

FIGURE 9.14 Apothecium of *Scutellinia scutellata*. (John C. Krug)

which, following meiosis, produces spores for sexual reproduction externally. These are called basidiospores. Sometimes the basidium is divided lengthwise, as in the jelly fungi. Different fungi have different structures producing the basidia. In mushrooms, the basidia are borne on the gills, but in other fungi they appear in pores or on spines and smooth surfaces. Release of spores is aided by a water droplet below the attachment of the spores to the basidium. Usually there are either two or four spores produced on a single basidium.

The Ascomycetes, or sac fungi, produce spores internally within an elongated, specialized cell, the ascus (Figure 9.13B). In one group of Ascomycetes, the spore-containing asci are borne within open cup-like structures, apothecia (Figure 9.14), while in another the asci are produced within closed, flask-like structures, called perithecia (Figure 9.15A). In certain genera the perithecia are embedded in compound sterile structures, called stromata (Figures 9.15B). The stromata usually develop from a separate origin, and frequently possess different tissue types, than the perithecia that form within the stromata. Spore-bearing fruiting bodies are sometimes collectively called ascocarps. Spore release is frequently aided by simple or complex structures at the tip of the ascus or by an expansive membrane inside the ascus wall. The spores, frequently called ascospores, can be either smooth or ornamented and can have one cell or many cells. Typically there are eight spores in a single ascus.

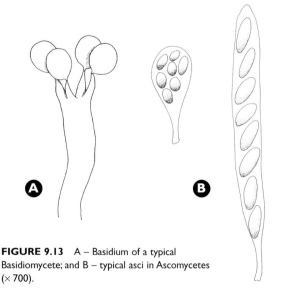

FIGURE 9.13 A – Basidium of a typical Basidiomycete; and B – typical asci in Ascomycetes (× 700).

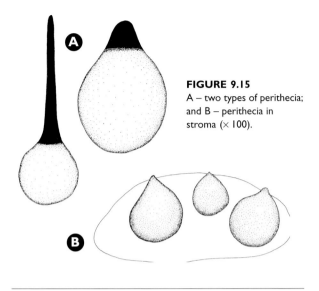

FIGURE 9.15
A – two types of perithecia; and B – perithecia in stroma (× 100).

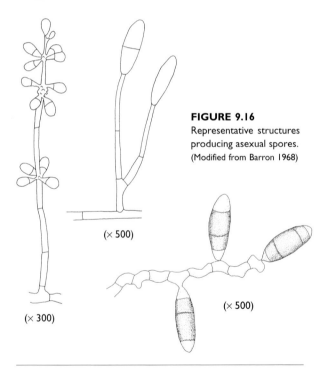

FIGURE 9.16
Representative structures producing asexual spores. (Modified from Barron 1968)

In both Ascomycetes and Basidiomycetes, reproduction can also occur asexually – that is, without sexual union – by production of asexual spores, called conidia. These spores can be borne directly on the fungal hyphae (filaments) or produced on specialized structures (Figure 9.16) developing on the hyphae, or within enclosed structures, called pycnidia, which sometimes superficially resemble perithecia. For illustrations of a range of conidia-bearing structures see Barron (1968) and Malloch (1981). A typical life cycle involving conidial formation is shown in Figure 9.17. In the life cycle the canker, which is sterile, can produce separate structures bearing asexual conidia, or sexual ascospores, both of which on germination form hyphae that give rise to a new canker. This in turn can develop structures bearing conidia or ascospores.

The Zygomycetes possess a mycelium – or mass of hyphae lacking cross-walls – from which develop stalked globular structures, sporangia (Figure 9.18). The sporangia bear asexual spores either internally (Figure 9.18A) or externally (Figure 9.18B), depending on the genus. Spore release depends largely on air currents. Spores are typically smooth, single celled, and lack flagella (motile appendages). Sexual reproduction occurs by the union of hyphae to produce sexual zygospores (Figure 9.18A).

Several additional groups formerly placed in the Zygomycetes are now considered to be only distantly related. Today, for example, Oomycetes are generally regarded as related to the algae rather than to the true fungi. Previously the downy mildews, which cause important plant diseases, were placed with the fungi but today are treated as Oomycetes. For the purposes of this chapter, however, they are considered with the fungi. The Oomycetes differ from the Zygomycetes in forming asexual spores with motile appendages, and having cellulose rather than chitin in the cell walls. Sexual reproduction is achieved from separate male and female organs; fertilization takes place via a germ tube and results in an oospore, which is the sexual spore. For further information on the classification, morphology, and life histories of fungi see Alexopoulos, Mims, and Blackwell (1996).

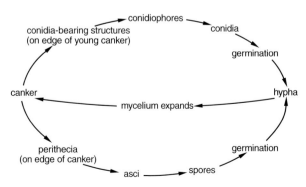

FIGURE 9.17 Life cycle of *Nectria galligena,* a characteristic Ascomycete.

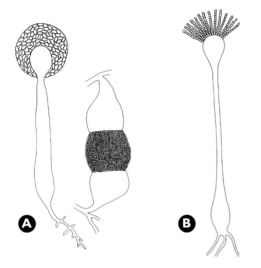

FIGURE 9.18 Two contrasting types of sporangia in Zygomycetes: A – spores produced internally, with zygospore at right (× 100); and B – spores borne externally (× 100). (Modified from Schröter 1897a)

HABITAT

The nature of the forest ecosystem that dominated southern Ontario prior to European colonization, and the various microhabitats and substrates, were discussed for bryophytes and lichens in Chapter 8. The loss of habitat and substrate is equally important for fungi but probably not quite so obvious. Fungi may be specific to certain substrates or hosts, infecting both living or dead tissues. They can also be quite cosmopolitan and sometimes can change hosts, especially near the limit of the host range. They can grow on the forest floor, in litter and on various substrates there, or in openings within the forest. It has been said that there is a fungus for every available habitat.

Unfortunately, we have little detailed information about fungi present in the original forest except for a few hints from Aboriginal medicine. There are a couple of lists of fungi for the greater Toronto region (Bell 1933; Langton 1911, 1913), but these are dominated by mushrooms, and are very incomplete, even for those fungi. Because of our lack of knowledge about fungi in the region, it is impossible in most instances to discuss changes in the overall fungal population. The discussion here therefore focuses on selected diseases and the possible influence of the introduction of agriculture and changing agricultural and forestry practices, with emphasis on changes in the fungal population where there is supporting information.

HOW PATHOGENIC FUNGI WORK

The very nature of decay in plants involves a succession of fungi on a given substrate. Specific fungi each play a vital role in this process. The order of this succession is complex and probably involves utilization and depletion of nutrients as well as competition among various fungi and other organisms. One of the major groups of fungi involved in this succession are the wood decay fungi, which for fallen woody substrates on the ground probably occur fairly late in the succession cycle. There are two types of decay, one affecting the inner heartwood of a tree, the other affecting the outer sapwood. These can be termed brown rot or white rot fungi, depending on the colour given to the wood by the decay.

Although we do not have a clear idea of the fungi present in the original forest, we can envisage a considerable amount of biodiversity. Certain species would be typical of these old forests. We know that pine and, in some places, spruce would have been a major component of the original forest. Cutting these forests would have reduced biodiversity.

One fungus that we find in old forests is a bracket type in which the basidia are borne in pores: *Phaeolus schweinitzii* (Figure 9.19) is generally confined to spruce and found only on old, mature trees. Thus, today it is found in the greater Toronto region only in those areas that contain very old spruce. *Grifolia frondosa* is confined to very mature oak trees. There are records of this fungus in the region from the 1930s and 1940s but no further records until the 1960s. The lack of records for the intervening years probably indicates that there were few trees of sufficient age at that time. Recently we discovered *Inonotus dryadeus* on very mature maple in the region. Previously the fungus was unrecorded for eastern Canada and is probably another example of a species confined to old-growth trees, in this case a hardwood. These three examples are all polypore types, but the mushroom *Pleurotus dryinus,* which is occasionally found in the region on mature hardwoods, may represent another example of a fungus found only on old trees. It is recorded by Langton (1913) as being rare. The infrequency of these fungi that are confined to very old trees, and thus would have been characteristic of the original old-growth forest, emphasizes the need to preserve what is left of our old-growth forests.

Cutting down forests and introducing agriculture appears to have allowed some fungi to exploit existing habitats. Langton (1913) indicates *Polyporus squamosus*

FIGURE 9.19 Bracket of *Phaeolus schweinitzii*, with lower fruiting body upturned to show pore surface. (John C. Krug)

FIGURE 9.20 Fruiting body of *Polyporus squamosus*. (Audrey B. Harris)

(Figure 9.20) as being rare, but today it is not infrequent on hardwoods, especially maple. The honey mushroom, *Armillariella mellea* (Figure 9.21), which is known to retain a viable population in the soil for centuries, is very pathogenic on various hardwoods. It is particularly devastating on young saplings, causing a serious root rot. In some young plantations almost the entire plantation is infected with this fungus. Another fungus causing a serious root rot is *Heterobasidion annosum* (= *Fomes annosus*). It is very prevalent on conifers, especially in plantations. It usually enters through a wound and spreads through the root system, which in plantations is interconnected. Thus, the stumps left from thinning operations are ideal sites for infection. Indeed, the fungus has become more prevalent in plantations than in native forest stands.

Chondrostereum purpureum (= *Stereum purpureum*) causes a decay of ornamental and fruit trees, especially well-established trees. It has become especially important in the urban environment but is also widespread in woodland forests. Infection occurs through open wounds, and the fungus causes a decay of the sapwood. It also causes the foliage to silver. It is particularly prevalent on mountain ash, crab apple, pears, and plums. For illustrations see Hiratsuka (1987).

The rust fungi are particularly important pathogens because of the economic losses they cause. They belong to the group of Basidiomycetes with a basidium divided lengthwise. Their life cycle is more complicated than most other fungi and consists of a number of distinct stages (Figure 9.22). The stages are: spermagonia that produce spermatia and receptive hyphae; aecia-producing aeciospores; uredinia-forming urediniospores; telia-bearing teliospores; and basidia that produce

FIGURE 9.21 Fruiting structure of the honey mushroom, *Armillariella mellea*. (John C. Krug)

basidiospores. The life cycle involves an alternation of uninucleate and binucleate spores. Usually the first two stages of the life cycle occur on one host, often referred to as the alternate host, and the uredinia and telia occur on a second host, the primary host. The basidia are not parasitic, as they are usually formed on fallen foliage. The two hosts are not necessarily closely related, and in most instances occur in distantly related plant families.

Not all rusts go through the same number of life cycle stages. Microcyclic rusts, for example, have lost one or several stages, and in some cases may complete their entire life cycle on one host. Although most rust species are restricted to a specific host, some may be similar morphologically but differ in their parasitic abilities. Such special variable forms are called *formae speciales*. Rusts can occur on a broad range of plants. For a general overview and further details of the rust life cycle see

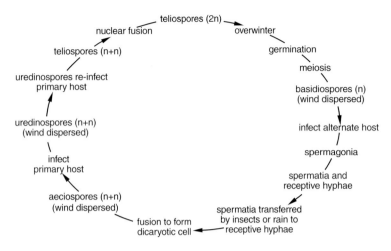

FIGURE 9.22 Typical life cycle of a rust fungus, n = haploid cell, n + n = dicaryotic cell, and 2n = diploid cell.

Alexopoulos, Mims, and Blackwell (1996) and Littlefield (1981).

With the introduction of cereal agriculture, the wheat rust – *Puccinia graminis* – became prevalent. Its primary host is wheat, while its alternate host is the European barberry, and it can also infect a number of different grasses. Since this barberry is not native, the rust was probably introduced from Europe. The rust has evolved various strains that infect different varieties of wheat. Because of the economic importance of wheat, barberry was banned from Canada in the 1930s (see Estey 1994). This legislation effectively controlled the rust in the greater Toronto region, as the alternate host disappeared. Nevertheless, the occasional outbreak can still occur by wind-borne spores from southern regions where urediniospores can survive. A number of *formae speciales* of *P. graminis* cause rust infections on other cereal grains.

With the introduction of tree plantations, a number of rust infections have become prevalent. One of these is by *Endocronartium harknessii*, the woodgate gall rust, a gall-forming, microcyclic rust on Scotch pine, *Pinus sylvestris*. It forms large galls within which aecia develop, and this can be a serious problem in pine plantations. Although commoner in western North America, it has become established in southern Ontario. For further information and illustrations of the galls formed by this rust see Ziller (1974).

Parmelee and Malloch (1972) reported the introduction of *Puccinia hysterium* on *Tragopogon pratense* in the greater Toronto area. This rust, which is microcyclic on *Tragopogon* (goat's beard), appears to have become firmly established, possibly introduced from Europe.

Another rust, originally introduced in western Canada but which has spread eastward, is the white pine blister rust, *Cronartium ribicola*. Here species of currants are the primary host and white pine the alternate host. Along with extensive logging of the white pine, this rust played a role in reducing these pine forests. Estey (1994) has given an excellent account of the history of the introduction of this rust. Although not of the same importance as the blister rust, *Chrysomyxa ledi* infects spruce (Figure 9.23) and alternates with *Ledum groenlandicum* (Labrador tea), which is the primary host. This is a native rust that has probably been present for many centuries.

A number of other introduced fungi have devastated elements in the original hardwood forest. One of these was the chestnut blight that infects the North American sweet chestnut (*Castanea dentata*). This fungus is a prime example of how an aggressive pathogen can completely destroy a susceptible host. It was introduced in the early part of the century on horticultural stock from Asia. The fungus is an Ascomycete, *Cryphonectria parasitica* (= *Endothia parasitica*), which causes foliage to wilt, cankers to form, and the tree to die eventually. By 1940, the disease had essentially eliminated the sweet chestnut, except for the occasional sapling. The fungus is able to survive on such saplings and

FIGURE 9.23 Infection of aecia on spruce by the rust *Chrysomyxa ledi*. (John C. Krug)

thus can infect secondary growth from old stumps. The ascospores and conidia are usually transmitted by insects or birds after being released by air currents. As a result of the spread of this disease the chestnut has disappeared from southern Ontario.

Recently the butternut (*Juglans cinerea*) has been dying throughout southern Ontario. The cause is another introduced pathogen, *Sirococcus clavigignenti-juglandacearum* – a conidial fungus. Initially it was introduced from Asia to the United States, but it has subsequently spread into Ontario. The spores are spread by air currents, infecting the trees and producing dark fruiting bodies. Rain washes the spores down the tree, where the fungus penetrates cracks and gradually forms a canker spreading around the tree. The resulting decline of the butternut has had an unfortunate secondary effect. With the loss of the nuts, there has been a serious decline of squirrels in infected forests.

Another introduced disease is *Ophiostoma ulmi* (= *Ceratocystis ulmi*), which is transmitted by insects. It was introduced during the 1940s on lumber from Europe and is spread by bark beetles. The fungus causes wilting and discolouration of the foliage (illustrated by Hiratsuka 1987) and finally death of the tree due to plugging of the vessels, especially those in springwood. It develops under the bark in the beetle chambers (Figure 9.24), and the conidia are spread by the beetles feeding on young elm shoots. This pathogen has largely been replaced by a more aggressive fungus, *Ophiostoma novo-ulmi,* also introduced. The net result has been the decline of the elm in southern Ontario, including the greater Toronto region. Those elms remaining show some degree of resistance to the fungus, whereas the general elm population was completely susceptible. The rock elm, found primarily in wet areas, also seems to be more resistant. The recent development of a protective hormone treatment promises to be helpful in saving the remaining elms.

With the development of orchards for apples and small stone fruits, a number of pathogens have become important. Prior to the development of orchards, these probably existed on wild members of the rose family (apples, hawthorns, cherries, plums), and, with increased urbanization, they probably became well established in ravines. One disease, caused by *Apiosporina morbosa* (= *Dibotryon morbosum*), is known as black knot of cherry because it forms large, black stromata (illustrated by Hiratsuka 1987). It infects both cherries and plums, and is well established in city ravines and in rural open and wooded areas. The spores are dispersed by the

FIGURE 9.24 Beetle chambers associated with infection by the Dutch elm disease fungus, *Ophiostoma ulmi*. (Martin Hubbes)

wind and readily infect other trees, eventually killing them. Early in the century it was not uncommon to find a few cherry trees in farm orchards, but black knot has caused the destruction of such trees except where protected by fungicides. Interestingly, Japanese plums appear to be resistant to this fungus.

An important disease of apples is apple scab, caused by *Venturia inaequalis*. It overwinters in fungal fruiting bodies on fallen leaves, and in the spring produces wind-borne ascospores that infect young apple leaves. These form a mycelium that develops into a stroma, the so-called scab. It forms conidia that are transported by rain to other leaves. The disease has become very prevalent in orchards and gardens, where it weakens the trees, but it can be controlled by fungicides.

Other diseases of fruit trees include *Monilinia fructicola,* the cause of brown rot of peaches. The fungus overwinters on shrivelled fruits (known as mummies) on the ground. Here ascospores are produced in the spring and are carried to blossoms and twigs, where infection is initiated. The developing mycelium produces conidia, which infect twigs and blossoms, forming cankers. Within these infections a new mycelium is formed, which produces air-borne conidia that infect the fruit and cause a brown rot. The infected fruit becomes penetrated with the fungus and falls to the ground, where it overwinters. The best control of the disease is orchard cleanliness. Another disease of peaches is *Taphrina deformans,* the cause of peach leaf curl. It has become more prevalent in recent years but can be controlled with fungicides.

With the development of ornamental gardening, especially in urban areas, roses of various types have

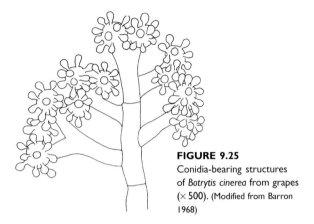

FIGURE 9.25
Conidia-bearing structures of *Botrytis cinerea* from grapes (× 500). (Modified from Barron 1968)

become very popular. One disease that has become established is black spot, so named because of the dark stromata that form on the leaves. The fungus, *Diplocarpon rosae*, overwinters on fallen leaves and produces wind-dispersed ascospores that infect the new leaves, where a new mycelium forms that produces conidia. These infect other leaves, especially when it rains and the conidia are splashed by rain drops to uninfected leaves. Later in the season a black stroma forms on the leaves, which finally fall to the ground. Diligence in cleaning up the leaves where the fungus overwinters will help control the disease, which probably originated from wild roses.

Another fungus that is important in gardens and orchards is *Botrytis cinerea* (Figure 9.25). It is an asexual fungus forming a greyish mould-like infection. It infects grapes, where it forms a growth in the developing fruit. The fungus is especially noticeable in wet seasons. Some particular wines, especially dessert wines, have been developed from such grapes, which are sweeter as a result of the infection. *Botrytis* also causes a number of vegetable diseases, such as the lettuce and onion mould; it is also responsible for infection of bulbs, especially tulips. Since the fungus also infects wild grapes and is present in soil, it was probably originally present in such habitats, from whence the spores infected agricultural crops.

Mildews are also important pathogens of a variety of hosts, including many wild plants, garden crops, and ornamentals. There are two types of mildews, the downy mildews, which belong to the Oomycetes, and the powdery mildews, which are Ascomycetes. Both are strongly parasitic, with individual species usually confined to a single host or limited number of hosts, and form a whitish growth on the leaves. The downy mildews infect a number of vegetable crops, but the best known is *Plasmopara viticola* (Figure 9.26) on grapes. This mildew is especially serious in wet seasons and can destroy the complete crop. It can be controlled with copper-based fungicides. The fungus originated on the native grape, where it does not seriously affect the host, but the wind-borne spores infect and ultimately destroy cultivated grapes.

In powdery mildews, the infection may be quite weak on some hosts but very destructive on others. The powdery appearance is due to conidia, which form masses on the leaves. Fungal fruiting bodies containing ascospores form later in the season on the leaves, and the fungus overwinters in this stage. Unsightly growths on ornamentals such as lilac and phlox are due to powdery mildews. A number of important cereals are also infected with powdery mildews. For illustrations see Hiratsuka (1987).

Many wild grasses are infected with *Claviceps purpurea,* which forms large, sterile structures called sclerotia in the flowering head of the grass (Figure 9.27). These hard, resistant masses of hyphae fall to the ground in the autumn, where they overwinter. They are protected from infection by other fungi because they contain a complex of mycotoxic alkaloids. The sclerotia, sometimes called ergots, germinate in the spring and produce stromata containing perithecia. These release ascospores that cause infection when susceptible grasses are in bloom. The infection destroys the ovary of the flower and eventually forms a sclerotium. The disease is especially prevalent on wild rye grasses but can also infect cultivated rye. If the grain is not thoroughly cleaned it can cause a disease known as ergotism, but with modern milling standards this is now rarely the case.

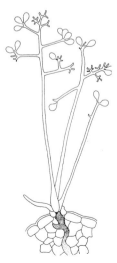

FIGURE 9.26
Spore-bearing structure of *Plasmopora viticola,* cause of downy mildew on grapes (× 300). (Modified from Millardet in Schröter 1897a).

FIGURE 9.27
Sclerotia of *Claviceps purpurea* in grass flowering head. (Modified from Lindau 1897).

Many fungi can cause leaf spots, though these are usually not too serious. We will mention only two of the more conspicuous ones. In the *Phyllosticta* leaf spot, infection is brought about by conidia that are splashed onto the leaf during rainy weather. A circular spot of dying leaf tissue becomes discoloured and is surrounded by a dark border. The infected area frequently falls out, giving the appearance of a hole in the leaf – thus the name 'shot hole of maple.' Another leaf disease, also on maple, is caused by species of *Rhytisma*, primarily *R. acerinum*. The disease is sometimes called tar spot, due to the black stroma that forms on the leaves during summer. Spores are released in spring from cup-like fruiting bodies that form on the stroma on fallen leaves. These diseases were probably present in the original forest.

Several fungi can cause problems on needles of conifers. These are especially important in artificial plantings and in plantations, such as those for Christmas trees. One of these is the needle cast fungus, *Lophodermium pinastri*, which causes the needles to discolour and die, particularly in red and Scotch pine. Another notable one is *Scirrhia acicola*, the cause of brown spot needle blight. This fungus causes banding or spotting on the needles, resulting finally in their death. *Dothistroma pini* was known as a minor pathogen of Monterey pine in California, but in the 1950s it became aggressive and destructive both in California and in plantations in Chile, East Africa, and New Zealand. In southern Ontario it is important primarily in young pine plantations and ornamental pine plantings, particularly of Austrian and ponderosa pine. It can also attack Douglas fir and European larch. The pathogen causes the needles to discolour and fall off, stunts the tree's growth, and ultimately kills it. In *Diplodia pinea*, infection can occur on both elongating shoots and needles. The disease appears on pines about 20 years old and kills current twig growth and generally stunts growth. This pathogen attacks pines, especially in ornamental plantings. Austrian, Scotch, ponderosa, and red pine are particularly susceptible.

A number of fungi cause die-back of twigs, and the infections are termed die-back diseases. Initially the twig is attacked at branch endings. The fungus then progresses until partial or entire branches are killed. A range of hardwoods can be attacked, including some woody shrubs such as alder, as well as some conifers. A number of fungi, including some species of *Dermea* – such as *D. balsamea* on hemlock – can cause die-back infections. Probably the most important die-back is caused by *Gremmeniella abietina* [= *Ascocalyx abietis*] on pines. The fungus causes yellowish discolouration of needles and shoot die-back. (For illustrations see Hiratsuka 1987.) However, the disease is really a canker disease, as stem cankers are often formed. Sometimes called the *Scleroderris* canker disease, it is another introduced pathogen and has been devastating in certain plantations.

Although a number of pathogens cause cankers, two important ones on hardwoods are the *Hypoxylon* canker on trembling aspen, caused by *H. mammatum*, and the *Nectria* canker on a range of hardwoods. The infection by *Nectria galligena* probably occurs through cracks and wounds, and the reddish perithecia fruiting on the margin of the canker are characteristic. This is sometimes confused with the canker caused by *Ceratocystis fimbriata* on aspen. In the *Hypoxylon* canker, infection is initiated through stem injuries by air-borne ascospores. A spreading mottled canker develops, on which fruiting bodies develop after several years, eventually causing death of the tree. It is illustrated by Hiratsuka (1987).

As well as the highly pathogenic fungi are others better considered as weak parasites. The bark fungus *Xylaria polymorpha*, for example, which is commonly known as dead man's fingers because of the shape of the stromata, is probably primarily saprophytic or a very weak pathogen. Members of both *Hypoxylon* and *Xylaria* are mentioned by Bell (1933) and have probably been present in the greater Toronto region for many years.

Although most fungi attack plant hosts, some parasitize other fungi. One of these is *Scopinella sphaerophila*, which grows on the stromata of *Apiosporina morbosa*, the black knot fungus, thus forming a sort of natural control. Likewise, the jelly fungus *Tremella mycophaga* attacks the fruiting layer of the bark fungus *Aleurodiscus amorphus* on balsam fir. Another fungus, *Hypomyces lactifluorum,* infects and distorts the fruiting body of species in the mushroom genus *Lactarius* (Figure 9.28). The stromata of *Cordyceps capitata* (Figure 9.29) grow on the subterranean false truffle, *Elaphomyces* sp.

We have mentioned a few of the many plant disease fungi and some weak parasites, but many others could be discussed, including the rusts on raspberries and certain roses that form conspicuous yellowish spots in mid-summer; the cedar apple rust that forms gelatinous growths (telia) on ornamental hawthorns and crab apples; and the smuts on corn and oats, where the kernels are replaced by masses of black spores. Some of the various cankers and galls on hardwoods, such as the *Diplodia* gall on aspen and balsam poplar and the *Eutypella* canker of maple, are also important. Illustrations for these two are provided by Hiratsuka (1987) and Manion (1981).

Many non-pathogenic fungi that would have been present in the original forest are still present and play important roles. Included here are many Ascomycetes and Basidiomycetes. Some of the more obvious of these saprophytic fungi are the morel (Figure 9.30), which is a modified cup fungus; turkey tail (Figure 9.31), a possibly mycorrhizal Basidiomycete; stinkhorns (Figure 9.32), where insects play a key role in spore dispersal; puffballs (Figure 9.33), where spores are released as a mass into the air; bird's nest fungi (Figure 9.34), where raindrops are involved in a special mechanism to trigger spore release; and jelly fungi (Figure 9.35), in which gelatinous tissue is present. Since most of these fungi are included by Bell (1933) and many are also listed by Langton (1911, 1913), they were all probably present in the original forest covering the greater Toronto region.

Although fire blight, caused by *Erwinia amylovora,* is a bacterial disease native to North America, the symptoms could be confused with some fungal infections. Initially the blossoms wilt, followed by twig blight and formation of cankers. The bacteria overwinter in the cankers from which they ooze in the spring and are disseminated by rain and insects. The disease is common on crab apples, hawthorns, and also originally on pears in the greater Toronto region. According to Manion (1981) fire blight was responsible for discontinuing major pear production in the northeast and establishing it in the northwest of Canada and the United States. Hiratsuka (1987) has good illustrations.

Some disease-causing fungi as well as many saprophytic ones were present for a long time, probably many centuries, in the original forests that preceded the arrival of Europeans in this region. From these forests the fungi were able to infect developing agricultural crops and plantations. Since many fungal spores can remain viable in the soil for years, they served as a ready source of potential fungal infection. Other fungi have been introduced, usually with devastating results. In still other instances, although the fungi have been well established, it is only in more recent years that the conditions have become ripe for serious expression and spread of disease.

This section has mentioned only a few of the fungi causing diseases in the greater Toronto region, and has focused on the most significant ones. Nevertheless, these illustrate a considerable biological diversity, again emphasizing the amazing biodiversity in the fungi.

FIGURE 9.28 Distorted fruiting body of a mushroom, *Lactarius* sp., infected with *Hypomyces lactifluorum*, which gives the characteristic red pigmentation. (John C. Krug)

FIGURE 9.29 Stroma of *Cordyceps capitata* parasitizing the false truffle, *Elaphomyces* sp. The arrows show the false truffle. (John C. Krug)

FIGURE 9.30 The morel, *Morchella esculenta*, is a modified cup fungus common in spring. (John C. Krug)

FIGURE 9.31 Fruiting body of the turkey tail, *Thelephora terrestris*, with lower fruiting body upturned to show lower surface. (John C. Krug)

FIGURE 9.33 *Calvatia gigantea*, a giant puffball. (John C. Krug)

FIGURE 9.35 *Dacromyces palmatus*, a jelly fungus. (John C. Krug)

FIGURE 9.32 *Phallus ravenelii*, a stinkhorn. (John C. Krug)

FIGURE 9.34 *Crucibulum laeve*, a bird's nest fungus showing the inner discs that contain the spores. (Mary Ferguson, FPSA, Royal Ontario Museum)

Invertebrates

D. Dudley Williams and Thomas R. Mason

From an environmental perspective, and apart from the sheer size of the present Metropolitan Toronto area conurbation, one of the region's most significant impacts has been the rapidity of its development. In contrast with similar-sized cities in Europe, for example, some of which have taken up to 2,000 years to reach their present size, this region has mushroomed from being virtually uninhabited in 1780 to today supporting a highly urbanized economy and millions of residents. As a consequence of this concentrated time frame, there have been fewer opportunities for harmonization of this urban mass, its human population, and their activities, with pre-existing and invading floras and faunas.

The purpose of this chapter is to examine some of the effects that this rapid growth of the human population has had on the populations of invertebrate animals that live in the region's ecosystems. We shall begin by looking at some of the important urban changes that have occurred over time and how these have promoted local species extinctions or invited invertebrate invaders. We shall then examine some case histories, and conclude with examples of problems that are likely to increase in the future.

CHANGES IN THE REGION THAT HAVE ACCOMPANIED URBANIZATION

Although there are differences in the rate of urbanization and its process among societies, there are also some similarities. With the enlargement of cities comes increasing distance of the human population core from natural environments. Greater pressures are placed on surrounding areas to support the needs of the population, for food, water, and recreation, for example. Changes in land use lead to changes in habitat suitability for other species. Increased transportation provides opportunities for species introductions via air, water, and land. Finally, there are problems of waste disposal and environmental toxicity, which may result in species extinctions.

Urbanization in the greater Toronto region is intense. But just as important as the coalescence of this region's human population is the fact that the area consumed possessed good quality soils together with some of the rarest environments in the country, including those that support large numbers of southern species for which the greater Toronto region is their most northerly limit (Scoggan 1978; McKay and Carling 1979; Soper and Heimburger 1982). Further, the adoption of the typical North American grid pattern for urban subdivision and transportation routing has limited future land use options (Jones 1966). Such a pattern differs markedly from the more haphazard 'organic' growth of many older cities in the Western world. Another important environmental pressure exerted by Toronto is the vast extent of its suburban sprawl, which has consumed even more land and has made natural habitats even more distant from the downtown core. Admittedly, preservation of some major 'green' valley corridors, such as the Don and Humber valleys, may have softened the impact of Toronto's expansion (Bailey 1973), but urban pressures have largely forced these into highly managed and degraded environmental fragments.

FEATURES OF EXTINCTION AND INVASION

Habitats that are most prone to invasion by non-native species exhibit the following characteristics: (1) they have a climate similar to that of the invader's habitat; (2) they have a low diversity of native species, which reduces competition; (3) they are disturbed, and thus are in the early stages of ecological succession; and (4) they have relatively few predators. Regions subject to heavy urbanization present just such features and are thus ripe for invasion.

The species that are likely to be able to take advantage of such habitats will have many of the following characteristics: (1) they have a large native range, combined with a high dispersal rate; (2) their reproductive rate is high, with reproduction occurring early in the life cycle; (3) their lifespan is relatively short; (4) they have high genetic variability to allow survival in a range of habitat types, combined with considerable physiological tolerance to adapt to changing local conditions; (5) they are opportunistic and/or generalist feeders; and (6) they are accustomed, or even adapted, to human-altered environments (Williams 1987, Lodge 1993).

Many of the same factors promoting invasion result in loss of resident species. Basically, local extinctions occur when a population experiences a death rate that cannot be compensated for by its birth rate or by immigration. Such circumstances rarely result from a species' own intrinsic failings but rather from changes to its environment. Foremost among the latter are disturbance, degradation, toxicity, and the introduction of predators and competitors, all of which accompany urbanization. Specialist species, those with a narrow range of environmental tolerance or those adapted to a single food type, are among the first to go. Eventually entire communities that have sensitive keystone species at their hearts will disappear. Some observers might see specialization as weakness, but in pristine (undisturbed) systems it is strongly selected for in order to reduce competition among species (Brewer 1979).

What must be made quite clear here is that local extinctions and species invasions are natural processes that have been proceeding since the beginning of life. These processes are important forces that promote the evolution of new species and new natural communities. At certain times in the earth's history, events of climate or geography have either accelerated or slowed the movement of species. Indeed, the present flora and fauna of Canada came into being only over the last 10,000 years. Before that time, virtually all of the country was covered by Pleistocene ice sheets many kilometres thick. With the extermination of the pre-Pleistocene biota, development of the present Canadian fauna and flora has been possible only as a result of invasion from ice-free refuge areas – largely in the United States but also in isolated sections of the Yukon, western Newfoundland, and the Arctic archipelago (Matthews 1979; Patrick and Williams 1990). The present-day biota is nowhere near as rich as that occurring in unglaciated regions of this continent, and must be regarded as primarily an interglacial one that can be expected to change in the near geological future as Canada's climate continues to change. What is of considerable concern, however, is the acceleration of extinction/invasion processes directly attributable to human activities. Concern has been expressed that this acceleration (beyond the rate of natural evolution), together with the relative uniformity of human-associated species, will lead to a much impoverished and homogenized global biota (Lodge 1993). Such a trend is detectable on much smaller geographical scales in the greater Toronto region.

SPECIFIC AND PERVASIVE INFLUENCES ON THE REGION'S INVERTEBRATE ENVIRONMENTS

Although it is difficult to find information that relates specifically to the immediate greater Toronto region, there are a number of studies that record broad environmental change in the southern half of Ontario. Some of the effects they document may also affect Toronto environments, although they may be difficult to discern due to prior and/or concomitant impacts of a more urban nature.

Acid Rain

The average deposition of free hydrogen ions in southern Ontario has been assessed at around 70 mequiv. per m^2, per year. This loading is about five to ten times that measured in remote areas of the province, and is of particular concern because the natural buffering capacity of lakes in exposed Precambrian Shield areas (beginning just north of Lake Simcoe) is low. A consequence of this is increased concentration (up to three times) of metal and other ions because they are mobilized by the acid (Dillon et al. 1980). Many freshwater invertebrates are known to respond extremely negatively, in terms of both numbers and diversity, to acidification of their environment (Hall and Ide 1987; Gunn and Keller 1990). For example, the snail *Amnicola limosa* has disappeared from Ontario lakes whose pH has fallen below 5.8 because such levels reduce fecundity and hatchling survival (Shaw and Mackie 1989). At present, however, because of the naturally high buffering capacity of the calcareous soils in the southern part of the province, fresh waters in the greater Toronto region have not been affected by elevated metal ions mobilized by acid rain. These waters do, nevertheless, receive substantial metal ions from industrial processing, and the emissions from vehicles further contribute nitrogen-based acidic substances. The influence that the latter may have on invertebrates in the region is largely unknown.

Increasing Temperature

Human activity has increased environmental temperature on local, regional, and global scales. Temperatures in the greater Toronto region have been increased, locally by nuclear power station cooling waters, which have influenced near-shore water temperatures of Lake Ontario, and by the heating of groundwater flowing through landfill sites, and more extensively through the heated air produced and retained within the city and its suburbs – the 'heat island.' In addition to these is the prediction, based on a climate modelling benchmark of a doubling of atmospheric carbon dioxide, that over the next 25 years global warming will result in a 2° to 3.5°C rise in the regional air temperature of Toronto (Hengeveld 1990; IPCC 1990). A recent, large-scale field experiment designed to examine the effects of such a rise on freshwater invertebrates showed the following changes in Valley Spring, Scarborough, over a two-year period: a decrease in total invertebrate densities; increased growth rate and precocious breeding in the amphipod crustacean *Hyalella azteca*; and smaller size at maturity and altered sex ratios in some species (Hogg and Williams 1996).

Water Pollution

This heading covers a multitude of contaminants, but only a few of the more than 350 toxic materials that have been measured in the Great Lakes Basin (DePinto 1991) can be summarized here. For example, the nearshore, benthic macroinvertebrate community of Lake Ontario is known to have been heavily affected by municipal and industrial inputs. One result of this has been a dramatic increase in oligochaete worm abundance (72 percent dominance of the nearshore community), more so than in any of the other Great Lakes (Nalepa 1991). Accompanying this has been a decrease in the numbers of the pollution-sensitive amphipod *Pontoporeia hoyi* (Hiltunen 1969). Toronto Harbour is particularly affected, with worm densities recorded at 200,000 m^{-2} and one species, *Tubifex tubifex*, dominating. At the mouth of the Don River, which receives heavy loads of organic and industrial wastes (Don Watershed Task Force 1994), the entire benthic community is composed of this and one other pollution-tolerant worm, *Limnodrilus hoffmeisteri* (Brinkhurst 1970). Oligochaetes, along with midge larvae (bloodworms), are known also to dominate sediments in the region contaminated with heavy metals. Concentrations of metals such as nickel, lead, zinc, and chromium have been shown to produce communities composed solely of pollution-tolerant forms. Even then, in the case of the midges, larval populations exhibited a high (27 percent) proportion of anatomical deformities known to result from these mutagens (Dickman, Yang, and Brindle 1990). Deformities in these larvae in Lake Ontario have also been linked to sediment contamination by radioactive wastes (Warwick et al. 1987).

Although the basic structure of the Lake Ontario zooplankton community has not changed significantly since the late 1960s, during the 1980s there was a decrease in density in nearshore areas, together with greater representation of smaller cladoceran species. It is believed that these changes may have resulted from increased predation by fishes (especially alewife) or by reduced food resources resulting from lower total phosphorus concentrations in the surface waters.

A quite recent pollutant that is affecting the Lake Ontario shore via surface and groundwater discharges from the greater Toronto region is road salt (Howard et al. 1993). Whereas the biological impact on the lake communities is largely unknown, there appear to be some species changes occurring in those low-order streams that receive much of the salt. A preliminary study has shown that the amphipod *Gammarus pseudolimnaeus* and turbellarian worms occurred only in springs with low chloride levels, whereas the larvae of tipulid and ceratopogonid flies were closely associated with high chloride waters (Williams et al. 1997). It is predicted that if road de-icing practices remain the same, the average chloride concentration in the Toronto region groundwater will reach around 426 ppm (Howard et al. 1993). This is nearly twice the drinking water quality objective of 250 ppm. That a considerable amount of this salt is also entering Lake Ontario is evident from winter measurements exceeding 10,000 ppm chloride in the lower Don River (Ontario Ministry of Environment and Energy, unpublished data).

INVERTEBRATE INTRODUCTIONS

Most of the foregoing examples were of environmental change diminishing the diversity of the *existing* invertebrate fauna in the greater Toronto region, with, in many cases, one or two highly tolerant forms becoming dominant as a result. This section deals with non-indigenous species that have been introduced or encouraged into the region as a consequence of human presence and activity. Since around the 1840s, the region has been subject to intensive colonization and commercial development, which have set the scene for a host of biological invasions, although species associated with the earliest European explorers and settlers precede this time (Ashworth 1986; Mills et al. 1993). We shall present example case histories for both aquatic and terrestrial environments.

Aquatic

Among the non-insect invertebrates, invading species of mollusc, crustacean, and worm have been best documented. Perhaps the best-known mollusc introduction to the region is *Dreissena polymorpha,* the zebra mussel (Figure 10.1A). It was first noticed in Lake Erie in 1988 and is likely to have been transported from Europe in the ballast water of ships. DNA comparison strongly indicates a Black Sea/Caspian Sea origin. It has

FIGURE 10.1 Three of the aquatic invertebrate species that have recently invaded the freshwater habitats in the Toronto region: A – a dense aggregation of the zebra mussel (*Dreissena polymorpha*) that has colonized the exposed (above the mud) shell surface of a unionid clam (W. Gary Sprules); B – the first instar (juvenile stage) of a female *Bythotrephes cederstroemi,* the spiny water flea (W. Gary Sprules); C – external and internal views of the shell of the Asiatic clam, *Corbicula fluminea.* (Kevin S. Cummings, Illinois Natural History Survey, Center for Biodiversity, Champaign, IL)

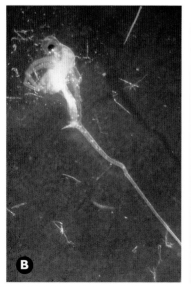

now spread throughout the lower Great Lakes and occurs at major ports on the upper lakes. Apart from adversely affecting native unionid clam populations, it has proved itself to be a major nuisance organism, fouling the intake pipes of Toronto water treatment plants. In addition, through its immense filtering capacity, it has begun to change water clarity, thus altering the energy and nutrient flow of local aquatic systems (Griffiths et al. 1991), and has caused shifts in the structure and function of benthic algal communities (Lowe and Pillsbury 1995). A second introduced species of *Dreissena* was confirmed from Lake Ontario in 1991 (May and Marsden 1992). It is predicted that from these Great Lakes port introductions, *Dreissena* has the potential to spread throughout most of North America (Strayer 1991).

Valvata piscinalis, the European valve snail, is thought to have been introduced into Lake Ontario in the straw and marsh-grass packing material that was used to transport chinaware from England and Eastern Europe. Although its first recorded occurrence was on the New York State shoreline around 1897, it was observed in packing debris on the Toronto lakeshore in 1912 (Latchford 1930). It is now quite common in the lower Great Lakes (Burch 1989), but its effects on endemic snail populations seem largely unknown.

The Asiatic clam, *Corbicula fluminea* (Figure 10.1C), has made incredible inroads into North America since its discovery in British Columbia in 1924 (Counts 1986). It has spread down the Pacific Coast, through the southern United States and entered the Great Lakes around 1978. However, as it cannot tolerate water temperatures of 2°C or less, its occurrence in Canada is limited to heated industrial effluent discharge areas. Its densities are often sufficiently high to clog water intake systems (Clarke 1981, Mills et al. 1993). This is a rather good example of an invading species surviving in an artificially created habitat. In other regions where it has become established, this clam has been responsible for triggering ecosystem-level changes in the indigenous flora (algae and macrophytes) and fauna (fish and birds) (Phelps 1994).

In total, some 14 species of mollusc are known to have invaded the Great Lakes through mechanisms ranging from bulk transport in ships' ballast or on their hulls to aquarium and bait releases, and from recreational boat traffic to the import of snail eggs on greenhouse plants (Mills et al. 1993).

Only six species of small to medium-sized crustaceans have invaded the Great Lakes. The largest of these is the amphipod *Gammarus fasciatus,* which probably came from U.S. eastern seaboard drainage systems via several human agencies (ships' ballast, aquatic plants, stocked fishes, and bait containers; Mills et al. 1993). Three non-native copepods occur in the area, although none seems more than locally abundant: *Eurytemora affinis* is a euryhaline species from the Atlantic Coast and was first recorded in Lake Ontario in 1958 (Anderson and Clayton 1959); *Skistodiaptomus pallidus* was first found in the lake in 1967 and probably came from the Mississippi drainage basin (Patalas 1969); and *Argulus japonicus,* a parasitic copepod native to Asia, was introduced with its host, the goldfish (Cressey 1978). In contrast, two exotic cladoceran species have had more impact. *Eubosmina coregoni* is a European species of water flea first recorded from Lake Michigan in 1966 (Wells 1970). By 1968 it was present in Lake Ontario and, today, it has spread throughout the entire Great Lakes to become one of the dominant species in the plankton. It was probably introduced to the region, and subsequently spread, in the ballast water of ocean- and lake-going vessels (Mills et al. 1993). *Bythotrephes cederstroemi,* the predatory spiny water flea (Figure 10.1B), is another European invader thought to have arrived in the late 1970s and spread in similar fashion. Its great abundance in the freshwater port of Leningrad, with which there was considerable trade through grain export from Great Lakes ports at that time, makes this region the most likely source (Sprules, Reissen, and Jin 1990). In Lake Michigan, its appearance produced a substantial decline in *Daphnia* populations, altered depth distribution patterns in surviving *Daphnia* populations, and reduced the abundance of *Leptodora kindti,* a native predatory cladoceran (Lehman and Caceres 1993).

A much larger crustacean, the crayfish *Orconectes rusticus,* has recently invaded the Duffins Creek watershed (Maude 1988). It has also invaded the Rouge River, where it now appears to be the dominant crayfish

> *The European valve snail is thought to have been introduced into Lake Ontario in the straw and marsh-grass packing material that was used to transport chinaware from England and Eastern Europe.*

(T.R. Mason, personal observation). A native of the midwestern United States, this species has probably invaded as a consequence of its use as fishing bait. In other parts of Ontario, this species has replaced native species such as *O. virilis* by virtue of its higher aggression (due to its larger claw size), its lower susceptibility to predation, and its interference with the reproductive success of residents (Garvey and Stein 1993).

Of the 139 non-indigenous aquatic organisms (plants, fishes, molluscs, and other invertebrates) that have become established in the Great Lakes since the 1800s, about one-third have been introduced in the past 30 years. This indicates a clear relationship with increasing human presence and activity. The opening of the St. Lawrence Seaway in 1959 was a particularly significant factor, as a majority of introductions have occurred via ship-related activities. Mills et al. (1993) report that although only 9 percent (13 species) of these invaders have had a significant effect on the local ecosystems, the true influence of the remainder is largely unknown.

Terrestrial

Molluscs, worms, and crustaceans also have been successful invaders of the land around Toronto. Alongside these have come the house centipedes, millipedes, spiders, and, possibly most important group of terrestrial invaders in this area, the mites.

The most obvious land snail to enter the greater Toronto region is *Cepaea nemoralis*, the European garden snail (Figure 10.2A). Unlike most native snails in the area, this species often will be found in the open, on walls and plants. The earliest known entry of this species to North America was via Burlington, New Jersey, in 1857 (Pilsbry 1939). This was a deliberate introduction, but several other accidental introductions occurred around that period in the eastern United States. The first Ontario record of this species was in the Owen Sound area in 1898, whereas the first records for Toronto were in 1944 (Oughton 1948). There appears to be no report of the species causing any damage to crops or garden plants. *Cepaea* is eaten by small mammalian predators (personal observation) and thus seems to provide a food resource for some native species. It may, nevertheless, also be a reservoir for parasites.

Slugs are a different matter. A total of 15 introduced species has been reported for North America (Chichester and Getz 1969). Introduction has occurred most likely through the transportation of agricultural products,

FIGURE 10.2 Three of the terrestrial invertebrate species that have recently invaded the Toronto region: A – the European garden snail, *Cepaea nemoralis*; B – several specimens of the parasitic varroa mite, *Varroa jacobsoni*, attached to the underside of a honey bee; C – dorsal view of *Cheiracanthium mildei* (Koch), the blackfooted sac spider. (*A and C:* courtesy of Mrs. C. Belyea; *B:* courtesy of Dr. S. Camazine, Penn State University)

as most of these species are residents of gardens and meadows. Some have a preference for soft plant growth and are considered agricultural and garden pests (Burch 1965). The largest species in the greater Toronto region is *Limax maximus,* the European striped slug. The first record of this species in Ontario occurred in Ottawa in 1904 (Chichester and Getz 1969). In Toronto, records date back to 1917 (Oughton 1948). Although this species may grow to 15 cm, it is considered to be less of a garden pest than introduced species of *Deroceras* (Baker 1939).

The genera *Limax, Arion,* and *Deroceras* have been introduced into many of the countries around the world that were colonized by Europeans. Although commonly regarded as garden pests, the various species appear to have limited their expansion mainly to greenhouses, gardens, and urban areas (Chichester and Getz 1969).

A discussion of introduced, non-insect, terrestrial invertebrates would not be complete without a brief consideration of earthworms. Of the 19 species of earthworm recorded in Ontario (Reynolds 1977), 17 are Palaearctic in origin, coming from Europe, Russia, and the Middle East, and are therefore non-native. Neither of the two endemic North American species of earthworm has been recorded in the greater Toronto region. The dates that the Palaearctic species were introduced are not known. They are assumed to have arrived with the earliest colonists from Europe. What changes these species may have brought about in Ontario ecosystems is unknown. However, a study done in the mountains of Kentucky showed that exotic earthworms typically occur only in areas that had been severely disturbed (Kalisz and Dotson 1989). It is likely, therefore, that such earthworms would not have invaded pristine local habitats.

Non-insect arthropods also contribute to the list of introduced biota in the greater Toronto region. Obtaining relevant information on the distribution and origins of these often common animals is surprisingly difficult. Entomologists and zoologists typically have not bothered making collections of these groups (Kevan 1983a, 1983b) since, in comparison with insects, with few exceptions they are of little economic importance.

The most frequently encountered terrestrial crustaceans known from this region are the Isopoda. Commonly known as sowbugs, woodlice, or slaters, they are observed most often under rocks, pieces of wood and bark, or any material that provides a damp crevice in which they can hide. The most common member of this order in the greater Toronto region is *Tracheoniscus rathkei,* which outnumbers all of the other isopods in this area (Walker 1927; Judd 1965). This species is believed to be introduced (Walker 1927).

Introduced myriapods in the region are represented by *Scutigera coleoptrata* and *Oxidus gracilis. Scutigera coleoptrata,* the house centipede, was first recorded in Canada in 1914 (Kevan 1983a). Thought to have originated in Mediterranean countries, the species has moved north into cooler climates by becoming a resident of houses. In the most northerly part of its range it is most commonly found in damp basements (Lewis 1981). Here it is seen most often running across walls in search of insects and other small arthropod prey. *Oxidus gracilis* is the common, flat-backed millipede most often found throughout the greater Toronto region in disturbed habitats and in greenhouses. This millipede has managed to travel with European colonists around the world and has become a cosmopolitan greenhouse pest. It does not appear to be able to survive far from human habitation. Of the known species of millipede found in Canada, 47 percent are introduced species, mainly from Europe (Shelley 1988).

Spiders are also among the species of introduced invertebrates in the region. Perhaps the most common spider in Toronto households is the European cellar spider, *Pholcus phalangioides.* It is most probable that this species arrived with the first settlers.

A more recent invader is the sac spider, *Cheiracanthium mildei* (Figure 10.2C), which has managed to take up residence in several parts of North America. A native of North Africa and southern Europe, the species was known to be present in the United States in 1953 (Kaston 1978). The first record of it occurring in the greater Toronto region was 1982 (Dondale and Redner 1982). Although a species of thickets and hedgerows in Europe, this type of sac spider has become a resident of houses and other human structures. Close proximity of this species to humans has led to numerous records of bites causing swelling and necrosis of tissue. Such bites often have been mistaken for those of the brown recluse spider, *Loxosceles reclusa* (Spielman and Levi 1970), although the latter is presently unknown in Canada.

The species of introduced, non-insect, terrestrial invertebrates that probably will have the greatest impact, both ecologically and economically, on the greater Toronto region, are two species of mite, *Acarapis woodi* and *Varroa jacobsoni* (Figure 10.2B). More commonly known as the honey bee tracheal mite and the varroa mite, respectively, both species are parasites of the honey bee.

To understand the impact of these mites, we must first consider the honey bee. More properly known as the European honey bee, *Apis mellifera* arrived in North America with the first waves of settlers. More recently, honey production and export have become big business in Ontario: by 1989, Canada ranked as one of the five top honey exporters in the world (Scott-Dupree and Otis 1989). It is therefore not surprising that the honey bee has also become the major insect pollinator for agricultural crops across Canada. Although it cannot be proven, it is quite probable that honey bees, as invaders, came into direct competition with native bees and eventually displaced many of them. Presently, honey bees are the major vector species for all plants in the greater Toronto region that require a non-specific pollinator for reproductive purposes. Thus, anything jeopardizing their effectiveness as pollinators must be considered a serious problem for many local plants and crops. The combination of these two mite species in the region represents the greatest threat to the honey bee and to the apiculture industry yet (De Jong, Morse, and Eickwort 1982).

The tracheal mite and the varroa mite are seriously threatening hives throughout the world, and Canada is no exception. These two pests are recent introductions: they were not reported in Canada before 1989 and 1991, respectively. Because of the high fatality rate of infected honey bee colonies, the progress of these two mite species into North America is being followed closely. Apiculturalists predict that feral populations of the honey bee probably will disappear from the region within a couple of years. The mites' effect on populations of honey bees goes well beyond the strictly commercial loss, as it leaves empty an important niche in the regional ecosystem. Because of the enormity of the consequences of the mite/honey bee relationship, we present a synopsis of the origins and life histories of the two species so that the biological complexities can be appreciated.

Acarapis woodi, the honey bee tracheal mite, is one of three species within the genus that associate with honey bees. The other two, *A. externus* and *A. dorsalis,* are not considered to be serious pests of honey bees. The honey bee is the only known host of the tracheal mite.

The tracheal mite seems first to have appeared in England around the turn of the century (De Jong, Morse, and Eickwort 1982), and possibly may have been derived from one of the external parasites of the honey bee. The mite's entire four-staged life cycle takes place within the bee's trachea, and both nymphs and adults obtain their food by piercing the tracheal wall and sucking their host's blood. Young bees become infected while still in the hive. When the mite load on an individual bee is high, its rate of activity becomes significantly impaired. Hives in which greater than 20 percent of the bees become infected are considered to be at risk, with hives infested in winter typically dying off. If the hive should survive, it is weak and the spring brood of bees is low in number. A recent importation of a strain of honey bee known as the 'Buckfast bee' has introduced a line of bees that are more resistant to the tracheal mite, which hopefully will reduce the severity of the mite's impact.

The tracheal mite and the varroa mite are seriously threatening bee hives throughout the world, and Canada is no exception.

The varroa mite is of much greater concern. This external parasite was originally a pest of the Asian honey bee, *Apis cerana*. With the introduction of *Apis mellifera* to southeast Asia, *Varroa jacobsoni* spread to this native species. They spread by, quite literally, jumping between bees. The varroa mite is large enough to be visible to the naked eye, and any bee coming within a centimetre becomes a focus for attraction. Healthy foraging bees raiding a weakened, mite-infested hive invariably carry a founding parasite colony back to their own hive.

Once in the new hive, gravid female mites leave and go in search of newly hatched bee larvae, upon which they lay eggs. The mite's life cycle is perfectly synchronized with that of its host, in that the mite nymphs feed on the bee larva and mature inside the sealed comb cell shortly before the bee is ready to emerge. Mating takes place inside the sealed cell, and mated female mites leave with the emerging bee. For some reason, drones are the preferred host. Hives in the final stages of infestation will have zero recruitment to the work force.

The varroa mite has spread extremely quickly. It was first observed on the European honey bee in the 1960s, in the Philippines. It had reached South America by 1971 and the United States by 1987. It was first detected in Canada in 1991. The mite was first observed in honey bee colonies at the Toronto Zoo in September 1996. At the zoo, hives that had appeared healthy in August were almost gone by September.

At this time, spread of the varroa mite has been slowed only by treating hives with chemicals specific to this species. However it is likely that feral European honey bee colonies in the region will not be able to survive. How long it might be before resistant strains of the mite develop within managed hives is unknown, but it is likely to happen.

The threat to the pollination of many crops and flowering plants is severe. Although it is hoped that the vacant niche likely to be created by the loss of the honey bee will be filled by other pollinating species, it is uncertain whether a large enough reservoir of alternative species still exists in the area.

PROBLEMS LIKELY TO PERSIST, ESCALATE, OR ARISE

As it is highly unlikely that, in the foreseeable future, growth of the human population in the greater Toronto region will either decline or level off, the impact on the region's ecosystems seems certain to escalate. This will be particularly so given the established trend of sprawling suburban growth into the surrounding countryside. The consequence will be further reduction and degradation of natural habitats, with the loss of native species populations on the local scale and perhaps of their genetic diversity on the larger scale. At the same time, more houses and other buildings will create more 'habitat' for those invertebrate species that thrive on human association. Invasion by exotic species will intensify as the importation of goods increases and will diversify as the number of Toronto's trading-partner countries swells. Some of these invading species (as in the case of the zebra mussel) will require expensive programs to correct their nuisance effect.

Doubtless there will be greater problems in the future with those invertebrates that affect our own health and that of our domesticated plants and animals. The health-threatening potential of most of these invertebrates is poorly known, but prominent among them are nematode worms that are of considerable concern to human health (Crofton 1968). In southern Ontario, a survey of three water treatment plants revealed 66 nematode genera in both untreated and treated water. Most of these passed through the sedimentation tanks and filters, and 50 percent emerged in the treated water, still alive (Mott and Harrison 1983). Some of the species present were plant parasites that, presumably, had the potential to infect greenhouse crops via irrigation.

Another example is the protozoan *Giardia lamblia,* a well-known inhabitant of the human intestine, producing cramps, nausea, diarrhoea, and fatigue. It is easily transmitted in human water supplies, and outbreaks occur in many countries. It can be killed by chlorination during water treatment, but the level required to eliminate its dormant cysts is considerably greater than that presently recommended by the treatment industry, which is aimed at destruction of coliform bacteria (Curds 1992).

A survey of three water treatment plants revealed 66 nematode genera in both untreated and treated water. Most passed through the sedimentation tanks and filters, and 50 percent emerged in the treated water, still alive.

Although human invertebrate pathogens have not yet produced major problems in the greater Toronto region, technological inadequacies may allow future exotic species direct access to the human population. Superimposed on such problems may be those that result from global warming, as even a one- or two-degree rise in the mean annual temperature of the greater Toronto region may be sufficient to allow species from more southerly climates to establish themselves. Based on the predicted changes in temperature for the lower Great Lakes over the next 25 years, combined with the urban 'heat island' effect, range extensions of terrestrial and aquatic invertebrates into the greater Toronto region will be inevitable.

Insects

Glenn B. Wiggins

No one knows how many species of insects live in the greater Toronto region. The first, and still the only, benchmark was set down in *The Natural History of the Toronto Region,* published by the Canadian Institute in 1913. The chapter on insects listed some 2,400 species, and that figure was certainly short of the real number because experts on many insect groups were not available at that time. The problem posed in understanding the insects of any region, then and still today, is that there are so many species and so few knowledgeable observers.

Even so, we know that the insect fauna of the greater Toronto region has been changing for a long time. Destruction of natural habitats has led to the disappearance of many native species, and to some extent these have been replaced by more resilient species, many of foreign origin. Changes observed in the distribution of a few of the native insects of the greater Toronto region over the first half of this century have been documented, revealing a gradual invasion by species from south of the Great Lakes, coupled with northward withdrawal by others (Walker 1957). These changes were attributed to a gradual warming of the climate of southern Ontario.

Some insects are agricultural or household pests, highly resilient cosmopolitan species able to take advantage of food resources provided through the activities of humans, especially the monocultures of farming and forestry. But by far the majority of insect species have important, often crucial, roles in ecosystems, and those of the greater Toronto region are no exception. Consequently, management of the natural environments of the region has to treat insects not as destructive foes to be thwarted, but mainly as fundamentally important parts of a healthy ecosystem. Consider the crucial ecological roles of insects:

1. Insects are intermediate links in food webs that transfer to vertebrate animals the nutrients and energy from the sun captured by plants. Multitudes of different insects feed on plants of all kinds, and they in turn may be consumed by predacious insects. Insects are food resources of many birds, mammals, reptiles, amphibians, and fish. For example, freshwater food webs produce fish and loons from algae and plants, but largely because aquatic insects are connecting links in the chain; and seasonal migrations of most birds are geared to an abundance of insects for food at more northern latitudes during their breeding season.
2. Two-thirds of all flowering plants depend on insects for fertilization and production of the fruits and seeds consumed by many birds and mammals. Nearly one-third of human food comes from plants pollinated by bees, including most cultivated fruit trees and many forage crops such as alfalfa. Through selective plant breeding, humans have made cultivated flowers larger and more colourful, but the underlying reason that plants produce flowers is to attract pollinators, which are chiefly insects. Broadly stated, flowering plants and insects are interdependent, and that came to be because they evolved together millions of years ago.
3. Insects of many kinds are involved in recycling nutrients from

Two-thirds of all flowering plants depend on insects for fertilization.

the decomposition of animal and plant matter. Nutrients from a dead mouse, for example, are transferred to a bird that captures a fly whose larva fed on the decomposing mouse. Insects are also part of the soil biota of invertebrate animals and fungi responsible for breaking down organic materials and sustaining the fertility of soil on which all plants depend.

4 Complex relationships of insect parasitoids with their hosts far exceed our present understanding; but these relationships have enhanced even further the ability of insects to permeate food webs, and also to hold in check population irruptions of host insects.

To comprehend the vast influence of insects on the world, it is helpful to understand a fundamental ecological truth: through millions of years of evolution, insects have permeated essential natural processes that make the terrestrial and freshwater worlds work as self-sustaining biological systems (see also Evans 1984; Wilson 1996). To view insects in this way is to be released from the anthropocentric notion that they can be assessed only in terms of people. The answers to common questions such as what good insects are, or flies, or mosquitoes, pertain not to humans but to competition for energy in the ecological processes that support all living things.

Entomology in the Greater Toronto Region

William Brodie at home in his library.
(Royal Ontario Museum Library no. AR001339)

The entomological pioneers in the greater Toronto region were naturalists fascinated by the diversity of insects. By 1863, interest was shared widely enough that a meeting was called for 16 April of that year in the rooms of the Canadian Institute 'for the purpose of taking into consideration the possibility of forming a Canadian Entomological Club.' The organizers were C.J.S. Bethune, a divinity student at Trinity College, and William Saunders, a druggist in London, Ontario, with the support of Henry Croft, professor of chemistry at the University of Toronto. This was pre-Confederation Canada West, but out of that meeting of nine persons was formed the Entomological Society of Canada, now a truly national organization with its head office in Ottawa, and the oldest entomological society in North America (see Wiggins 1966).

Among the Toronto entomologists at the turn of the century was William Brodie, a dentist by profession but a dedicated naturalist and entomologist. Following his formal retirement, Brodie was appointed in 1903 as the first provincial biologist of Ontario, and he added his sizeable personal collection of insects to the Provincial Museum, then maintained in the Normal School Building in Toronto. Brodie studied Hymenoptera, and in particular gall-forming wasps. He was a valued source of entomological information and encouragement for others, including the young Edmund Walker, who later became professor and head of the Department of Zoology at the University of Toronto.

The opening of the Royal Ontario Museum of Zoology in 1914 provided the opportunity to consolidate the institutional insect collections of the Provincial Museum, the University of Toronto, the Canadian Institute, and many individuals. The ROM collection now holds the main body of data on the insect fauna of the greater Toronto region; some of that information has appeared in various ROM publications (e.g., Walker 1957; Riotte 1992). Permanent collections are crucial for documenting the history of faunal changes; this is especially true for insects, where the number of species far exceeds the number of knowledgeable workers at any given time.

For some years, the Toronto Entomologists' Association has provided a focus for amateurs interested in the study of insects; results from a shared interest in butterflies have been published in *The Ontario Butterfly Atlas* (Holmes et al. 1991), which includes data on species from the greater Toronto region.

The enormous numbers of insect species now living confirm that insects have been highly successful in penetrating food webs. Approximately 1 million species of living animals are now known in the world, and some 750,000 of them are insects – thus, three-quarters of all animal species now known are insects. Moreover, estimates for all living things fall somewhere between 10 million and 100 million species (Wilson 1992). This astonishing gap reflects a lack of understanding about the biological support systems of the planet; and a high proportion of the unknowns are insects. So, if you want to see the real world and how it works – watch insects.

Watching insects is well suited to excursions in the greater Toronto region because just about everywhere one goes there are more insects than all other animals combined. An inquiring observer of insects never runs out of interesting things to see; and because insects are in fact difficult to avoid, everyone has a good deal of experience with them. This chapter offers a framework for organizing personal experience and knowledge about insects within the larger context of the evolution and ecology of insects worldwide. My objective is to encourage an understanding of these animals in nature, which involves watching living insects and connecting what they do with the way the world works.

THE ORDERS OF INSECTS

As insect faunas from all parts of the world are explored and named, it turns out that their enormous diversity can be encompassed in a system of about 30 major categories known as orders. This underlying pattern of consistency amidst incredible variety has been developed over the past 250 years as the basis for organizing knowledge through biological classification: orders are subdivided into families; families into genera; and genera comprise from one to hundreds of species that are the discrete reproductive groups of individuals in natural ecosystems.

Some observations are offered below about each of the insect orders that could be encountered in the greater Toronto region, drawing together a few of the features that make them distinctive and remarkable animals. At this level, insects of the region are a microcosm of the insects of the world. Illustrations are helpful in visualizing the insects discussed here, and are available in field

Three-quarters of all animal species now known are insects.

guides (e.g., Borror and White 1970). Because there are more species in even a restricted area than can be treated in a single book, general field guides to insects are based at the level of orders and families. Even so, learning to recognize families of insects is a satisfying way for observers to appreciate a great deal about the diversity and biology of these creatures. For the most part, the orders of insects are intuitively distinct sorts of animals; for example, beetles, dragonflies, cockroaches, moths and butterflies, flies, and wasps all belong to different insect orders. At the family level, among beetles, for example, there are carpet beetles, carrion beetles, click beetles, diving beetles, fireflies, fungus beetles, blister beetles, hister beetles, riffle beetles, rove beetles, sap beetles, spider beetles, weevils, and many, many more. In identifying small insects, we can see a surprising amount of their structural detail with a 10×-magnifying lens.

I. INSECTS FROM AN AGE BEFORE INSECTS HAD WINGS: APTERYGOTA

Palaeontologists know a great deal about dinosaurs – knowledge derived from studying fossilized skeletons, placing them in geological time, and making comparisons with living reptiles. In a similar manner, palaeoentomologists have traced the ancestry of insects back into past ages of the earth. Fossilized remains of insects have been studied from far back in time, including forms that, from as early as the Devonian Period nearly 400 million years ago, have never had wings. These primitively wingless insects are assigned to the Apterygota, from the Greek *a* (without) and *pteron* (wing).

1.1 Silverfish: Order Thysanura

In our society, school children know that dinosaurs ranged the continents from about 245 to 65 million years ago, when they became extinct, and mammals and birds became the dominant vertebrate animals on land. But how many people know that the silverfish in their houses are far more ancient creatures than dinosaurs? The ancestry of silverfish can be traced to the Carboniferous Period, approximately 300 million years ago. Their secretive life style worked so well without wings that silverfish already had a pedigree of some 50 million years by the time the first dinosaurs appeared; they were contemporaries of dinosaurs for 180 million years and still persist 65 million years after the dinosaurs became extinct. In our area, silverfish survive in buildings where temperature and humidity combine to provide suitable living conditions throughout the year. Their simple food requirements make silverfish undemanding animals, able to live as scavengers on such products of human society as the sizing and glaze on paper and glue in the bindings of books. Silverfish grow to about 12 mm in length; their common name derives from the shiny scales that cover much of the body and which are readily shed, leaving predators such as ants with jaws full of scales while the owner runs away. The name Thysanura is also descriptive, deriving from the Greek *thysanos* (fringe) and *oura* (tail).

A few groups of tiny arthropods traditionally considered to be insects without wings, and so treated in most books, are now believed to have had a separate ancestry. Among these are the springtails (order Collembola), which are abundant in the greater Toronto region in damp sites under logs and in soil; they are often encountered in masses on snow on sunny days – hence the popular name 'snowfleas.' Although only 1 to 3 mm long, springtails can be recognized because they jump, propelled by the muscular release of a hinged furca on the underside of the abdomen – giving rise to their name. Springtails are incredibly abundant in soil (see Evans 1984), and have an important part in sustaining the fertility of soil through degradation of organic materials.

2. INSECTS WITH ANCIENT WINGS: PALAEOPTERA

Vertebrate animals took up flying several times in the history of life; bats and birds became successful fliers through modification to different parts of the fore limbs, and some extinct reptiles also could fly. Ancient insects evidently developed wings only once, but that was one of the momentous events in the history of life because flight proved to be an enormous advantage for insects. The wings of insects didn't just happen, of course, but evolved from simple beginnings over a long period. Two types of wings are recognized among the insects now living. They differ fundamentally in the mechanism of the hinged joint articulating the wings with the body (see Evans 1984). Two orders of living insects have a more primitive form of wing articulation than the others, and are assigned to the Palaeoptera (from the Greek *palaios* (ancient) and *pteron* (wing).

2.1 Mayflies: Order Ephemeroptera

Immature stages, or nymphs, of most mayflies graze algae and fine organic particles from rocks on the bottom of streams and lakes; a few are predacious on other aquatic insects. Adult mayflies have two or three long filaments at the end of the abdomen, and their wings at rest are held together vertically above the body. Mayflies are famous around the world for two reasons. First, adults rarely live longer than two or three days, and some species are indeed ephemeral in completing their entire winged reproductive existence within 24 hours, hence their name, from the Greek *ephemeros* (short-lived) and *pteron* (wing). Second, mayflies are the only living insects that moult one more time after their wings first appear; the insects are not sexually mature until the second winged stage. The second moult of winged mayflies appears to have been part of an ancient life style that has not changed for 300 million years since mayflies lived in the Carboniferous Period.

Adult mayflies rarely live longer than two or three days, and some species complete their entire winged reproductive existence within 24 hours.

Mayfly adults seldom stray far from the water where they lived as nymphs, and in the greater Toronto region they can be found along unpolluted headwater streams and in ponds. Up to the end of the first half of this century, before pollution levels in the Great Lakes became troublesome, adults of burrowing mayflies emerged in hordes every year from Lake Erie, and for a few days were shovelled up every morning from beneath the street lights of shoreline towns. These insects provided a significant food resource for fish and in recent years have been returning to some parts of Lake Erie. Mayflies live in Lake Ontario, too, but not in such large concentrations.

2.2 Dragonflies and Damselflies: Order Odonata

Dragonflies at rest hold their wings outstretched at right angles to the body; most damselflies hold their wings together in line with the body, and they are also more slender than dragonflies. Even with their ancient wings, dragonflies are marvellously adept in flight, and often range far from water, where the immature stages live. In both adult and nymphal stages, dragonflies and damselflies prey on other insects. The name Odonata, from the Greek *odon* (tooth), alludes to their strong mandibles.

For reproduction they return to ponds and slow streams, where males of some species defend breeding territories along the shoreline. Reproductive behaviour in the Odonata is richly varied (see Evans 1984); it is not difficult to observe, because the brighter the day, the more activity there is. Close observation of plants and partly submerged tree branches along a shoreline where dragonflies are active will often reveal empty nymphal skins still fastened where the claws grasped the rough surface while the adult insect crawled out of a split in the back. Few people have seen the emergence of adult Odonata, but those who have remember the event for the rest of their lives as an amazing and beautiful phenomenon (Figure 11.1). Ancestors of these dragonflies were doing the same things along marshy shores 200 million years ago as dinosaurs lumbered through the shallows.

3. INSECTS WITH NEW WINGS: NEOPTERA

Because all other winged insects have a form of wing articulation that bestows greater flexibility on wing movements, they are placed in the Neoptera – from the Greek *neos* (new) and *pteron* (wing). They differ from each other in a multitude of other features, however, and on this basis are assigned to many different orders.

In some groups of winged insects the wings have been lost through specialization for particular life styles.

Insects and Ecological Sustainability

All ecosystems change over time, but in areas such as the greater Toronto region, which has been subjected to accelerating human pressure for some 200 years, the environment has been degraded and natural habitats for plants and animals reduced. Although some 2,400 species of insects were recorded by 1913 (Walker 1913), the composition of the insect fauna of the greater Toronto region has never been investigated intensively. There are so many species of insects and so few insect taxonomists that the biological changes taking place in insect communities can be neither documented nor interpreted. And that is an unfortunate gap in biological knowledge because insects are so deeply integrated into food webs that they are a sensitive barometer of even small distortions in biological communities.

A special case for monitoring environmental change can be made for aquatic insects – species of the orders Ephemeroptera (mayflies), Odonata (dragonflies), Plecoptera (stoneflies), Trichoptera (caddisflies), and for several families of Diptera (flies) and Coleoptera (beetles). The network of streams draining any region reflects changes in the biological condition or health of the watershed from deforestation, erosion, organic pollution from sewage, runoff from application of fertilizers and pesticides, toxic chemical pollution from industrial effluents, and other consequences of the indifference of people to the natural environment in which they live. These disturbances alter the composition of natural stream communities, which largely comprise aquatic insects. A growing body of knowledge on the identity and biology of the larval stages of aquatic insects enables freshwater biologists to monitor changes in water quality through detection of changes in the communities of stream insects. Biological monitoring of this sort provides an insight into disturbances in the aquatic community year-round, an advantage over the specific-time data from chemical analysis of water.

The greater Toronto region is richly endowed with wooded ravines, each with its stream. Some habitats in peripheral areas still support diverse insect communities, but for how much longer depends on the effectiveness of measures for protecting natural environments. Fundamental improvement or further degradation in the natural environment would be mirrored by changes in the communities of stream-dwelling insects. Research programs of this kind require facilities to house the collections in order that comparisons can be made from one survey to another. An Ontario-wide program of biological monitoring using aquatic insects was discontinued some years ago, and the collections deposited for safekeeping in the Royal Ontario Museum's Centre for Biodiversity and Conservation Biology. For many years the museum has had a leading role in the international biological community concerned with basic research on the identification and biology of aquatic insects.

Some moths are wingless, for example, as are fleas and lice, but all were derived from winged ancestors.

3.1 Grasshoppers and Crickets: Order Orthoptera

Orthoptera derives from the Greek *orthos* (straight) and *pteron* (wing). These are familiar insects to everyone, and not least because most of them communicate by sound – but here there are intriguing differences between grasshoppers and crickets that are not so familiar. Grasshoppers produce sound (not all do) by striking their fore wings against a ridge on the hind legs; and grasshoppers hear each other with sound receptors on the first abdominal segment. Crickets and their relatives make sounds by rubbing the fore wings together; and their sound receivers are located on the first pair of legs. The function of the sounds is to attract mates, and in most species only males can call. Grasshoppers are active during the day, and call while flying; they have short antennae. Crickets and their relatives are active mainly at night and call from a resting position; these insects have long thread-like antennae, and the large green katydids or so-called long-horned grasshoppers are really related to crickets.

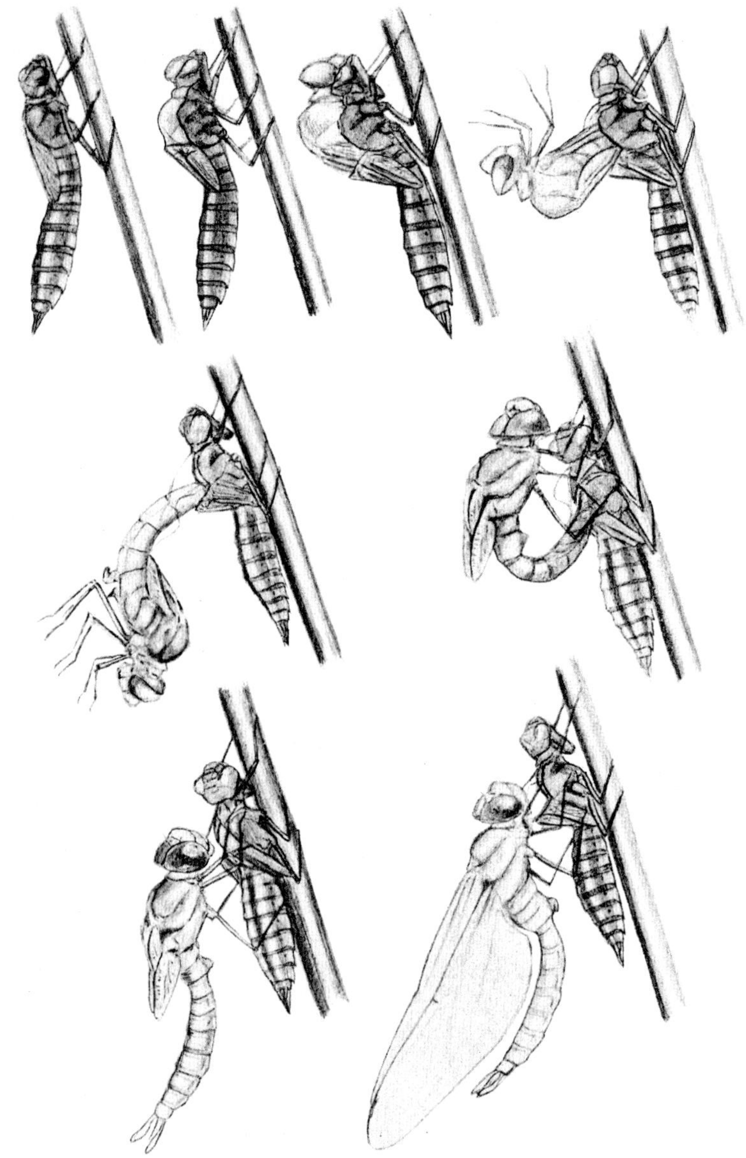

Numerous species of crickets and grasshoppers occur in the greater Toronto region and, to a perceptive listener, the males have calls as distinctive as the songs of different species of birds or frogs. The subtle natural music on summer nights is really the communication network of crickets and their relatives (see Evans 1984; Dethier 1992). With a flashlight you can track down nocturnal singers and watch their wings vibrate as the sound is produced: greenish-white tree crickets perched on low shrubs, and green katydids on goldenrods and other plants. Wingless, and therefore silent, camel crickets are sometimes encountered in damp cellars of old houses, or under logs on the ground. Typical grasshoppers include the cracker locusts of hot dry roadsides; when disturbed, they fly a short distance crackling as they go and settle to the ground, obscured by cryptic colouring. The hordes of plague locusts that devastate agricultural crops at tropical latitudes are true grasshoppers.

The musical cricket-on-the-hearth in the writing of Charles Dickens was a house cricket, an introduced cosmopolitan species that originated in a warm climate; this species lives in heated buildings in the greater Toronto region. House crickets have been known in

FIGURE 11.1 Emergence of a dragonfly adult from the nymphal stage.
(Drawn by E.M. Walker, University of Toronto Studies, Biological Series no. 11, 1912)

North America for more than 200 years, and are believed to have crossed the ocean on cargo ships.

In some books on insects, mantids, cockroaches, and walkingsticks are included in the order Orthoptera; in others these groups are treated as separate orders, as here.

3.2 Mantids: Order Mantodea

The European, or praying, mantid is often seen in the greater Toronto region, having become naturalized in Ontario at least as early as 1914, perhaps originating from introductions in New York State. It is a large and fascinating insect, wholly dependent for food on capturing other insects. This species overwinters as eggs embedded in a spongy mass fastened to plants or to buildings. Some European mantids are green, others are brown, but they are all the same species. A larger species, also introduced, is the Chinese mantid, which has reached southern Ontario and might be encountered in the greater Toronto region. The name mantid is derived from the Greek *mantis,* meaning soothsayer or prophet, given because in their resting posture mantids give the appearance of praying.

3.3 Cockroaches: Order Blattodea

When *The Natural History of the Toronto Region* was published by the Canadian Institute in 1913, eight species of cockroaches (whose name comes from the Latin *blatta,* meaning cockroach) were included. Two of these species are native to southern Ontario, the other six are introduced. One native species (*Parcoblatta virginica*) was cited as 'rare,' the other (*Parcoblatta pennsylvanica*) as 'not uncommon.' If either of these native species still occurs in the greater Toronto region, it has to be regarded as extremely rare; the absence of recent records suggests that both are probably extinct in the region. However, *P. pennsylvanica* is common in Muskoka and in eastern Ontario. The native cockroaches of Ontario live under logs and rocks, and enter houses occasionally, but they have never become established in urban areas. The cockroaches of Toronto are introduced species shared with most other northern cities; they thrive under disturbed conditions and live in heated buildings because they cannot survive winter conditions outdoors. These urban species came originally from Africa and Asia, and have been spread around the globe through commerce. Most cockroaches are scavengers, feeding mainly on decaying plant materials with associated fungi and micro-organisms (see Evans 1984).

3.4 Termites: Order Isoptera

The name Isoptera comes from the Greek *iso* (equal) and *pteron* (wing), a reference to the similar size of fore and hind wings, which is an uncommon condition among insects. Most termites are tropical, but the eastern subterranean termite, a native species ranging from New England to Minnesota and southward, occurs in southwestern Ontario and has been known in the greater Toronto region since 1938. This species has become notorious in central urban Toronto for the damage caused to buildings. All termites eat wood, and under natural conditions they have an important role in the breakdown of dead wood and the recycling of nutrients. Damage to buildings occurs when support timbers are in contact with the ground and termites enter the wood directly from colonies in the soil. They can also bridge a masonry foundation by constructing thin mud tubes or galleries from the soil to the wooden structure of the building; these galleries are one of the few clues that termites are present in a building. Termites are able to digest the cellulose in wood because of specialized protozoans that live in their digestive tract. Termites are social insects living in colonies that often comprise thousands of individuals, representing at least three distinctive castes: winged reproductive males and females, wingless workers, and soldiers.

3.5 Grylloblattids: Order Grylloblattodea

Grylloblattids are confined to mountainous terrain of western North America, Siberia, China, and Japan. Although they do not occur in the greater Toronto region, they do have a close connection with this city: it was Professor Edmund Walker of the University of Toronto who discovered grylloblattids in Banff, Alberta, in 1913. He first interpreted their unique combination of features of both crickets and cockroaches, conveyed in the composite name that he created for them (from

---

Plants and partly submerged tree branches along a shoreline where dragonflies are active will often have empty nymphal skins still fastened where the claws grasped the rough surface while the adult insect crawled out of a split in the back.

the Latin *gryllus* [cricket] and *blatta* [cockroach]). This was the last of the insect orders to be discovered. Grylloblattids are an interesting chapter in the history of entomology in Canada; and *Grylloblatta campodeiformis* E.M. Walker is enshrined in the emblem of the Entomological Society of Canada (see Wiggins 1966).

3.6 Walkingsticks: Order Phasmatodea

Walkingsticks are extraordinary creatures, fully deserving to be called apparitions (Phasmatodea derives from the Greek *phasma*, for apparition or spirit). The order is mainly tropical, and includes the longest insect now living, which attains a length of 30 cm. Since the wings of many species are reduced or lacking, these insects gain protection from predators through camouflage to resemble twigs. In some species, expansions of the legs and other body parts resemble leaves, including even veins in the leaves. Walkingsticks are herbivores, and the single species of the greater Toronto region feeds on leaves of oak, basswood, and several other trees including black locust, and on a variety of shrubs. Eggs are scattered on the ground, and remain in the leaf litter through the winter.

E.M. Walker

Edmund M. Walker, mid-1940s.
(Courtesy Conrad E. Heidenreich)

Grylloblatta campodeiformis
(Edmund M. Walker, photographer.
Courtesy Conrad E. Heidenreich)

Edmund Murton Walker was born on 5 October 1877 in Windsor, Ontario. In his youth Walker developed a keen interest in insects, which he attempted to classify, describe, and sketch. In 1900 he graduated in natural science from the University of Toronto and was persuaded by his father to obtain a degree in medicine, which he completed in 1903. Following his internship, he returned to his real interest, insects, and in 1905 went to the University of Berlin for a year of postgraduate work in invertebrate zoology. Here he met his future wife, Eleanora Walzel. They were married in her native Austria in 1909 and had four children.

In 1906, Walker was appointed lecturer in invertebrate zoology at the University of Toronto. He remained at that institution until he retired in 1948, serving as chair of the Department of Zoology from 1934 until his retirement. In 1914 he began to develop the invertebrate collection at the Royal Ontario Museum, which became a lifelong pursuit. He was editor of the *Transactions of the Royal Canadian Institute* from 1924 to 1945.

Walker's first interest was in the Orthoptera (grasshoppers and their relatives), in which area his major contribution was the discovery of *Grylloblatta campodeiformis*, a relict orthopteran link between cockroaches and crickets. His most extensive contribution was to the Odonata (dragonflies), which culminated in a three-volume work, *The Odonata of Canada and Alaska*. The third volume was completed by Dr. Philip S. Corbet and published after Walker's death. Walker was a gifted teacher and a fine artist equally proficient in landscape painting and drawing of the anatomical details of insects. He had a wonderful ear for music and would delight his audiences by mimicking the calls of various insects and birds. His great strength was as a field biologist combining botany and all facets of zoology into an integrated view of the natural environment. Although he travelled and collected over much of Canada, the province of Ontario, and especially the greater Toronto area, remained of special interest to him. He died in Toronto on 14 February 1969.

3.7 Stoneflies: Order Plecoptera

Plecoptera – from the Greek *plecos* (folded) and *pteron* (wing), in reference to the hind wings that are folded like a fan when the insect is at rest – is a small order of aquatic insects whose nymphs occur in the greater Toronto region in unpolluted headwater streams. Nymphs of some stoneflies graze algae and fine organic particles on rocks of the streambed, others feed on decaying leaves and the associated fungi, and others are predatory on aquatic insects. Adult stoneflies are secretive and, because they are weak flyers, do not venture far from the margins of their streams. By drumming their abdomen against the ground, male and female stoneflies are able to make sounds that aid in attracting mates; the signals are distinctive for each species. Some adult stoneflies are conspicuous in winter and in early spring when they emerge from the nymphal stage to crawl over snow-covered banks on sunny days. As with mayflies and dragonflies, the ancestors of present-day stoneflies lived long before dinosaurs appeared on earth.

3.8 Earwigs: Order Dermaptera

The name Dermaptera, from the Greek *derma* (skin) and *pteron* (wing), refers to the coarse texture of the short fore wings that cover the folded, membranous hind wings. A few species of earwigs are native to southern Ontario and some could be encountered in the greater Toronto region, but the one known to most people is the larger, introduced European earwig. As for many introduced species, this one can be abundant, and often seeks shelter in buildings. Gardeners encounter earwigs under the petals of flowers or the leaves of cabbage and lettuce, where the insects lie concealed during the day. At night, earwigs move about in search of other insects, which are their principal food. When earwigs are abundant they will also eat garden plants, but on balance earwigs are biological controls on herbivorous insects. Earwigs make nests underground, where they pass the winter; after driving males out of the nests in the spring, females deposit eggs. In behaviour highly unusual for insects, the female remains with the eggs and tends the newly hatched immatures as they forage during the night and return to the underground nest during the day.

In behaviour highly unusual for insects, the female earwig remains with the eggs and tends the newly hatched immatures as they forage during the night and return to the underground nest during the day.

3.9 Booklice and Barklice: Order Psocoptera

Lice is a misleading name for these small insects, for none of them is parasitic. Booklice are greyish-white, wingless insects, approximately 2 mm long. They are common in houses, especially among items left undisturbed under humid conditions. Barklice have wings and are about 5 mm long; they are often found under loose bark of dead trees. All of these insects feed on fungi associated with organic materials. The name Psocoptera is derived from the Greek *psocho* (to grind) and *pteron* (wing).

3.10 Chewing Lice: Order Mallophaga

Chewing lice are external parasites, chiefly on birds but also on mammals. These insects are secondarily wingless – they have lost the wings their ancestors had – and range in length up to 5 mm. They feed on feathers or hair; the name Mallophaga comes from the Greek *mallos* (wool) and *phagein* (to eat). Different species of lice are specific to different hosts; they perish when removed from the host. Chewing lice do not occur on humans, although they are common on domestic animals, especially poultry and pigeons.

3.11 Sucking Lice: Order Anoplura

These wingless external parasitic insects suck blood from mammalian hosts. The head and body lice of humans belong to the Anoplura – from the Greek *anoplos* (unarmed) and *oura* (tail). They are vectors of certain diseases, chiefly epidemic typhus. Lice become a problem mainly when people live under crowded, unsanitary conditions, but infestations of head lice among school children are sometimes persistent. In some recent references, the Mallophaga and Anoplura are combined in a single order with the marvellous name, Phthiraptera, from the Greek *phtheir* (louse), *a* (without), and *pteron* (wing). All lice are derived from winged ancestors.

3.12 Thrips: Order Thysanoptera

Thrips are common insects but are seldom seen because they are so small, ranging in length from 0.5 to 2 mm. They are usually associated with flowers and can be found by tapping wild flowers over a white pan or sheet of paper. Under magnification, the characteristic slender fringed wings from which they derive their name – *thysanos* (fringe) and *pteron* (wing) (Greek) – are evident. The mouthparts of thrips are adapted for sucking juices from plants, although some feed on fungal spores or are predacious. Some species are destructive pests of horticultural plants.

3.13 Bugs: Order Hemiptera

These are the insects to which the name 'bug' is correctly applied. The fore wings are distinctive because the basal part is thickened and opaque and the apical part is a typical translucent membrane with veins – hence the derivation for the name of this order, from the Greek *hemi* (half) and *pteron* (wing). Bugs also have a prominent beak formed from mouthparts modified for piercing and sucking out the juices of plant hosts or animal prey on which they feed. Most, such as plant bugs, chinch bugs, and squash bugs, feed on plants; ambush bugs, assassin bugs, and some stink bugs feed on other insects. Most aquatic bugs are predators, such as water striders, backswimmers, waterscorpions, and the giant water bugs that are often attracted to lights near water. Bed bugs are also members of the Hemiptera; as bloodsucking parasites of birds and bats, and of humans too, they do not have wings (see Evans 1984). Many species of bugs emit a powerful odour. One species of assassin bug is introduced and occurs in houses where it preys on invertebrates such as spiders, carpet beetles, and bed bugs; the immature nymphal stages look like animated dust balls because bits of debris adhere to a sticky secretion on their bodies, providing camouflage and leading to the name 'masked hunter.' The western conifer seed bug, a rather large insect that turns up in buildings in the greater Toronto region, is an unusual example of a native insect expanding its range.

3.14 Cicadas, Aphids, and Relatives: Order Homoptera

These insects are closely related to the Hemiptera, but their wings are uniformly membranous, signified in the name Homoptera, from the Greek *homos* (uniform) and *pteron* (wing). They also have sucking mouthparts, and include some of the most serious pests of plants: aphids, scale insects, and whiteflies. Among the largest insects in this order are cicadas; the males are responsible for the loud, shrill calls from high up in trees during hot summer days. Nymphal cicadas live in the soil, sucking plant juices from roots; when they emerge as winged adults, the cast nymphal skin is often found attached to the trunk of a tree. Related species occurring south of the Great Lakes are the periodical cicadas, notable because they pass 13 or 17 years underground as nymphs. Most cicadas, including those in the greater Toronto region, have a life cycle of about two years. Froghoppers are much smaller than cicadas, and are best known from their nymphs, which are called spittlebugs because as they feed on grass and other low plants, a foamy mass is exuded as a protective cover.

4. INSECTS WITH COMPLETE METAMORPHOSIS: ENDOPTERYGOTA

These are neopterous insects with a resting pupal stage in which the worm-like larva undergoes complete transformation or metamorphosis to a winged adult of entirely different form and habits. The other insects discussed previously lack the resting pupal stage. The name Endopterygota (Greek: *endo* [inside], *pteron* [wing]) alludes to the internal development of the wings, in contrast to all other winged insects, where the wings can be seen as buds developing externally with each moult, the exopterygote condition.

When complete metamorphosis arose in the evolution of insects, a new frontier was opened in the exploitation by insects of energy from terrestrial and freshwater ecosystems. The ecological impact is evident in the numbers of species: roughly 80 percent of all living insect species have life histories with a pupal stage. The underlying explanation for this strong ecological penetration is that complete metamorphosis enables the larval stage to become a highly specialized feeder – e.g., as a leaf-eating caterpillar, or as a legless fly maggot in a heap of decomposing organic material – but through the pupal stage the insect is transformed into a winged adult dependent on wholly different food, such as the nectar of flowers. Consequently, in each endopterygote species, different life stages exploit different sources of energy and nutrients during the course of a single life cycle, reducing competition between adults and larvae. For comparison with incomplete metamorphosis, the original life style of insects, walk through a meadow and you will see both adult and wingless immature grasshoppers living together and using the same food resources. Examine a colony of aphids on a plant stem and you will see the young feeding together with winged adults.

4.1 Fishflies, Lacewings, and Relatives: Order Neuroptera

Several rather disparate families are assigned to this small order, but they share conspicuous wing venation with many interconnecting cross branches, accounting for the name Neuroptera (Greek: *neuron* [nerve], *pteron* [wing]). The most common members in the greater Toronto region are lacewings, exquisite green or brown insects 2 cm long. They are common on plants, where they prey on aphids. Larvae in some other families, such as fishflies, dobsonflies, and alderflies, are aquatic. Larvae of dobsonflies are hellgrammites, very large freshwater predators often used for bait by anglers. These aquatic families are treated in some references as a separate order, Megaloptera, a name based on the large size of some species.

4.2 Beetles: Order Coleoptera

Beetles are distinguished by heavily armoured bodies, much of the armour derived from the first pair of wings, which have become thickened, convex plates called elytra. Each elytron hinges forward to expose a second pair of folded membranous wings used in flight, hence the order's name, Coleoptera, from the Greek *coleos* (sheath or scabbard) and *pteron* (wing). A fact of life not widely appreciated is that more than 25 percent of all living animal species are beetles – nearly 300,000 known species (Wilson 1992), plus vast numbers still to be discovered. Consequently, a substantial part of the ecological success of insects overall can be attributed to beetles, and this is as true for the greater Toronto region as it is for any other part of the world. Diversity in beetles even in the greater Toronto region is nearly inexhaustible, and approximately 1,100 species were recorded by 1913. Why beetles are so extraordinarily successful is one of the fundamental questions about life on earth. The best answer available so far is found in the advantages of complete metamorphosis, described above, and in the protective armoured body of adult beetles. The armoured body also reduces moisture loss; consequently, beetles are more successful than most other insects in very dry habitats.

Beetles are cryptic animals, usually carrying on with their lives in concealment. Close inspection of the flowers of native plants will reveal some beetles; milkweed and goldenrod, for example, are usually dependable, and leaf beetles occur on many plants. Others can be found under the bark of dead trees. The black beetles under logs and rocks are ground beetles; on dry sites the iridescent beetles that fly in quick, short bursts are tiger beetles; both prey on other insects. Our fauna of ladybird beetles has begun to change in recent years as native species are replaced by introduced species (Marshall 1995). Several families of beetles live in water, including diving beetles and the whirligig beetles that cluster together on the surface and have divided eyes for vision above and below the water surface. A field guide to beetles (White 1983) is available for those who wish to probe more deeply into beetle life.

4.3 Twisted-Wing Parasites: Order Strepsiptera

This is a small order of tiny, obscure insects that are mostly internal parasites of others, chiefly Homoptera, Orthoptera, and Hymenoptera. Although Strepsiptera – from the Greek, *streptos* (twisted), *pteron* (wing), in reference to the aberrant wings – do occur in the greater Toronto region, they are likely to be found only by rearing infected host insects.

More than 25 percent of all living animal species are beetles.

4.4 Scorpionflies and Allies: Order Mecoptera

Insects of the small order Mecoptera – from the Greek *mecos* (length) and *pteron* (wing) – occur in moist wooded areas of the greater Toronto region. Adults have a prolonged snout, and the wings of the common scorpionflies have conspicuous brown blotches. The name scorpionfly is derived from the bulbous apex of the abdomen in males of the most common family, which resembles, in form only, the stinging apparatus of scorpions. Larval and adult scorpionflies are predators or scavengers of other insects.

4.5 Caddisflies: Order Trichoptera

Adult caddisflies are found most often around lights, as are moths and other nocturnal insects; they resemble small moths, but the wings are held roof-like at rest, rather than the more flattened position of most moths. Their wings are covered with tiny hairs, as implied by the name Trichoptera, from the Greek *trichos* (hair) and *pteron* (wing). However, larvae of caddisflies are aquatic in streams and ponds, and some are distinctive for the cases they carry about, constructed of twigs or rock fragments fastened together with silk. These case-making larvae feed mainly on decaying leaves. Larvae of other caddisflies construct stationary shelters on submerged rocks and logs, often supplemented with a fine-meshed silken net to filter food materials carried by the current. Unpolluted streams in the outlying parts of the greater Toronto region usually have abundant populations of larval caddisflies, and they can be readily seen by turning over submerged rocks and logs.

4.6 Moths and Butterflies: Order Lepidoptera

Lepidoptera, from the Greek *lepidos* (scale) and *pteron* (wing), in reference to the microscopic scales covering the wings and body, are one of the most successful groups of insects, the world total now numbering over 112,000 known species (Wilson 1992). In both larval and adult stages they are intimately associated with plants because the major diversification of both flowering plants and Lepidoptera occurred together during the Cretaceous Period some 100 million years ago. Lepidoptera

Flower flies gain protection from predators by mimicking stinging bees and wasps.

exploited the irruption in flowering plant diversity by specializing to a large extent on particular groups of larval food plants, thereby partitioning these resources among more species. Butterflies are a visual delight as they fly from flower to flower sipping nectar on sunny days, but butterflies are the day-flying contingent representing only about 10 percent of Lepidoptera (see Evans 1984). The remaining 90 percent of the species are moths, mostly active as part of the nocturnal world that humans rarely see. Some sense of the rich diversity of moths can be gained by watching for them around lights. Among the few exceptions to the predominantly nocturnal moths are clear-winged sphinx moths, which are active during the day. Like other sphinx moths, they fly rapidly and hover over flowers like hummingbirds, uncoiling a long, tubular proboscis to extract nectar from deep inside the flower, while transferring pollen from one flower to another – the plant's price for giving up its nutritious nectar. Field

---

The Monarch Butterfly

Everyone is familiar with the Monarch butterfly. Rich brownish red with contrasting black veins on the wings, these insects grace the summer landscape. But Monarchs are unable to live through cold winters. The first Monarchs to appear every year in the greater Toronto region have flown in from areas to the south. The females seek out patches of native milkweed and deposit eggs. The caterpillars feed only on milkweed, and in a few weeks enter the pupal stage, enclosed in a transparent integument suspended from a leaf or other support. About 10 to 15 days later, a Monarch butterfly emerges, expands its compressed wings, and takes to the air. Towards the end of summer, adults of the new generations from much of eastern and central North America fly south, eventually congregating in trees of fir forests of the Transvolcanic Range in central Mexico. There they remain from November through March, and with return of spring conditions, the butterflies begin to mate and to move northward. Monarchs reappearing in Ontario and the greater Toronto region in May arose from a wave of generations produced to the south as the returning migrants follow the northward advance of spring (Malcolm and Zalucki 1993).

The migration of the Monarch butterfly is more like that of a bird than of an insect. Much of the story was documented by Professor Fred Urquhart, now retired from Scarborough College of the University of Toronto. He was fascinated by the Monarchs that disappeared from his boyhood insect-collecting haunts in east Toronto, only to reappear every spring; later as a university professor, he devised a method of fastening a numbered tag to the wing. Recapture of tagged butterflies proved that they had flown enormous distances southward in the autumn; later in the 1970s Urquhart and his wife, Norah, discovered overwintering sites in Mexico, and undertook efforts with the Mexican government to protect those areas in the mountain forest. In recognition of their work the Urquharts were jointly awarded the Order of Canada in 1998.

The return of the Monarch butterfly to the greater Toronto region every year depends on successful overwintering of the adult butterflies to the south. Succeeding generations depend on milkweed, which is the only plant

Stages in the life cycle of a Monarch butterfly. (left, courtesy A. Joan Bennett; below, courtesy Frank Parhizgar)

eaten by the caterpillars. Milkweed is regarded as an invasive and noxious plant in Ontario and often eradicated; but, without it, our enjoyment of Monarch butterflies and their incredible journey would come to an end. The Committee on the Status of Endangered Wildlife in Canada has placed monarchs as vulnerable on the endangered species list.

The Monarch-milkweed connection has another twist. The Viceroy butterfly, which also occurs in the greater Toronto region, resembles the Monarch in colour pattern. The two species are easily confused, but the Viceroy can be distinguished by a curved black line intersecting the longitudinal veins of the hind wing. The conventional explanation for the close similarity between Monarchs and the unrelated Viceroy is that Monarchs are unpalatable to predators (mainly birds) because of poisonous cardenolides (plant-derived steroid compounds that include glycosides used in treatment for congestive heart failure in humans) sequestered from the milkweed food of the caterpillars. The conspicuous colour pattern of Monarch butterflies reminds potential predators not to touch. Viceroy butterflies, which usually do not feed on milkweed and are usually palatable, have gained protection from potential predators through evolution of a mimicking colour pattern. However, the story is proving to be rather more complex (e.g., Malcolm and Zalucki 1993). For example, studies in chemical ecology reveal that milkweed species differ in their production of emetic cardenolides; the milkweeds eaten by caterpillars of the spring arrivals on the Gulf Coast are highly toxic, and as the butterflies of the next generation move northward, predators are conditioned to avoid Monarchs, even though subsequent reproduction in the north occurs on milkweeds of low toxicity.

The biological bottom line is that milkweed species produce variable concentrations of poisonous compounds such as cardenolides as protection against all herbivores, and a few insects have overcome this chemical defence. Among them, Monarch butterflies, through natural selection, have added the poisonous compounds to their own predator defence, and successfully enough to warrant advertising their unpalatability by conspicuous colours. Monarch caterpillars are outlandishly banded with black, white, and yellow. Close examination of milkweed plants in the greater Toronto region will yield other insects that have broken through the chemical defence: milkweed bugs suck juices from the seeds, a rounded leaf beetle feeds on foliage, and elongate long-horned beetles with prominent antennae feed on the roots as larvae and as adults eat the leaves and flowers. All of them are conspicuously marked with red and black and are also evidently meant to be seen.

Viceroy butterfly adult. (W.A. Crich, Royal Ontario Museum)

guides are available for butterflies (Opler 1992; Layberry, Hall, and Lafontaine 1998) and moths (Covell 1984).

4.7 Flies: Order Diptera

The name Diptera (Greek: *di* [two], *pteron* [wing]) describes a distinguishing feature of true flies: they have only the first pair of wings; the second wings have become reduced to tiny stalked knobs called halteres (see Evans 1984). The impact of flies on the natural world is immense (see Oldroyd 1966). Somewhere around 100,000 species of flies are now known (Wilson 1992), which means that one in every ten species of animals now known to science is a fly of some kind – and exploration of the flies of the world is still far from completion. Observers in the greater Toronto region can expect to see representatives of a substantial segment of the world's families of flies. Sunny days are best, especially around flowers; lapping up the flowers' offering of nectar, the flies are dusted with pollen, which they transfer to other plants of the same kind. Among the insects attracted to flowers are flower (or hover) flies, manoeuvring like helicopters, with their abdomens banded in yellow and black. Flower flies gain protection from predators by mimicking stinging bees and wasps, but their single pair of wings and short, bristle-like antennae give the ruse away. Robber flies are large, aggressive predators of other insects; they capture even beetles, taken in flight when the usually protective elytra are extended.

Other flies exploit proteins in the blood of birds and mammals as food: mosquitoes, black flies, deer flies, and biting midges are a distressing nuisance at our latitude; but in warmer parts of the globe blood-feeding insects transmit malaria, yellow fever, onchocerciasis, leishmaniasis, and other diseases of humans. Black flies live in unpolluted running waters and are much less troublesome in Toronto than in cottage country. Mosquitoes are a problem even in populated areas because larvae live in standing waters. In the greater Toronto region, temporary snowmelt pools are principal habitats; larvae hatch from eggs that have lain on dry ground for much of the preceding year. Larval habitats for other species are provided by water in such unlikely places as discarded tires and flower containers in cemeteries. Native anopheline mosquitoes are capable of transmitting malaria, and did so in pioneer times until the disease was eradicated in Canada. A few species of culicine mosquitoes are vectors of arboviruses (arthropod-borne viruses), including various forms causing encephalitis in birds and horses, and also humans (McLintock 1976). Eradicating mosquitoes and black flies is difficult. The shallow pools where larvae of some mosquitoes live are also the breeding habitats of many other aquatic insects and several frogs and salamanders. Consequently, application of chemicals to the pools to eradicate mosquito larvae is ecologically unsound. Moreover, insect electrocution traps intended to reduce locally the numbers of mosquitoes and other biting insects are ineffective because few blood-feeding insects are attracted to lights. Identification of the corpses in these traps reveals a very high proportion of beneficial insects.

Through the marvel of complete metamorphosis, flies, superbly agile in flight, develop from worm-like, legless larvae, or maggots, that are not often seen because they are deeply involved in the biological affairs of the terrestrial and freshwater worlds. Consumption of decaying organic material and recycling of nutrients are major roles of larval flies. Larvae of marsh flies are predators of snails; larvae of tachinid flies are parasitoids inside other insects. Cluster fly larvae are parasitoids of earthworms; the adult flies hibernate in wall spaces of buildings where, warmed up to activity, they congregate around windows. Goldenrod gall flies are seldom seen, but feeding of their larvae inside the stem stimulates growth of the common bulbous gall on goldenrod, providing a localized food supply. The larva pupates inside the gall in spring, and the adult emerges through a tiny hole – if it has escaped attack by parasitoid wasps or winter foraging by downy woodpeckers. Gall gnat larvae cause the formation of the bud-like pine cone galls so common on willow twigs.

4.8 Fleas: Order Siphonaptera

Fleas are tiny insects, living as external parasites on mammals and birds. The name of this order, Siphonaptera – from the Greek *siphon* (tube), *a* (without), and *pteron* (wing) – alludes to the tubular structure of the mouthparts, which facilitates blood feeding, and also to the absence of wings, which were lost during the course of the specialized evolution of fleas and other external insect parasites such as lice. Fleas are flattened laterally, facilitating movement through the hair or feathers of their hosts. Lacking wings, fleas move mainly by jumping, often from one host to another. Resilin, a highly elastic protein in the thorax, and one that stores energy when compressed, enables fleas to broad-jump up to 31 cm. The tiny worm-like larvae of fleas feed on bits of organic debris in the nests of their hosts, and they spin a silken cocoon for metamorphosis. Rat fleas were responsible for spreading bubonic plague, which is mainly a disease of rodents; but the Black Death that swept through medieval Europe followed when the numbers of rats and fleas rose to epidemic proportions and humans were bitten, too. In the greater Toronto region, personal experience with fleas is usually restricted to occasional introductions into the household by the family dog or cat.

4.9 Wasps, Ants, and Relatives: Order Hymenoptera

To an inquiring observer, the Hymenoptera, from the Greek *hymen* (membrane) and *pteron* (wing), of the greater Toronto region can reveal a great deal about how the world works. Around 100,000 species are known worldwide, and penetration of these insects into biological affairs is vast. Sawflies and wood wasps, whose larvae feed on plants, are living representatives of the ancestral Hymenoptera; larval sawflies are destructive defoliators of coniferous trees such as pines, and several introduced species are widespread in southern Ontario. Other wasps and their relatives such as bees and ants were derived later in the history of the Hymenoptera, and all of them have a wasp waist – strongly constricted basal abdominal segments that enable a female to manoeuvre its abdomen and ovipositor while stinging or depositing eggs on host insects. Solitary wasps deposit their eggs individually in chambers provisioned with grasshoppers, caterpillars, or other prey immobilized through stinging, but still alive. These wasps can be seen digging nest chambers in sandy soil, and returning with their prey. Highly developed social behaviour, unique among insects other than termites, led to cooperation and division of labour among the more highly evolved Hymenoptera and thus to huge colonies of honey bees, paper wasps, and ants (see Wilson 1996).

With the decline of introduced honey bees in North America through disease and parasitic mites (see Chapter 10), pollen-feeding native bees are becoming more widely appreciated as pollinating agents; they were, after all, the bee pollinators of native North American plants before honey bees were introduced. Unlike the social honey bees, these are solitary nesting bees; they are smaller and more obscure than honey bees, but many different species of native bees can be seen on flowers as they gather pollen to provision their individual nesting cells where the larvae develop.

Parasitoid Hymenoptera are not social insects, and hence their activities are rather obscure, but their impact on other insects is enormous (see Evans 1984). Parasitoid larvae feed inside or on the body of a host insect, ultimately causing its death, as distinct from parasitic insects such as lice and fleas that feed on a host but do not kill it. The parasitoid life style exploited so very successfully in the diversification of Hymenoptera, and also in Diptera, was made possible through complete metamorphosis; larvae that pass their entire development in the body of a host are simplified to the essentials of a feeding machine, lacking the head or legs of conventional insect larvae. Adults of parasitoid Hymenoptera range in size from large ichneumons to tiny fairyflies less than 1 mm long. The smallest species are parasitoids of insect eggs, and some of them are secondary parasitoids of primary parasitoids. Because their host relationships are often fairly specific, parasitoid Hymenoptera are used in biological control programs, where they reduce infestations of particular insect species destructive in agriculture. Parasitoid Hymenoptera present a vast new frontier for discovery in biological science.

12

Fish

Henry Regier and Gordon Wichert

Viewed from above, the river valleys appear to be the most natural or least modified part of the Toronto region landscape. But within the valleys the river channels and aquatic species associations have mostly been highly modified. Only a few small streams in the moraine and escarpment parts of the landscape far upstream from the Lake Ontario shore are still quite natural.

Where wooded and grassy slopes of the valleys have persisted, they have acted as buffers and filters to protect a stream from some of the harmful effects of development on the flatter, drier lands immediately adjacent. But bad effects of careless development often obliterated such natural protection. Some valleys were cleared to the stream edge, and bordering wetlands were filled. Stream courses were deliberately altered with dams and channelling, usually in ecologically degrading ways. Harm done upstream is inexorably transmitted downstream. These issues were discussed briefly in Chapter 3, and are developed further below.

The ecosystems within pristine streams in old-growth forests are themselves in an old-growth state. The numerous fish, beaver, otter, mussels, crayfish, amphibians, beetles, bugs, leeches, and water-related birds of such streams belong to species in which individual organisms become comparatively large and old. Older individuals of such species have a 'sense of place.' Each relates possessively, at least during the reproductive season, to habitat features that it identifies as important to itself, using instincts and learning capabilities that have emerged in the course of evolution.

Reproduction is one of the most risky and difficult duties of a living creature. Fish are particularly vulnerable to capture by anglers during the spawning season. For centuries perceptive anglers have sensed that a species' reproductive process is choreographed in time and space to increase the chance of successful reproduction. While preoccupied with the rituals of spawning, fish are not wary of such predators. Fish reproduction can be disrupted through careless or greedy fishing practices. Hence legal measures have long been taken to protect fish, especially during a spawning season.

Even when and where fishing was not allowed during spawning, the fisheries usually intensified in other seasons and suppressed the populations of valued, old-growth species to the advantage of less valued species. Fish were increasingly barred from spawning areas by dams. Spawning and rearing habitat became fouled because of pollution and eventually by pesticides and hazardous contaminants. They also became silted due to other abuses. Fish hatcheries were then created in attempts to mitigate the excessive reduction in large, spawner-size fish, and to compensate for the inaccessibility and degradation of natural spawning habitat. As a mitigative measure, in the absence of rehabilitative action directed at the exploitive fishery and habitat degradation, hatchery programs usually failed to achieve their purposes.

A natural stream has an intricate socio-ecological organization that includes a 'negotiated' system of sharing within and among the various species to satisfy the needs for habitat resources and spaces that are special to each. Much of this socio-ecological organization, based on information, some of which was instinctive and some learned, disintegrated following human abuse of Toronto's aquatic ecosystems. Sensitive, long-lived, old-growth native species waned in number while insensitive, short-lived opportunists, including numerous non-native species, thrived. We generally prize the ecologically sensitive more than the insensitive species.

Complex organization in streams is possible where the flow regime does not fluctuate erratically, especially during warmer months. But almost all sectors of economic development in the region have had the effect of greatly increasing the fluctuations of stream flow. Increasingly intense but shorter floods alternated with unnaturally extended periods of low or no flow during summer and winter droughts. The larger the volume and faster the rate of flow during these intense floods, the bigger the wobbles or sinuosities that the stream carves into its substrate, and the wider and shallower the stream in cross-section. Landscape artists have painted such abused streams in periods of calm, with exposed and eroded banks, presumably not knowing their ugly moods in stormy times and their degraded ecological state even in calm seasons. Only the ecologically untutored perceive these as pleasant country scenes. Only the most adaptive aquatic species can tolerate such debased environments.

During the past two centuries of industrial, agricultural, and urban development, flooding became progressively worse. Each flood can be likened to manoeuvring an ever more ungainly bulldozer down a stream to demolish parts of the existing stream banks and bed, fill in some deeper stretches and wetlands, and carve a new, less stable stream channel. Over the decades the bulldozer-like floods became larger and more forceful, and more destructive not only of natural features but also of structures that humans had built in the streams and on the floodplains. Mill dams built to withstand last year's flood were breached by next year's bigger flood, with a mass of the liberated water careening downstream. Engineering attempts to constrain the effects of floods locally with concrete and steel usually exacerbated the problem – locally and elsewhere. One result was a transformation of hidden, mysterious, pristine streams into open, ugly, and dangerous waterways that then became excessively warm and often dry during droughts. Special habitats of many native fish species were obliterated by such transformations.

Early developers needed to dispose of wastes, including sewage from privies, sawdust from lumber mills, mash from breweries, washings from woollen mills, and whey from cheese factories. Such wastes were often directed into nearby streams to be carried away inexpensively from an entrepreneur's property. Bacteria decomposing these wastes used dissolved oxygen that was needed by other creatures, which then suffocated. Other wastes were directly poisonous to aquatic organisms. Again the species we value most highly tend to be the most sensitive to such pollutants. Altogether pollution had disastrous effects on the ecological associations of pristine streams.

Each flood can be likened to manoeuvring an ever more ungainly bulldozer down a stream to demolish parts of the existing banks and bed, fill in some deeper stretches and wetlands, and carve a new, less stable stream channel.

TWO HUNDRED YEARS AND COUNTING

Early in the nineteenth century, much of the Toronto region landscape was logged, burned, and converted to primitive agriculture. Stream courses and flow regimes were changed rapidly. By mid-century the migratory Atlantic salmon were becoming scarce, largely because of hydrological changes. They disappeared from Lake Ontario by the end of the nineteenth century.

A hundred years ago, the larger streams, wetlands, and Toronto Bay near urbanizing Toronto had already become ecological slums. Only the most tolerant fish species, usually small in size, were to be found in summer in such ugly waters. Anglers had no interest in them. People who could afford to do so turned their backs, built homes farther inland, and stayed away from such waters. Dirty industries, railways, and highways were built on these devalued shores. People who worked or travelled in these ecoslums, or lived near them, risked their health.

In the twentieth century in rural parts of the region, agriculture became progressively more mechanized and chemicalized, in ecologically careless ways. The general process of intensifying degradation of the waters of the Toronto region continued until the 1950s. Following assessments of the damage done in each of the different watersheds in the Toronto region, corrective action then got under way and has progressed with increasing vigour for four or five decades. Regeneration is now well under way. But new kinds of environmental abuse, such as the release of newly invented harmful chemicals of many kinds and the further evolution of an urban smog and heat island, have offset some of the benefits of corrective action directed toward earlier kinds of abuses.

In what follows we examine how the species of three fish families have adapted, or failed to adapt, to two centuries of increasingly intensive effects of development-oriented humans on these fishes' habitats. Then we will briefly review a new kind of progress, with a dozen different kinds of corrective or rehabilitative programs undertaken in recent decades.

FAMILIES OF FISH SPECIES AS INDICATORS OF HABITAT CONDITIONS

In the Toronto region, three families of fish species – salmonids, percids, and centrarchids – have always been of special interest, whether to Aboriginal fishers, recreational anglers, or commercial harvesters. We have selected the three fish families here to help to tell a complicated story about ecological degradation and partial recovery. Some species in each of these families are well known to the public. Also, much scientific research in Toronto has been directed particularly at species of these three families. All are sensitive to the ecological degradation that has been such a frequent consequence of 'modern progress.'

Centrarchids include the sunfishes and the smallmouth and largemouth basses. Species of this family thrive in moderately large, shallow, and slow-moving waters with aquatic plants, rocks, and big woody debris. They prefer cool water, about 18°C in spring, to spawn, and warm temperatures near 30°C in summer; they become relatively inactive in winter at temperatures below about 5°C. They once thrived in pristine Toronto Bay, coastal wetlands, lower river reaches, and beaver ponds. There are no non-native centrarchids in the Toronto region.

Percids include yellow perch, walleye, and small darters. These species show less preference than centrarchids for large structural features such as water plants and boulders, and are found on or near a gravelly bottom swept by moderate currents. They prefer cooler temperatures than the centrarchids, spawning in spring and then in summer occupying waters between 20° and 25°C; they remain active in winter even in near-freezing water. In the pristine state, they were found along the Lake Ontario shore and Toronto Bay and up the larger streams to areas of rapids and riffles.

Salmonids include species of the salmonine subfamily – i.e., the 'chars' (brook trout and lake trout), the 'trouts' (the Atlantic salmon and brown trout), and the 'salmons' (rainbow trout or steelhead, chinook, and coho). The first three of these seven species are native here. The last four were introduced, have become naturalized here to a minor extent, and are maintained at unnatural levels with hatchery stocking to please anglers. All these salmonines prefer temperatures below 20°C year-round, so they are to be found in summer either in the colder spring-fed streams or in the deeper offshore waters of the lake, where they are vulnerable to lamprey. In fall the spawners migrate to clean gravelly rapids or reefs, where the eggs incubate over winter. Salmonids may be quite active in winter, as ice fishers know. The smaller of these species and the juveniles of the larger species that migrate upstream tend to stay in the spring-fed streams over summer, when the larger species and older individuals of the migratory species are in deeper lake waters.

Bacteria decomposing the wastes dumped in streams used dissolved oxygen that was needed by other creatures, which then suffocated.

All our freshwater fish are cold-blooded in that they have little physiological capability to regulate their internal temperature independently from external water temperatures. But they can sense temperature and the difference between their current internal temperature and their genetically fixed preferred temperature. Their behavioural and anatomical capabilities then permit them to seek habitat with temperature close to their preference. These fish also have genetically fixed tolerance limits for low oxygen concentrations, and behavioural capabilities to avoid suffocation, if possible.

A CONCENTRIC MODEL OF PREFERRED HABITATS

In summer, the geographic, hydrographic distribution of centrarchids, percids, and salmonids in a watershed

Sea Lamprey

Lake trout with lamprey wounds and a lamprey attached, in a fishnet. (Great Lakes Fishery Commission)

Sea lamprey is a primitive species in an evolutionary sense, with cartilage instead of bone in its skeleton and no paired fins or jaws. But that does not mean it is ecologically unsuccessful. With its funnel-shaped, toothed mouth it attaches to the side of an individual of another fish species using suction. It then rasps through the skin of the fish and sucks the blood and other liquids from its prey.

The first account of the sea lamprey in the Great Lakes Basin was by the naturalist Charles Fothergill, who described one taken from Duffins Creek in 1833. The part of the Erie Canal that linked an upper tributary of the Hudson River of the Atlantic seaboard to a tributary of Lake Ontario had been completed a few years previously. The sea lamprey apparently entered the Great Lakes Basin via this canal connection. It is well-known that this species attaches to the bottoms of boats and hitch-hikes along.

Sea lamprey are often erroneously called eels. Eels evolved from fish with fins and still possess some internal vestiges of those fins. Eels have normal jaws and ingest their prey whole or in pieces that they tear from their victims. Eels migrate from our freshwaters into deep equatorial waters of the Atlantic Ocean to spawn, and the almost-transparent elvers then may take years to migrate back into the freshwaters. In the Toronto region, eels are now much less common than they were once, apparently because of the dams in the St. Lawrence Seaway system.

Adult sea lamprey run from deeper lake water up into clear, clean tributaries in spring to spawn in nests in gravelly riffles. The spawners then die. The young emerge as tiny creatures something like aquatic earthworms and burrow into soft mud bottoms. They filter the water that flows by their burrow openings for organic particles and use the digestible parts as food. Depending on how warm the water is in summer and how much food comes floating by, this larval existence may last more than ten years. When a larva, also called an ammocoete, reaches a length of 10 to 12 cm, it transforms into an adult, leaves its burrow, and migrates into deeper waters of lakes or of the ocean in coastal streams. Thereafter it acts like an external parasite or sucking predator on species of fish that it prefers. In this adult stage it may take a year or two to mobilize the anatomical and physiological resources to reproduce. To grow 100 g it may destroy 2,000 g of prey fish.

Anglers detest the sea lamprey because it is particularly destructive of some of the fish species that they prefer most. The soft-scaled species of the Salmonid family are easy prey to the sea lamprey. The lake whitefish, a salmonid, and lake trout are particularly vulnerable to lamprey attack, in part because they all prefer clear, clean water of low temperature in summer. These salmonids appear to have no behavioural defences against lamprey attack.

Fishery managers have fought a costly war against sea lamprey in Lake Ontario since the 1970s, and in the upper Great Lakes since the 1950s. Streams in which larval sea lamprey are common are treated with relatively selective toxins or lampricides. Such chemicals need to be used very carefully – in fact, it is impossible to use them without adversely affecting some other stream organisms. Certain species of caddis fly of interest to fly fishers appear to be harmed by one of the lampricides, as are the large aquatic amphibians called mud puppies. Because sea lamprey are not strong swimmers, they can be stopped from running upstream by low-head barrier dams less than 2 m in height. In the greater Toronto region three such lamprey barriers have been constructed: one several kilometres above the mouth of each of the Humber and Credit Rivers and another halfway to the source of Duffins Creek.

Sea lamprey are caught commercially in Europe, smoked, and sold in delicatessen shops. Such an enterprise was attempted with sea lamprey from Lake Huron waters years ago with the expectation that if the venture were successful, lamprey fishers would probably overexploit the sea lamprey, to the advantage of the anglers who competed with the sea lamprey. Unfortunately, it was commercially unsuccessful.

and the nearby lake is a bit like a target in archery. The bull's-eye is at and near the mouth of the stream, where the waters become warmest in summer and most suitable for the centrarchids. Farther upstream into the mid-reaches of the larger streams and farther out into the bay and along the lakeshore, the waters are cooler in summer and thus appropriate for the percids. Still farther upstream into the moraine and escarpment and offshore into the deeper parts of the lake, the cold waters in summer sustain the salmonids. Much more than just the temperature of their habitat is important to these species, of course, but water temperature is a key feature of their ecological niches. With these families in particular, all species require a high quality and adequate quantity of moving water and healthy ecosystems generally. They do not thrive where rapid siltation occurs, where much organic matter decomposes, or where mats of filamentous algae proliferate. That is why they may be selected as integrative indicators of the quality of such ecosystems.

HISTORICAL DEGRADATION AND PARTIAL RECOVERY

Some major changes in the aquatic ecosystems of the Toronto region can be illustrated by relating the concentric distributional pattern of our three families to the changes in temperature and water quality regimes due to human activities. All of the following human influences have had the effect of warming the water in summer, when temperatures play a major ecological role, especially with species of the families selected here:

- removing bushes and trees that shaded the stream from the sun, to speed the flow of water during the spring or to build a structure on the banks
- draining and ditching low-lying lands and wetlands to farm and develop the land, and incidentally to reduce infiltration and flow of waters through aquifers, which unintentionally diminishes cold flows from springs and seeps into the stream
- changing the cross-sectional dimensions of a stream from narrow and deep to broad and shallow, resulting in greater irradiation of the water because of greater surface area
- building mill dams and reservoirs, which exposed large areas of water to the sun

- reducing water flows from aquifers in summer, resulting in a smaller volume of water flowing at a reduced rate and becoming warmer because of the sun's irradiation
- releasing partially purified waste water from sewage treatment plants at temperatures above the ambient stream temperatures
- releasing cooling water from industries, mostly into Toronto Bay
- polluting the air with particulates, resulting in a heat island with regional average air temperatures eventually reaching some 2°C above that of the pristine state
- venting heat into the atmosphere and loading it with greenhouse gases, causing a continental increase of air temperature of about 1°C.

In recent years, corrective measures have been undertaken against some of these practices. Such programs may eventually lead to reduction in the intensity of warming of the water. We judge that, to date, cooling due to such rehabilitative effects is minimal.

Figures 12.1, 12.2, and 12.3 show where populations of different species of the three families of fish that we have selected as integrative indicators of ecosystem quality were found in two major surveys of streams in the Toronto region, in 1946-54 and 1984-5. Many less complete surveys were done over the years, and the data from these generally support the findings sketched below.

We would expect that the warming effects of the nine stresses identified above, when taken together, should have been most pronounced in parts of a stream basin that are fed only by minor aquifers and are most fully 'developed' toward an urban form. Thus much of the Etobicoke Creek, Don River, and Highland Creek basins, and the lower half of the Humber River basin and Toronto Bay should show warming effects. Small streams flowing from springs in the Niagara Escarpment and Oak Ridges Moraine and larger streams that gather cold waters from such spring-fed streams should show less warming effect. A similar pattern relates to water quality factors not related to temperature because the escarpment and moraine have not yet been 'developed' intensively.

What implications might we expect in terms of changes through time of the distributions of centrarchids, percids, and salmonids? Wherever water quality is poor we would

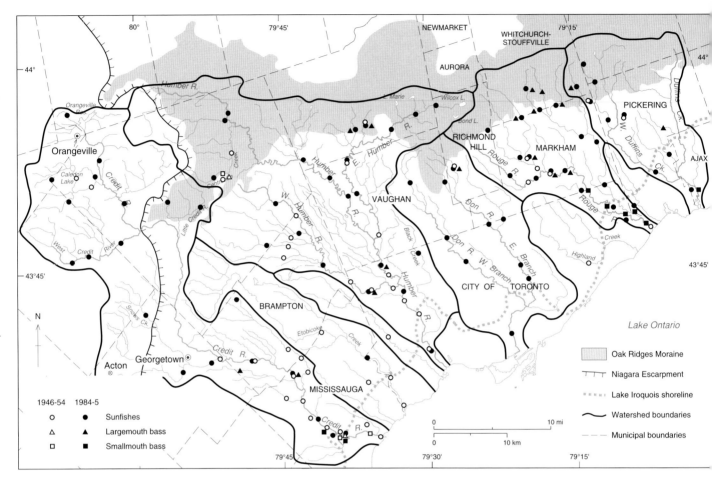

FIGURE 12.1 Comparisons of 1946-54 and 1984-5 fish surveys for bass and sunfishes (centrarchids).

expect to find few individuals of these families. Generally water quality has been most degraded toward the bull's-eye of the imaginary target mentioned above – i.e., in the waters of the lower basin and immediately adjacent lake waters. Water quality has been least degraded near the upstream sources of these stream systems and in the deeper offshore waters. This pattern should have had several consequences on the spatial distribution of our indicator species and families in summer. The greatest adverse effects should have appeared with centrarchids, like smallmouth bass, native in and near the mouths of the streams. Moderately adverse effects could be expected with percids like the darters in the mid-reaches of larger streams. Even lesser effects should have been felt by headwater salmonids like the brook trout. Thus the sampling data for 1946-54 should reflect few centrarchids in the more urbanized downstream reaches of the streams of the Toronto region.

Where water quality has always been acceptable, the distributional pattern of increases in water temperature should have led to an increase in the distributional area of warm-water centrarchids (sunfishes and basses) and extension of the limits of their ranges upstream in the basins and along shore in the lake. The distributional band of the cool-water percid darters should have been shifted outward from a centre of downtown Toronto. The distribution in summer of the salmonids like brook trout should have shrunk into the spring-fed tributaries.

The waters that were most degraded in 1948-1954 have become partially rehabilitated. These waters are in the lower reaches of the Don and Humber Rivers and

Fish

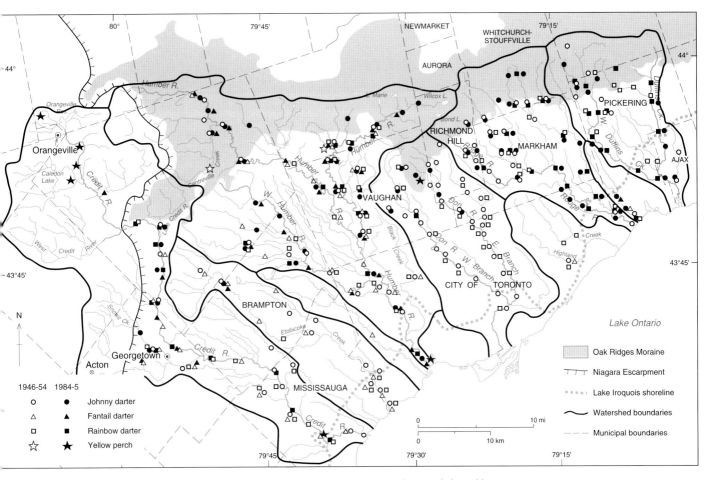

FIGURE 12.2 Comparisons of 1946-54 and 1984-5 fish surveys for darter and yellow perch (percids).

Petticoat Creek. Here some centrarchids should have reappeared.

Urban development between the earlier and later dates was not managed well with respect to adverse effects on mid-reaches of the larger streams. In these areas the percid darters should have become less prominent.

To determine the accuracy of these predictions, we can compare them to the data in Figures 12.1, 12.2, and 12.3. Each of these maps shows the boundaries of the traditional urbanizing municipalities in the Toronto region. For the earlier period, 1948-54, relatively small urban areas existed toward the centre of each of the boundaries. The boundaries as shown are more relevant to the later period, 1984-5.

Comparison of the data in the maps with the hypothetical expectations sketched above shows that the data generally support the expectations. Note that in 1948-54 few centrarchids and percids were found in the headwaters of the Credit River that lie above the Niagara Escarpment at the northwest end of the Toronto region. By 1984-5 species populations of these two families were found in numerous locales above the escarpment. Presumably they were transported there by humans, perhaps through release of bait fish by anglers.

Note also that the non-native salmonid rainbow trout and brown trout occur in more locales of the larger streams in the later period. Some of the individuals of these species taken in the surveys may have been from hatchery plantings. These species are somewhat more tolerant of human effects than is the native brook trout, and may displace the latter from its marginal habitats.

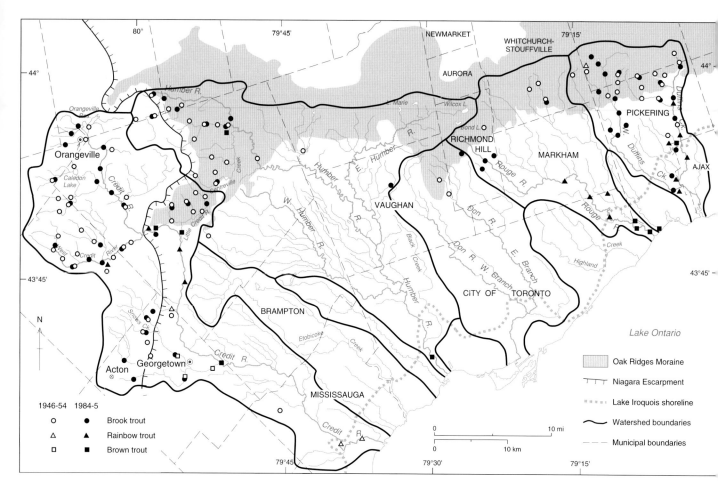

FIGURE 12.3 Comparisons of 1946-54 and 1984-5 fish surveys for trout families (salmonids).

CONCLUSION

Much rehabilitative work has occurred in the Toronto region streams since the last area-wide survey in 1984-5. This work has been directed more toward chemical water quality than to water temperature, water flows, and structural habitat features. We might expect, as a consequence, that salmonids, percids, and centrarchids may all have become somewhat more prominent members of numerous local fish associations.

Species and families of aquatic insects have long been used in the Toronto region for indicator purposes similar to those for which we have used fish. Molluscs, crayfish, and aquatic plants can also be used in this way. Perhaps fewer data are available for these other forms of aquatic life, but this shortage could be remedied for the future. An informed naturalist can accurately diagnose many of the ills of aquatic ecosystems from a consideration of the presence and abundance of a selection of different species, families, and orders. Over time, these organisms necessarily integrate some adverse effects of whatever humans are doing in a watershed upstream to the sampling locale.

Water runs downhill, above and below ground level. Water is a

The organisms that live in flowing waters are subjected to the consequences of almost everything that happens upstream in the basin.

special kind of chemical and physical substance. It carries with it some of the features of almost anything it encounters in its flowings. So the organisms that live in flowing waters are subjected to the consequences of almost everything that happens upstream in the basin.

The adaptive capabilities of individuals are often limited, especially for species that have evolved a strong sense of place. Mostly they adapt through time by changing their geographical hydrographical distributions, or disappearing locally.

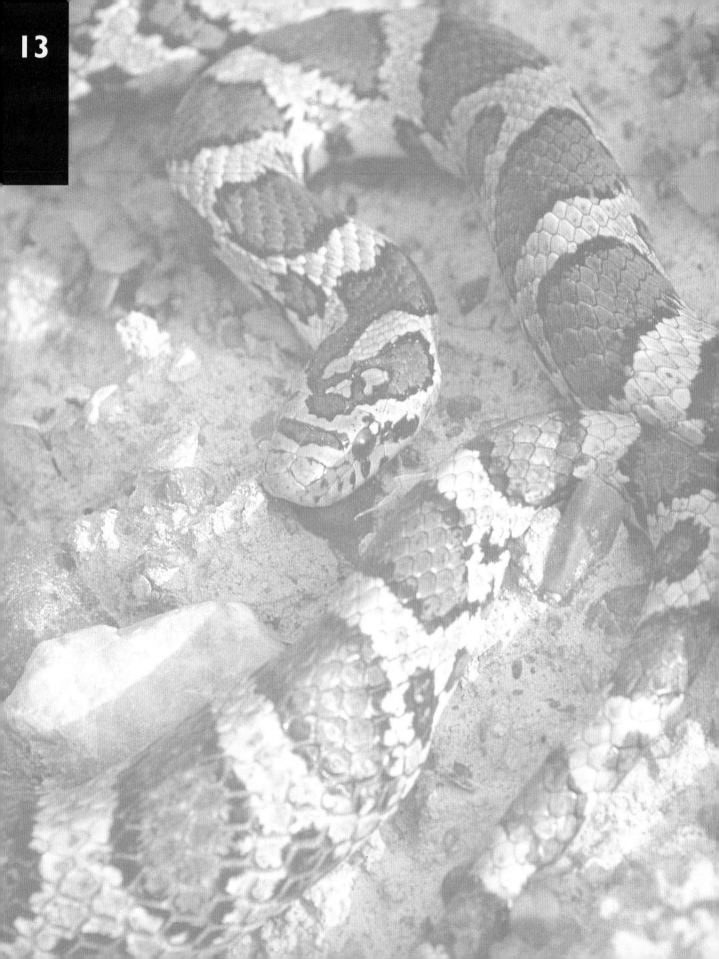

Amphibians and Reptiles

Robert Johnson

The first and last complete faunal survey of the greater Toronto region was published in the Canadian Institute's *The Natural History of the Toronto Region* (Faull 1913). Since that time only one publication has described the distribution of amphibians and reptiles in Toronto (Johnson 1982). Because of this lack of data, amphibians and reptiles were rarely considered in determining environmentally sensitive area (ESA) designations. This is unfortunate because the presence of amphibians and some reptile species often indicates wetland type and the availability of groundwater. For example, some species require permanent water deep enough to avoid freezing; others breed in vernal pools without aquatic predators such as fish; and some reptiles and amphibians overwinter in woodland leaf litter. The number of amphibian species may change as a result of larger landscape changes in groundwater supplies before other species of animals less sensitive to moisture changes respond. Changes in species may indicate changes in the water resources that support wetlands. A high number of amphibian and reptile species at one location would reinforce ESA designations or flag areas that do not strictly qualify as ESAs (Eagles and Adindu 1978).

This chapter outlines past and current distribution of amphibians and reptiles in the greater Toronto region. It suggests what changes have altered ecosystems and had an impact on amphibians and reptiles in the region. As well, it describes the species currently found in this urban area, identifies significant habitats, and discusses factors affecting the sustainability of amphibian and reptile populations.

THE URBAN LANDSCAPE IN 1913

Growth of the City of Toronto was guided by local geological conditions, with the Lake Iroquois shoreline escarpment initially deflecting settlement along the Lake Ontario shoreline. Toronto was laid out in rectangular form, superimposed over the river systems that still dissect the area. As Keys writes in the 1913 Canadian Institute volume, 'Neither hill nor dale, creek nor river, bluff nor ravine has been allowed to deflect the monotonous straight lines of [Toronto's] streets' (Keys 1913).

The 1913 publication provided a number of options for people who wanted to see local amphibian and reptile species, recommending excursions to various habitats that could support such animals. These included Bear Swamp, occupying 120 to 160 ha stretching 5 km from Lake Ontario on Highland Creek; numerous bogs along the Oak Ridges Moraine; a wide strip of woodland and several small ponds near the beach at Lorne Park; and Milne's Swamp, about 2 km southwest of Markham Village. In addition, sphagnum bogs were located at Highland Creek, west of High Park, and until 1905 there was a bog at Swansea (Humber Bay) with pitcher plants and sundew (Graham 1913). The damp woods and ravines that supported ferns and equisetums (Ivey 1913) provided microhabitats in which amphibians could seek refuge. Several large wetlands survived behind barrier beaches of sand and gravel along the Lake Ontario shoreline. As wetlands were drained, filled, or used to dispose of garbage and sewage, these large wetlands, meadows, woodlots, and river valleys provided source populations for increasingly isolated wetland habitats.

Numerous plant species – including *Potamogeton, Sagittaria, Cyperus, Carex, Calla, Pontederia,* and *Symplocarpu* – were, and still are, indicators of a variety of wetland areas. These species could be found at Ashbridge's Bay, Howard Lake, Grenadier Pond, and the Toronto Islands (Scott 1913). Ashbridge's Bay and the Don River delta were cut up into a continuous strand of ponds, weedy lagoons, large bogs, and dense cattail stands (Barnett 1971).

Early photographs of Toronto provide insights into what the urban landscape was like. Although most of the upland forests had been felled, many streams supported tangles of vegetation along their course. Similarly, roadside verges were overgrown. This vegetation provided cover for reptiles and amphibians to move from the river valleys into and across the cityscape. The gradual loss of continuity in undeveloped areas of the tablelands, as well as the intolerance of city residents, many carrying old-world fears and mistaken ideas about amphibians and reptiles, led to the isolation of reptiles and amphibians in patches of remaining habitats.

If the number of amphibians and reptiles found within the city was declining, there were still areas close to the city, like Ashbridge's Bay, with its 'large marshy area,' that supported a variety of species (Williams 1913). Thus, in 1913, Williams counted 11 reptiles and Piersol found 16 amphibians in the vicinity of Toronto. The lists that follow use the original species spellings with modern equivalents in parentheses. Piersol's list included the mudpuppy, spotted salamander, salamander (now known to be two species, the Jefferson salamander and the blue-spotted salamander), black newt (dusky salamander), red-backed salamander, salamander (four-toed salamander), newt (red-spotted newt), common toad, swamp tree frog (striped chorus frog), common tree frog (gray treefrog), spring peeper, leopard frog, pickerel frog, wood frog, green frog, and bullfrog. In addition, Piersol suggested that cricket frog, northern frog (mink frog), and Cambridge frog might occur near Toronto. Williams' list of reptiles included nine snakes – the red-bellied (redbelly) snake, DeKay's snake (brown snake), riband snake (ribbon snake), garter snake, water snake, grass snake (smooth green snake), ring-necked (ringneck) snake, milk snake, hog-nosed (hognose) snake (or blowing adder) – and two turtles, the snapping turtle and mud or painted turtle.

DISTRIBUTION OF AMPHIBIANS AND REPTILES IN 1913

Piersol and Williams provide insights on the distribution of amphibians and reptiles respectively. Mudpuppies were common in the Don and Humber Rivers and in the Toronto Island lagoons. The spotted salamander was once found 'on all sides of the city in woods.' The Jefferson salamander, however, was not common except in woods to the east of Toronto and on Scarborough Heights. The black newt, *Desmoganthus nigra,* was known from a single specimen in the Biological Museum. The specimen was collected in Toronto but no data were available from any other locality. It is possible that the species was a dusky salamander, *Desmognathus fuscus*. The dusky salamander is found in the habitat described by Piersol for the black newt, 'usually in or near clear, cold springs or brooks in rocky localities.' Today the Niagara Gorge is the closest location to Toronto where the dusky salamander can be found. Although Piersol described the eastern red-backed salamander as appearing 'on all sides of the city,' he noted that the four-toed salamander was known from a single specimen in the Humber valley. This report is consistent with records of a remnant bog near the Humber. The red-spotted newt was found on 'all sides of the city in pools, bays of larger bodies of water (e.g., Howard Lake and Island Lagoons), and in the quieter portions of streams where there is considerable vegetation growing in the water' (Piersol 1913) – just where it is found now.

The natural history of nine species of toads and frogs included the common toad, which was, as the name suggests, common, and was found in woods, along hedges and fences, and in gardens, as it is today. Striped chorus frogs were found in low, marshy land and the Humber and Don River valleys. The gray treefrog was 'found on all sides of the city,' as were the wood, green, and leopard frogs. The spring peeper was found in woods only in the west of the city. The area west of the city was also the only location for the pickerel frog, which frequented marshes and brooks. The bullfrog was found only 'in the marshy margins of bays, rivers or large ponds.'

The habitat of snakes was diverse. The redbelly snake was considered not very common (Williams 1913), although this is most likely a comment on its secretive nature rather than on its true abundance, especially in woodlots. The brown snake was often found on wasteland at the side of railroads. The ribbon snake was found at Woodbine and Balmy Beaches. The garter snake was found in woods and grassy fields and was considered to be the most common of the 'large' snakes. Smooth green and milk snakes were fairly common, as was the water snake, which was found near streams and ponds. The ringneck snake was considered rather rare. So was the hognose snake, or blowing adder, once fairly common in High Park, where it undoubtedly fed on the large toad populations that the ponds and sandy soils support. The nine species of snake are all quite harmless.

The two species of turtle were common. Snapping turtles were especially common at the Toronto Islands, where they are still found today.

IMPACT OF URBANIZATION ON AMPHIBIAN AND REPTILE POPULATIONS

Amphibians and reptiles depend on specific habitats for critical periods of their life history. Breeding sites, overwintering sites, and egg-laying sites may limit the distribution of amphibians and reptiles. Many species rely on past success and site fidelity, and return to specific sites for overwintering or egg laying. Ponds, especially ephemeral ones, swamps, wet meadows, and moist woodlots with connections to upland or open summer habitat are threatened habitats in urban ecosystems. Although many of these areas were previously considered to be waste or of low value, they are increasingly likely to be used for recreation or modified to support development. Disruptions in these crucial habitats may have major impacts on the amphibian and reptile fauna, often before impacts on other species are recognized. For example, biotoxin accumulation or changes in hydrology that affect survivorship in moisture-dependent species may signal slow deterioration or disruption of local ecosystems before this impact is identified in other species. Even more dramatic may be the loss of whole populations if hibernation sites are destroyed or altered while occupied.

Naturally occurring local extirpations are compounded by barriers to dispersal, such as motorways, channelized streams, or fences. Water-land and meadow-woodlot interfaces may be altered by mowing, loss of shrub layers, and herbicide or pesticide use. The invasion of shrubs onto wet meadows may result from natural succession or induced changes in hydrology. These changes can affect habitat structure, microclimate, thermoregulation behaviours, and prey base.

Superimposed on these specific habitat needs are longer term or annual trends in climatic variables, such as early or late frosts or freezing temperatures that affect adult survivorship and that of developing eggs in spring ponds or fall turtle nests. Droughts can reduce the number of young of amphibian urban populations already affected by the cumulative impacts of low-level stressors. Local extirpations may occur more frequently when migrants from source populations cannot recolonize or survive in isolated or marginal habitats. Thus, the importance of identifying critical habitats, source populations, and dispersal routes to ensure the survival of existing populations of amphibians and reptiles.

Wetlands and areas prone to flooding were the last habitats to be drained and filled. As the size of wetlands was reduced and hydrological systems altered, the dynamic changes that sustained ecosystem diversity and succession were buffered or eliminated. We can only speculate on any positive aspects of the many millponds that dotted the river valleys to provide the energy for milling lumber and flour. Nevertheless, the loss of forest cover and subsequent shift to an agricultural landbase reduced the diversity of habitats and their continuity along the Lake Ontario shoreline. Some predator species, such as raccoon, found favourable habitats in the changing landscape. Along with predatory fish, often introduced to wetlands and streams, they had a negative impact on amphibians and reptiles, especially in isolated habitats.

Green frogs, toads, northern brown snakes, and garter snakes may have expanded their ranges. Not only can they tolerate disturbed habitats but they would have benefited from the increase in permanent ponds and an increased prey base in the agricultural landscape. Tailed amphibians, however, declined as the forests were cleared and as drainage lowered water tables from urban woodlots. Snakes, especially if aggressive or brightly coloured, were persecuted by those who feared them, as were snapping turtles, also harvested for human food. This persecution continues to the present day.

The immediate impact of urbanization was the increasing impairment of hydrological functions. Ditching and draining gradually reduced the number of wetlands, and wet prairies disappeared as ditching

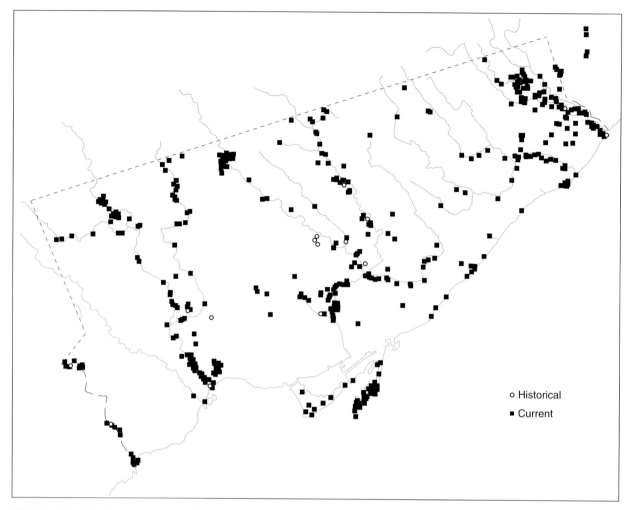

FIGURE 13.1 Distribution of reptiles in Toronto, 1982-97.

lowered water tables. The vernal woodland pools and ephemeral wetlands that supported a diverse community of breeding amphibians no longer dotted the spring landscape.

Although spring flooding was still a concern in river valleys, most of the bottomland wetlands were drained or isolated from river courses. The urban landscape became more homogeneous, and the course and pattern of hydrological change became more predictable. As wetlands aged, and with increased sedimentation in the stabilized wetland ecosystems, the diversity of wetland types and ages was reduced.

Eight of the 27 species of reptiles and amphibians catalogued in 1913 are no longer found in the greater Toronto region.

Changes in water quality have either an immediate or a cumulative impact on aquatic organisms. Although early-twentieth-century observers had no record of the effects of water pollution on amphibians and reptiles, they identified it as the reason for brook trout declines (Nash 1913). Certainly lack of alternatives for waste disposal or effluent treatments led to the use of wetlands for this purpose – affecting all wetland species.

Given all of these factors, it is perhaps no surprise that eight of the species catalogued in 1913 have been extirpated from the greater Toronto region: the dusky salamander, the four-toed

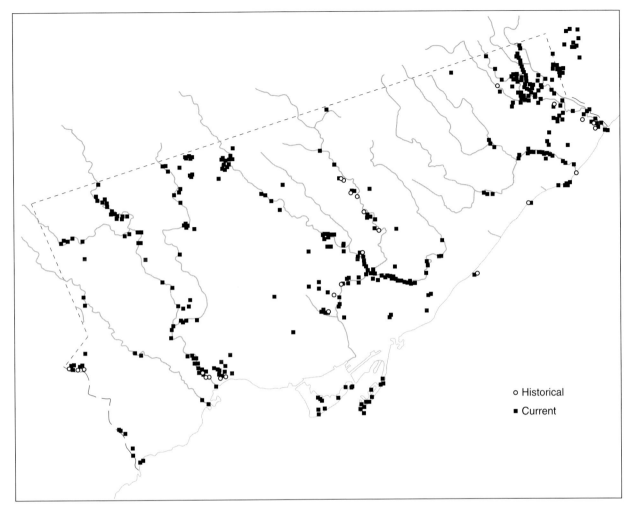

FIGURE 13.2 Distribution of amphibians in Toronto, 1982-97.

salamander, the bullfrog, pickerel frog, spring peeper, ribbon snake, ringneck snake, and hognose snake. Species once more widespread but now rare include the spotted salamander, blue-spotted salamander, red-spotted newt, northern water snake, and smooth green snake.

DISTRIBUTION OF AMPHIBIANS AND REPTILES 1982-97

Of 45 species of reptiles and amphibians currently found in southern Ontario (Johnson 1989), 21 (46 percent) are found in the Toronto area. Of these 21, there are 5 species of salamanders, 6 of frogs and toads, 4 of turtles, and 6 of snakes. Another 8 species – the Jefferson salamander, four-toed salamander, pickerel frog, bullfrog, spring peeper, eastern spiny softshell turtle, northern ribbon snake, and northern ringneck snake – are found in the surrounding area of the Toronto-Hamilton bioregion, which extends from the Niagara Escarpment in the west, along the Oak Ridges Moraine in the north, to Kingston in the east (Johnson 1982; Plourde et al. 1989; Royal Commission 1992; Lamond 1994). Not only do these geographical features define the boundary of the bioregion but they also support source populations of amphibians and reptiles and are reservoirs of biodiversity.

The locality data for the 11 amphibians and 10 reptiles from 1982 to 1997 in the Toronto region demonstrate the importance of valley ecosystems. Following an initial study of amphibian and reptile distribution

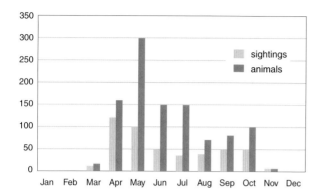

FIGURE 13.3 Total number of amphibian sightings, by month, 1982-97.

(Johnson 1982), over 15 years I searched all river systems in Metropolitan Toronto, parks, permanent wetlands, woodlots, the Lake Ontario waterfront, and open areas of fields and ephemeral ponds. I also recorded all observations submitted by members of the Toronto Field Naturalists. It is no surprise that amphibian distribution depends on wetlands or moist microhabitats. The distribution of reptiles (Figure 13.1) correlates with that of amphibians (Figure 13.2) and is, in part, indicative of the quality of the shared habitats. Figure 13.3 indicates the total number of amphibian sightings over the course of the study, for each month.

Table 13.1 shows the number of sightings of amphibians and reptiles in the Toronto bioregion from 1982 to 1997. Of the species listed some, especially if found together, may be used as indicators of habitat disturbance in an urban ecosystem. The Rouge River, for example, the least disturbed river system in the study area, has species similar to those found in non-urban and less disturbed habitats outside the region. A lower number of species and loss of more sensitive or specialized species occurs in more disturbed habitats. As well, over the period the study made consistent recordings of some species from disturbed sites, indicating their tolerance of disturbance, ability to recover from it, or 'generalist' life history (if they opportunistically eat a variety of foods).

TABLE 13.1

Number of records of native amphibians and reptiles in the Toronto bioregion, 1982-97

Amphibians			Reptiles		
Species	Number 1982-97	(Number since 1989)	Species	Number 1982-97	(Number since 1989)
American toad	260	(91)	Eastern garter snake	286	(105)
Green frog	95	(32)	Midland painted turtle	179	(74)
Northern leopard frog	90	(26)	Northern brown snake	168	(71)
Striped chorus frog	69	(28)	Snapping turtle	106	(48)
Wood frog	60	(28)	Eastern milk snake	58	(26)
Eastern redback salamander	56	(21)	Northern redbelly snake	36	(14)
Gray treefrog	25	(16)	Blanding's turtle	26	(14)
Red-spotted newt	10	(2)	Map turtle	16	(9)
Spotted salamander	9	(3)	Smooth green snake	11	(1)
Mudpuppy	4	(1)	Northern water snake	5	(1)
Blue-spotted salamander	3	(1)			
Unconfirmed			*Unconfirmed*		
Spring peeper			Northern ribbon snake		
Pickerel frog			Northern ringneck snake		
Bullfrog			Musk turtle		
			Eastern spiny softshell turtle		
			Wood turtle		

Note: A record refers to site locality. In many cases more than one specimen may have been recorded at a single site.

On the basis of these criteria the following species can be considered tolerant of altered urban ecosystems: the American toad, green frog (if permanent water is available), striped chorus frog, eastern garter snake, northern brown snake, midland painted turtle, and snapping turtle. Other species may indicate more pristine sites or remnant ecosystems in urban environments: the eastern redback salamander, spotted salamander, northern leopard frog, gray treefrog, wood frog, northern redbelly snake, and eastern milk snake.

Rare species in the Toronto urban ecosystem, or species that depend on rare or declining habitats in the area, include the mudpuppy, blue-spotted salamander, spotted salamander, red-spotted newt, water snake, smooth green snake, Blanding's turtle, and map turtle. One introduced species, the red-eared slider, has been recorded on 17 occasions. This released pet turtle, native to the central and southern United States, overwinters in 5 ponds in Toronto.

Amphibians

Tailed Amphibians

Eastern redback salamander (*Plethodon cinereus*)

The eastern redback salamander (Figure 13.4) is the most common tailed amphibian in Toronto. It is found under logs and rocks on moist soils along the wooded slopes of all major river valley systems and wooded ravines within the city. There are two colour morphs: the typical red-backed or red-striped morph, and a black or grey variety. Perched water tables and moist soils may result in some surprising extensions onto tablelands and may account for the presence of this species in High Park. This is the only species of salamander in the Toronto area that has no lungs; all gas exchange occurs across the skin. They lay their eggs within cavities in moist logs. Larval development occurs within the egg, from which hatches a fully formed salamander. Although independent of free-standing water, they may be extirpated from areas where improved drainage for development lowers water tables to the extent that surface moisture is unavailable to the salamander or the logs they inhabit.

Red-spotted newt (*Notophthalmus v. viridescens*)

The adult red-spotted newt lives and breeds in ponds and slow-moving backwaters of streams. Both habitats are utilized in Toronto, although most ponds are threatened by landfill for industrial development or by deteriorating water quality. Although there is a brightly coloured terrestrial 'eft' stage in its life cycle, no efts have been encountered.

Spotted salamander (*Ambystoma maculatum*)

Along with the blue-spotted salamander, the spotted salamander is one of the rarest and most exciting finds in the Toronto area. The only recent sightings of this species were made by school groups in the Humber marshes and West Pond (west of High Park) areas in the City of Toronto. The presence of this species, usually associated with rich woodlands, indicates the quality of these isolated habitat patches. In Scarborough, extirpations have occurred as a result of pond contamination and the filling of wetlands in an industrial area, with modification of the surrounding habitat.

Spotted salamanders are encountered only in early spring or late fall under logs or rocks. They spend the summer underground where it is cool. For this reason it and the blue-spotted salamander are often called mole salamanders. The conspicuous globular egg masses of the spotted salamander may be used to determine presence in early April after the salamanders have left breeding ponds.

Mudpuppy (*Necturus maculosus*)

The mudpuppy, also known as the water dog, is the largest tailed amphibian in Ontario; it can grow up to 25 or 30 cm. The mudpuppy (Figure 13.5) is often accidentally caught by anglers using worms or minnows as bait and usually through winter ice. With permanent gills, this species is totally aquatic and restricted to permanent bodies of water. Because it is aquatic and

FIGURE 13.4 Eastern redback salamander. (Robert Johnson)

FIGURE 13.5 Mud puppy. (Robert Johnson)

FIGURE 13.6 American toad. (Robert Johnson)

secretive, surveying for this bizarre-looking creature is difficult, and all sightings are from larger river systems. There appears to be no reason why this species is no longer found at the Toronto Islands, where it was recorded as being abundant in 1913.

Blue-spotted salamander (*Ambystoma laterale*)

The Rouge River valley may be the only remaining area where this species can be found. The only sighting of a blue-spotted salamander since 1989 was by a school group on a nature hike, who found and photographed the specimen. The Rouge River area represents some of the best woodlands in Toronto, with many vernal ponds in which the salamander may breed.

Frogs and Toads

American toad (*Bufo americanus*)

The American toad (Figure 13.6) is the most common amphibian found in Toronto. Its success is in part a result of its preference for shallow and temporary bodies of water for breeding and its ability to hibernate underground. It is well adapted to the disturbed habitats of urban ecosystems. Any area that floods temporarily with spring melt water and rainwater may be utilized for breeding. Although toads usually return to their birth ponds to breed, there are always newly mature toads to exploit newly formed ponds. Nowhere is this adaptability more evident than in their use of freshwater 'pools' formed on top of in-ground swimming pool covers after the snowmelt and spring rains accumulate. These 'ponds' warm faster than natural groundwater ponds, and toads readily lay their eggs in these locations. This is also indicative of the likelihood of successful colonization of backyard ponds constructed for wildlife.

Many recent subdivisions have begun to mature, and backyard gardens provide summer foraging opportunities and overwintering sites for the toad. In one Scarborough community school children have documented the last remaining pond that supports the local toad population and have become crusading pond guardians. The popularity of backyard ponds provides the breeding habitat that will allow this species to persist in even the most urbanized habitats. In areas where natural ponds remain, road mortalities of both adults moving to and away from breeding ponds and juveniles leaving their birth pond, are very high. There are opportunities to document these road crossings and to try to protect the toads with signs. In other urban areas, mortalities have been reduced by closing roads during the breeding season or constructing tunnels under roads.

One of the largest populations of toads occurs on the Toronto Islands, where residents have become fond of their backyard amphibian. The ideal conditions on these islands, and the consequent longevity of the toads found there, may be responsible for the large body mass of toads in this location.

Toads migrate across roads from winter hibernation areas the first week of April. Eggs are laid the first two weeks of May and the first toadlets metamorphose at the end of June. They bury themselves underground for the winter in early October, when the soil temperature is 6°C.

Green frog (*Rana clamitans*)

This typical pond amphibian, which requires – and is an indicator of – permanent ponds as a tadpole, typically takes two years to develop. If the pond is shallow and dries or freezes to the bottom, the tadpoles will die. Winterkill also has an impact on adult frogs that overwinter on the pond bottom. The green frog is widely distributed in the typical eutrophic marshlands that persist in urban areas, and its distribution reflects closely the few remaining large ponds, marshes, river backwaters, and oxbows. Without disturbance to inundate invasive plants, or lacking water level fluctuations to provide germination beds for a diversity of wetlands plants, many wetlands become shallow, cattail-choked marshes due to sedimentation or succession. Although these marshes become too shallow for green frogs, the American toad thrives under such conditions.

Newly metamorphosed green frogs will travel some distance from their birth ponds to small ponds, streams, and backyard ponds. Although mortality of this age class is high, overland dispersal results in the colonization of new ponds and stream habitats. This species is most commonly confused with the bullfrog, accounting for the occasional report of bullfrogs in the greater Toronto region.

Northern leopard frog (*Rana pipiens*)

Leopard frogs are found in the larger ponds or stream backwaters and riverfront marshes that support open meadows. The natural succession from wet meadow to a shrub/tree community reduces available foraging habitat for these frogs. In urban areas, lowered water tables may result in the loss of wet meadows. Wet meadows are also the easiest habitats for well-meaning homeowners to eliminate. One can fill in or channelize a wet area at the back of a garden without realizing the impact that this 'improvement' may have on the meadow community.

Leopard frogs first call the second and third week of April, and froglets appear the first week of August. They begin to enter ponds to hibernate the second week of October, when the air temperature is 6°C.

Striped chorus frog (*Pseudacris t. triseriata*)

This frog is found in larger marshes and wetlands or flooded ditches within the vicinity of wetlands. It is tolerant of more open or lightly wooded sites that form an ecotone between older woodlots and young field or meadow communities. It overwinters on land, often in old fields or adjacent woodlots. Chorus frogs begin calling in the first two weeks of April. For reasons unknown, and from anecdotal accounts, it appears that the places in the Toronto region that once supported spring peepers now support chorus frogs.

Wood frog (*Rana sylvatica*)

A good indicator of the few remaining flooded woodlands or swamps in Toronto, the wood frog hibernates on land in woodlots. Again the Rouge River valley and adjacent tablelands are the stronghold habitats for this species. Wood frogs call in the first week of April and lay their eggs shortly thereafter.

Gray treefrog (*Hyla versicolor*)

Relict populations of gray treefrogs remain in those areas characterized by ponds fringed with shrubs or willow and nearby woodlots for overwintering. The Rouge valley is a stronghold for this species. Although once found in valley wetlands of the Don River, sedimentation and invasion by a monoculture of cattail resulted in the extirpation of gray treefrogs from this habitat. Gray treefrogs begin calling the second week of May.

The American toad is the most common amphibian found in Toronto.

Unconfirmed Populations

There are three species of amphibians not recorded in Toronto but for which there are anecdotal records and nearby populations.

Spring peeper (*Pseudacris crucifer*)

Although there were unconfirmed reports of the spring peeper during the time frame of the study, it is considered to be extirpated from the Toronto urban area. Its call is often confused with the early spring call of the chorus frog. The spring peeper is found within colonization distance of the greater Toronto region in the Oak Ridges Moraine wetlands in the north and in wetlands east and west of the city. Although flooded woodlands – the preferred habitat of the spring peeper – remain in Toronto, the spring peeper is no longer to be found there. In many areas, observers suggest that spring peepers have been replaced by the chorus frog in habitats from which it was formerly reported.

Pickerel frog (*Rana palustris*)

There are three unsubstantiated reports of this species from habitats that might sustain it. Anecdotal records exist for Etobicoke Creek, a small tributary stream and adjacent pond of the Rouge River, and the north Ajax area.

Bullfrog (*Rana catesbeiana*)

Much like the spring peeper, the closest existing populations for this large-pond and lake species are in the Oak Ridges Moraine and east and west of Toronto. The species is often mistaken for the green frog, which is similar in colour but has a lateral ridge along each side of its back.

Reptiles

Turtles

Snapping turtle (*Chelydra serpentina*)

The snapping turtle is the most abundant turtle in the Toronto area. It is found in streams, marshes, ponds, and the warm shoreline bays of Lake Ontario. It is exhilarating to see this primitive-looking creature lumbering out of the primeval swamp to climb the steep valley slopes and lay 20 to 40 eggs in a nest dug in urban gardens and lawns. One urban family has been watching snapping turtles lay their eggs in the sandy soil of their front lawn for over 15 years.

One urban family has been watching snapping turtles lay their eggs in the sandy soil of their front lawn for over 15 years.

Raccoons may claim most of these nests, but there are still enough to delight people walking in High Park, for example, who come upon a nest erupting with tiny turtles. Persecuted to this day, the snapping turtle is blamed for killing waterfowl. While it is true that snapping turtles will take young geese or ducks, more ducklings are predated by raccoons, foxes, large fish, and owls. It is often mistakenly assumed that snapping turtles have killed what they chanced upon already dead or dying. Snapping turtles are also feared because of their bite, which they will demonstrate when threatened on land. But in the water these turtles are good swimmers, and they avoid or flee from any threat. There are no records of anyone bitten by a snapping turtle in the water but there are many records of anglers bitten by large fish.

Recent landfill and wetland creation projects along the Lake Ontario shoreline will improve foraging and overwintering habitat for the snapping turtle, and warmer water will facilitate movement along the Toronto waterfront. Sadly, recent research has shown that turtles feeding on contaminated fish accumulate toxins that may affect the fertility of eggs.

Snapping turtles have been observed moving under ice at Ward's Island in February but they usually emerge from hibernation the last week of April. Eggs are laid during the first two weeks of June and hatch in September. Hibernation begins in October.

Midland painted turtle (*Chrysemys picta marginata*)

The midland painted turtle is the typical 'pond' turtle, often seen basking on logs and rocks of Toronto wetlands. Its numbers often indicate large productive wetlands or river marshes. In separate years, over 50 were seen basking in Centennial Swamp, Scarborough (now drained for housing), and 35 in old gravel quarry ponds (Amos Ponds on the Scarborough/Pickering border). Painted turtles often inhabit the same wetlands as snapping turtles.

Painted turtles first emerge from hibernation and bask on logs the last week of March and lay eggs the first two weeks of June.

Map turtle (*Graptemys geographica*)

Map turtles, like painted turtles, are considered to be 'basking' turtles most often observed perched on a rock in larger bodies of water. Over the 15 years of the study, map turtles have been documented expanding their range from the protected marshes of the Humber River along the Lake Ontario waterfront by utilizing the newly constructed and sheltered waters of lakefront landfills and marinas – first colonizing East Point Park, then the Toronto Islands, and finally the Leslie Street Spit for the first time in 1989.

Blanding's turtle (*Emydoidea blandingi*)

Although found at a number of locations in Toronto, the Blanding's turtle may have been introduced at all but the Humber marshes site. When encountered in other areas of Ontario, this large turtle is often picked up off roads by well-meaning people and returned to Toronto, only to escape or be released into nearby and often landlocked wetlands in the city. Blanding's turtles are easily recognized by the high domed shell and bright yellow neck.

Unconfirmed Species

There are in addition three species of turtles not recorded recently in Toronto but that are either within migratory distance or for which there are anecdotal records.

Musk turtle (Sternotherus odoratus)

The musk turtle is a small, secretive turtle that scrambles about the bottom of wetlands. It is often mistakenly identified as a young snapping turtle.

Eastern spiny softshell turtle (Apalone spinifera spinifera)

Although the closest current population is in Cootes Paradise, near Hamilton, the softshell turtle is also found in Lake Erie and the Thames River drainage. There have been two tantalizing reports of this species by local anglers. Both sightings were from habitats that are ideal for this species, and both had good views of this unmistakable turtle. In fact, one person caught the turtle on hook and line, also a frequent occurrence in areas where the turtle is found. We must consider the Rouge River mouth and Lynde Shores, Ajax, sightings as anecdotal.

Wood turtle (Clemmys insculpta)

There is one record for the wood turtle in the Rouge River valley. Although the terrestrial and gravelly stream habitats in the valley are ideal for this species, this record has not been substantiated. Although the sighting was by a reliable observer, it was most likely of a released or escaped captive turtle.

Snakes

Northern brown snake (Storeria d. dekayi)

The northern brown is one of the most widespread snakes in Toronto ravines and valleylands. This species prefers open areas and wooded valleylands but survives well in the altered habitats and waste lands of urban Toronto. It is often found under garden refuse, stumps, rubble, old boards, and cardboard.

The northern brown snake basks on warm roads – 21 of 32 at one location were killed by cars – and pathways near or at hibernation sites, particularly in early spring and late fall. The peak of this snake's movement from foraging to hibernation sites occurs between 6 September and 24 October.

Northern redbelly snake (Storeria o. occipitomaculata)

Although widespread in Toronto, the northern redbelly snake is secretive and is restricted to moist sites in wooded valleys and tablelands. It also forages in overgrown fields during the summer. The snake shelters under logs and rocks. As many as 25 snakes have been seen dead on roads at one location near the Rouge valley, indicating a movement from summer foraging in fields to woodland overwintering sites. Such movements in one year peaked between 16 and 23 October.

Northern water snake (Nerodia s. sipedon)

The northern water snake is rare in Toronto. All sightings were from a rocky section of the river or rivermouth marshes of the lower Rouge valley. There are populations on the outskirts of Toronto within distance of the Oak Ridges Moraine headwaters of rivers that flow though the city.

Eastern garter snake (Thamnophis s. sirtalis)

The commonest snake in Toronto, the garter snake feeds on toads and frogs and is often associated with habitats where such prey are found. It will tolerate human disturbance and often ranges from valleylands into urban wastelands, railway and hydro corridors, and backyards. Garter snakes have been observed hibernating in old wells, groundhog holes, concrete walls under bridges, wooden railway retaining walls, rock piles, and gabion baskets that line streams or bridges. Frequent reports of snakes around homes result from their using crumbling foundations of houses and garages, or cavities along in-ground swimming pools, to gain access to underground or frost-free hibernation sites. At one location, over 30 snakes were counted as they basked around a groundhog hole they used as an overwintering site.

In 1983, the garter snake was the first reptile encountered on the newly formed Leslie Street Spit landfill site, feeding on the toads and frogs that also appeared that year. The concrete rubble and rock that was used as a base for the spit provides ideal underground hibernation sites. In a simplified predator/prey relationship, garter snakes are found in abundance on the Toronto

> *In 1983, the garter snake was the first reptile encountered on the newly formed Leslie Street Spit landfill site.*

FIGURE 13.7 Eastern milk snake with its eggs. (Robert Johnson)

Islands, where they prey on a sizeable toad and leopard frog population.

Garter snakes emerge from hibernation during the first three weeks of April, breed shortly thereafter, and give birth the first week of August. They enter hibernation at the end of October.

Smooth green snake (*Opheodrys vernalis*)

A beautiful bright green snake, the smooth green snake is well camouflaged in the meadows it prefers. This snake feeds exclusively on invertebrates. Populations surviving in the many open areas of a rapidly expanding Toronto in the 1950s and 1960s may have been significantly affected by the extensive use of pesticides, particularly DDT. This snake persists in isolated localities in fields or meadows surrounding a pond.

Eastern milk snake (*Lampropeltis t. triangulum*)

A species from isolated and unconnected areas of the greater Toronto region, the milk snake (Figure 13.7) is usually indicative of relict farms and outbuildings, such as Don Valley Brickworks, Valley Halla farm, Todmorden Mills, or horse stables at York Mills and Don Mills. The young mice that this snake preys upon are abundant in these areas, as they undoubtedly were when the farms were operating. However, the milk snake does not restrict its diet to mammals; a milk snake in the Humber Marshes was observed consuming a northern brown snake in 1986.

The milk snake is often persecuted because its reddish colour is associated with venomous snakes; its habit of vibrating its tail when disturbed is reminiscent of rattlesnakes; and it tends to overwinter in the crumbling foundations of old homes or outbuildings, where it comes into contact with people.

Unconfirmed Species

There are only two snake species not recorded recently in Toronto but which may be encountered nearby or for which there are anecdotal records.

Northern ringneck snake (*Diadophis punctatus edwardsi*)

The northern ringneck is not common in the area, and there have been no recent reports after sightings along Etobicoke Creek and nearby western creeks. This species feeds on salamanders and small frogs, and frequents the habitats that its prey prefers.

Northern ribbon snake (*Thamnophis sauritus septentironalis*)

Ribbon snakes have not been reported in Toronto but populations are found associated with nearby wetlands. Although often confused with garter snakes, a close look will reveal a whitish, half-moon-shaped scale in the front of the eye. The yellow side stripes are on scale rows 3 and 4 up from the belly, whereas the garter snake has stripes on scale rows 2 and 3.

IMPLICATIONS FOR THE FUTURE OF THE REMAINING ECOSYSTEMS

If we use the data provided by Williams (1913) and Piersol (1913) to summarize important sites in the Toronto region in the second decade of the twentieth century, we can identify significant amphibian and reptile populations in High Park (Howard Lake), the Toronto Islands, Ashbridge's Bay, Scarborough Heights, the Humber valley, and the Don valley. Important amphibian and reptile habitats identified by the distribution data from 1982 to 1997 include, by number of sightings: the Rouge River valley and marshes, the Humber valley and marshes, G. Ross Lord Park, the Toronto Islands, West Pond, Taylor Creek Park, High Park, and Grenadier Pond. Wetlands within these areas are important habitats for urban biodiversity. For example, the Humber River marshes support 17 species of reptiles and amphibians.

The habitats that characterize these areas – and because of changes in urban hydrology, perhaps the most threatened amphibian and reptile habitats – are flooded woodlands (swamps), wet meadows, moist woodlands (often valley slopes), and marshes. Wetlands are important breeding and overwintering sites, as are urban woodlands. Woodlands provide summer shelter, moisture, prey, and winter hibernation sites for salamanders, frogs, and snakes. The wood frog and gray treefrog, for example, hibernate only in the surface leaf litter of woodlands. The redback salamander, which lays its eggs in moist, rotting logs, survives in High Park. It depends on a forested slope with a cool spring nearby. Its continued presence is a testament to the ability of even some sensitive species to live in a heavily used urban park surrounded by millions of people. Amazingly, two large salamander species are still found in the Rouge and Humber River valleys.

There currently exists within the Toronto bioregion good amphibian and reptile habitat. Yet the sustainability of amphibians and reptiles in the region depends on the north-south corridors, which provide connections to sizeable source populations in the hinterlands, and the availability of east-west links that facilitate movement into and across the city. The survival of amphibian and reptile populations is most uncertain as they move out of the valley systems and into the east-west linkages along rail and hydro corridors or parks and cemeteries. Recent attempts to naturalize parks and service corridors have been compromised by the perception that parks, shorelines, corridors, and woodlots should be tidy, or that human safety is jeopardized by areas undergoing natural succession. Nonetheless, trends toward re-establishing forested links between tablelands, restoring valley wetlands, and using landscape plans that protect or create wetland and upland habitats (Gosselin and Johnson 1995) along rail and hydro corridors all reduce the impact of cyclical loss of species or populations that colonize marginal (or sink) habitats. These measures also help the movement of amphibians and reptiles from our more pristine and diverse ecosystems to those valleylands that have been most affected by urbanization.

Marginal urban habitats may be sinks for valleyland populations of amphibians and reptiles. For this reason, every effort should be made to maintain existing wild places and their linkages within the bioregion. These wild places may be the only source of amphibians and reptiles to colonize tablelands and newly restored habitats.

There are two trends in urban ecosystems: ecosystem dynamics are altered when urbanization reduces the uncertainty and impact of environmental perturbations (e.g., flooding and erosion are controlled or eliminated); and succession is fixed in time and has to be managed in order to persist. There is a loss of habitat diversity and of the mosaic of successional stages: successional patches tend to be of the same age or structure. Habitats may slowly deteriorate due to the impact of and responses to stresses in the urban environment. These stresses include increased sedimentation and pollutants, lowered water tables and altered hydrological cycles, and the impact of edge disturbance and invasive species on small, isolated habitat patches.

Some species manage to adapt to the modified habitats around them. Nowhere is this more evident than the toads that breed in swimming pool covers or the garter snakes that use foundations or retaining walls of bridges or in-ground swimming pools to overwinter.

In the last 15 years there has been a shift toward habitat creation, although concerns over safety and mosquitoes still limit wetland-creation projects, particularly on land under public ownership. Habitat-creation projects should be designed within bioregional constraints and an ecological context, and should recognize local hydrological and successional changes that do not require continued inputs of energy to maintain.

Other than human persecution, loss or marginalization of valleyland habitat is the greatest threat to the sustainability of healthy amphibian and reptile populations.

One of the benefits of a recent increase in beaver populations in the Toronto bioregion and across Ontario has been the number of new wetlands it has created. Once beavers consume their food supply they move on, leaving behind wetlands that undergo succession. Thus, a mosaic of wetlands of different ages and structures are available for many species of wildlife. Beavers are therefore today a keystone species and one of the few components of wetland ecosystem creation that we do not control. Beaver ponds sometimes swell to flood roads, but techniques are available to maintain water levels that allow beavers to live without creating flooding problems. Other than this legitimate concern, beaver ponds in cold-water streams where there are fisheries sometimes bother anglers.

Other than human persecution, loss or marginalization of valleyland habitat is the greatest threat to the sustainability of healthy amphibian and reptile populations. The existing river valleys in the greater

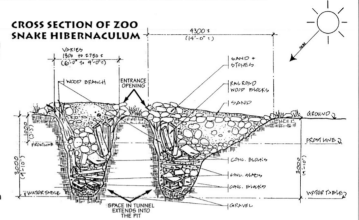

FIGURE 13.8 Snake hibernaculum. (Toronto Zoo)

Toronto region serve as models of which habitats can be sustained in the urban context. Thus the Rouge River valley and associated wetlands might be the model for restoring the Don River, with a realistic expectation of the species assemblage that might be sustained there.

Recent habitat modification projects may facilitate the movement of amphibians and reptiles onto tablelands or between watersheds. For example, habitat-creation projects developed and demonstrated at the Toronto Zoo include the design and guidelines for creating underground snake hibernacula (Figure 13.8), turtle basking logs, turtle nesting beaches and platforms, and amphibian breeding ponds (Gosselin and Johnson 1995). The Zoo also constructed a snake hibernaculum for Metro Works as an alternative to an adjacent underground maintenance chamber that many brown and garter snakes were using to overwinter.

The Zoo also provides information on living with snakes around homes and on the danger that released pet turtles and frogs pose to native wildlife. Zoo materials are available to explain why turtles are on roads, why snapping turtles should not be feared, why frogs and newts should not be purchased for backyard ponds, and why native wildlife should not be transported to stock created habitats.

These examples of habitat creation projects are site specific, and similar efforts elsewhere may not always have the desired results. They do not replace natural areas that are part of sustainable ecosystems, and we do not always understand the importance of crucial habitat in the context of landscapes. The discovery by outdoor education students of a snapping turtle digging into a mud-bottomed spring can teach us a lesson here. In the spring the turtle emerged from the mud – with five others. To survive the winter these huge beasts were totally dependent on a single muddy spring located over 50 m from the Rouge River site where they spent the summer. Without the accidental discovery of the turtle about to hibernate, no one would have selected the stream as an important area for a sizable portion of the Rouge River snapping turtle population. It is unlikely that the site would have been considered worthy of protection in the face of development.

This simple event points out the importance of landscape conservation. All habitats function as part of an interconnected landscape, permitting the creation and succession of a mosaic of habitats, the importance of which we may not fully understand. We still have much to learn from the wildlife that surrounds us. The habitats we create may never function in the same way as those we mimic, and are best used as adjuncts to natural systems rather than as replacements.

The City of Toronto recognizes the importance of diversity in its sustainability and contribution to continued economic development. This diversity extends not only to its cultural mosaic but also to its landscape mosaic. The city has become more and more aware of the value of ecological functions and biodiversity – if not for biological reasons, then certainly in terms of impact of altered ecosystems on economic development, quality of life, and cost of clean air, water, and soil. The presence of amphibians and reptiles in the greater Toronto region indicates habitats supporting a diverse collection of plants and animals. Amphibian and reptile populations, particularly in combination with other types of plants and animals, may be good indicators of ecosystem dynamics and the impacts of changes in UV light, acid rain, and chemical contaminants. The challenge for the future is to maintain both the quantity of urban habitat and air and water quality and connectivity of habitats.

Most conservation programs have focused on identifying and in some cases protecting habitats with rare species. Unfortunately, rare urban habitats such as flooded woodlands, wetlands, and wet meadows that do not support rare species, but that have significant numbers of locally common or widely distributed species, are frequently overlooked. However, the sustainability of a great blue heronry on the borders of the city, for example, is based on the scatter and biomass of frogs that inhabit a mosaic of small wetlands. As these wetlands are lost the flight distance for these birds increases, so that increased amounts of energy are expended to feed young in the nest. No one wetland may be noteworthy. But in the context of the whole landscape and successional patterns, the relationship of these wetlands to one another and their upland habitats increases biodiversity and sustainability of the landscape as a whole.

More important, the opportunity to encounter and tolerate such diverse life forms contributes to the diversity of urban experience. When Faull edited the Canadian Institute's volume on the natural history of Toronto, there was opportunity to experience wildlife in our backyard. Diversity of experience was but a carriage ride away. This remains our goal today: a snapping turtle laying eggs, a spotted salamander under a log, and the blurting trill of a gray treefrog just a bus ride away.

14

Mammals

Jenna M. Dunlop and M. Brock Fenton

Mammals are excellent indicators of the impact of human changes on the environment. The sensitivity of mammals to human impact is reflected by the changes in species composition in the Toronto region in the last 100 years. While some species are extremely tolerant of human-wrought changes to habitat, others are much more intolerant of even minor habitat disruption. Raccoons are good examples of the first category: they are opportunistic and quick to exploit feeding and housing situations presented by human habitation. In contrast, water shrews quickly disappear in the face of habitat disruption.

Mammals have an enormous impact on terrestrial environments, even when the impact that humans have on their surroundings is excluded. Mammals' quest for food and shelter affects people. Squirrels and skunks, as well as raccoons, are often uninvited housemates, while these and other mammals such as house mice and rats commonly exploit garbage and other food sources. Although the composition of mammal faunas in the Toronto region has changed, most species of mammals are relatively invisible to people, making it difficult for us to assess their presence and population levels, both in the past and in the present.

The fossil record shows that Toronto's tradition of being a place where mammals from elsewhere gather goes back a long time (Harington 1978). Each year, fossil remains of previous mammalian inhabitants of the Toronto region erode out of the Scarborough Bluffs or are found during excavations. Fossil deposits from as early as 40,000 years ago illustrate that the mammals of the Toronto region included species that had originated in Asia (the gigantic elephant-like mammoth and mastodon) along with others that had came from the south (a species of extinct giant beaver). In the next 35,000 years, the fossil record for this area includes muskoxen-like creatures apparently specialized for more arctic conditions, and large, ferocious species such as grizzly bears that now live farther west in North America.

By 5,000 years ago, a visitor would have recognized many of the mammals of the Toronto region, including some species that today live in surrounding regions that are less densely populated by humans. At that time, there were muskrats in the rivers as there are today, and although the giant beavers are gone, the smaller Canadian beavers remain, as do the grey squirrels, deer mice, and meadow voles. The fossils suggest that 5,000 years ago it was somewhat warmer than today: Toronto's list of mammals then included both grey foxes and pine voles (Harington 1978), species that today occur only in warmer areas to the south (Peterson 1966).

WHAT ARE MAMMALS?

There are approximately 5,000 species of modern mammals living in the world, and the vast majority share several signal characteristics. Mammals give birth to live young and feed their young milk that is produced by mammary glands, which are specialized sweat glands. Mammals tend to be warm blooded, and most of them have a covering of fur. Most mammals have teeth specialized to perform different functions such as piercing, cutting, crushing, or grinding.

Mammals are classified into different orders, families, and genera. A list of the mammals known from the Toronto region over time (Table 14.1) also shows their position in this classification.

TABLE 14.1

Mammals known from the greater Toronto region, 1800 to the present

	Species	Latin name	Recording	Introduced species
Marsupial	Virginia opossum	*Didelphis virginiana*	First recorded 1950	N
Insectivores	Hairy tailed mole	*Parascalops breweri*	Present	N
	Star-nosed mole	*Condylura cristata*	Present	N
	Big short-tailed shrew	*Blarina brevicauda*	Present	N
	Common shrew	*Sorex cinereus*	Present	N
	Smoky shrew	*Sorex fumeus*	Present	N
		Sorex hoyi	Last recorded 1960	N
	Water shrew	*Sorex palustris*	Present	N
Bats	Big brown bat*	*Eptesicus fuscus*	Present	N
	Silver-haired bat	*Lasionycteris noctivagans*	Present	N
	Red bat	*Lasiurus borealis*	Present	N
	Hoary bat*	*Lasiurus cinereus*	Present	N
	Eastern small-footed bat	*Myotis leibii*	Last recorded 1950	N
	Little brown bat	*Myotis lucifugus*	Present	N
	Northern long-eared bat	*Myotis septentrionalis*	Last recorded 1950	N
	Eastern pipistrelle	*Pipistrellus subflavus*	Last recorded 1960	N
Carnivores	Raccoon*	*Procyon lotor*	Present	N
	Domestic dog	*Canis familiaris*	Present	Y
	Coyote*	*Canis latrans*	Present	N
	Arctic fox	*Alopex lagopus*	Last recorded 1960	N
	Red fox*	*Vulpes vulpes*	Present	N
	Ermine	*Mustela erminea*	Present	N
	Long-tailed weasel	*Mustela frenata*	Last recorded 1960	N
	Mink*	*Mustela vison*	Present	N
	Otter	*Lutra canadensis*	Present	N
	Striped skunk*	*Mephitis mephitis*	Present	N
	Domestic cat	*Felis catus*	Present	Y
	Canada lynx	*Lynx canadensis*	Last recorded 1960	N
	Bobcat	*Lynx rufus*	Last recorded 1800s	N
	Black bear	*Ursus americanus*	Last recorded 1950	N
Rabbits	Snowshoe hare	*Lepus americanus*	Last recorded 1970	N
	European hare	*Lepus europaeus*	Last recorded 1960	Y
	Domestic rabbit	*Oryctolagus cuniculus*	Present	Y
	Cottontail rabbit*	*Sylvilagus floridanus*	Present	N
Rodents	Northern flying squirrel	*Glaucomys sabrinus*	Present	N
	Eastern flying squirrel	*Glaucomys volans*	Last recorded 1800s	N
	Eastern grey squirrel*	*Sciurus carolinensis*	Present	N
	Red squirrel	*Tamiasciurus hudsonicus*	Present	N
	Eastern chipmunk*	*Tamias striatus*	Present	N
	Woodchuck*	*Marmota monax*	Present	N
	Red-backed mouse	*Clethrionymus gapperi*	Last recorded 1920	N

▶

◀ TABLE 14.1

	Species	Latin name	Recording	Introduced species
Rodents	House mouse*	*Mus musculus*	Present	Y
	Woodland jumping mouse	*Napaeozapus insignis*	Present	N
	White-footed mouse*	*Peromyscus leucopus*	Present	N
	Deer mouse	*Peromyscus maniculatus*	Present	N
	Meadow jumping mouse	*Zapus hudsonius*	Present	N
	Meadow vole*	*Microtus pennsylvanicus*	Present	N
	Southern lemming	*Synaptomys cooperi*	Last recorded 1930	N
	Norway rat*	*Rattus norvegicus*	Present	Y
	Roof rat	*Rattus rattus*	Present	Y
	Common muskrat*	*Ondatra zibethicus*	Present	N
	Beaver*	*Castor canadensis*	Present	N
	Nutria	*Myocastor coypus*	Last recorded 1960	Y
	Porcupine	*Erethizon dorsatum*	Present	N
Hoofed mammals	Elk	*Cervus elaphus*	Last recorded 1950	N
	White-tailed deer	*Odocoileus virginianus*	Present	N
	Caribou	*Rangifer tarandus*	Last recorded 1950	N
	Domestic horse	*Equus caballus*	Present	Y

* Species discussed in detail in the text.

THE LIFE STYLES OF MAMMALS: THEIR VALUE AS INDICATORS

Humans are among the most versatile of mammals, using technology to live and thrive under a wide range of climatic conditions. Other species of mammals share our eclectic diet and some our proclivity for warm, sheltered places. The range of mammals' dietary and other habits is one element making them useful as indicators of habitat quality. While some species are strictly herbivores, others eat mainly insects or other meat. Still others, like humans, are omnivorous. Species that are mainly herbivorous include the hares and rabbits; rodents such as voles, beavers, muskrats, and porcupines; and ungulates such as deer. Mammals that feed mainly on invertebrates such as insects include shrews, moles, and bats, as well as some of the smaller carnivores. The larger carnivores, from weasels to wolves and lynx, eat mainly meat, and larger species such as wolves, bobcats, and wolverines come into conflict with people when they eat domestic animals.

Omnivorous mammals have been quick to exploit human behaviour. Smaller species such as chipmunks and squirrels, medium-sized ones like groundhogs, raccoons, and skunks, and even the larger black bears, all commonly feed on the by-products of human activity, from crops to garbage. While smaller and medium-sized omnivores can easily coexist with humans, even when our population density is high, bears are usually relatively quickly excluded from our cities and towns because they can injure people and may even treat us as prey (Herero 1985).

While most mammals live mainly in terrestrial habitats, others such as water shrews, muskrats, beavers, mink, and otters are amphibious and their life cycles are tied to water. Other mammals are arboreal, living in trees, while others are fossorial, living underground. Some mammals live in the forest litter or beneath the snow, according to the season. This range of habitat preferences allows us to use mammals to assess a wide range of habitat conditions. For instance, in the Toronto region, some tracts of land support white-tailed deer, which are often seen in town, and this means that other, less conspicuous woodlands species such as red fox, coyote, moles and shrews, and rodents can also survive here. Although there has been impressive urbanization of the Toronto region in the last 100 years, many parks and protected areas such as Mount Pleasant Cemetery and High Park, as well as various golf courses, are sanctuaries for mammals.

SOME GREATER TORONTO REGION MAMMALS

The Carnivores

The Raccoon *(Procyon lotor)*

One of most distinctive and familiar of the mammals living in the Toronto region, the raccoon is easily recognizable by the black mask on its white face, and the dark rings on its bushy tail. Raccoons have stocky bodies, broad heads, and pointed snouts. Their fur is usually grey or black on the back and paler on the belly. The versatility of raccoons contributes to their success as denizens of our city.

The most common social group in raccoons is a mother and her young of that year. Raccoons can live a long time, but probably only 1 percent survive longer than seven years, and the average lifespan is about three years. Although larger carnivores will prey on raccoons, the largest cause of raccoon mortality is human action, either active hunting, habitat destruction, or collisions with vehicles.

Almost everyone who lives in and around Toronto will have seen raccoons, most often at night or in the early morning. In a suburb just west of the Humber River and north of Dundas Street, on hot summer days, raccoons sprawled in the crotches of trees are a familiar sight. In less clement weather raccoons spend the day in more sheltered sites such as hollow trees or culverts and, all too often, in attics. Raccoons eat both animals and plants, so that they quickly learn to exploit garbage as a food source. Raccoons are very abundant in the Toronto region, where they may reach densities of over 100 per km^2.

Like other mammals, raccoons are susceptible to rabies, a viral disease. Rabies in raccoons in Ontario has been relatively uncommon, so the progression of a raccoon-specific strain of rabies north from the vicinity of Washington, DC, has been a matter of great concern for biologists and public health officials in Ontario. By 1994, raccoons with rabies had been recorded very close to the Ontario-New York State border. In New York State, the arrival of raccoon rabies saw an increase in the overall number of rabies cases from 54 in 1989 to 2,750 in 1993. Expenses related to rabies more than doubled with the arrival of raccoon rabies in New Jersey, while they quadrupled in New York State. The high density of raccoons in urban areas such as greater Toronto makes this an important issue. By vaccinating raccoons against rabies in a buffer zone along the New York-Ontario border, wildlife biologists have succeeded in stemming the northward movement of the disease. For more about the raccoon rabies situation in Ontario, see Rosatte et al. (1997).

Raccoons receive mixed reviews from people living in the Toronto region. While some enjoy watching their antics, others complain about overturned garbage containers and holes dug in lawns by raccoons foraging for grubs. Animal control companies are kept busy dislodging raccoons that have moved into attics.

The Red Fox *(Vulpes vulpes)*

Red foxes are another common denizen of the greater Toronto region. In April 1997, you could often see a pair of these foxes in a backyard near Royal York and Dundas, within 10 m of the back door. Early risers are more apt to see these mammals, which are mainly nocturnal.

Red foxes are very successful small carnivores, weighing about 3 to 7 kg, with long, erect pointed ears and a sharp muzzle. Their distinctive bushy tails serve both for warmth and for balance. Red foxes vary in colour from black to silver, as well as red. In an urban situation, foxes range widely, from the parklands that are so prominent in much of the Toronto region, to developments where the density of humans and their dwellings is much higher. One key factor determining the distribution of red foxes is the availability of suitable hiding places to spend the day. Foxes usually dig their own dens, but they may use an abandoned woodchuck den.

Like raccoons, the versatility of red foxes is a large factor in their success at colonizing urban areas (Figure 14.1). These nimble predators adjust their hunting strategy according to the prey they are pursuing (Henry 1986). This flexibility gives them access to a wide range of food, including the garbage produced by humans. Red foxes often store food in caches for later eating.

Male and female red foxes will mate and remain in one area with the young pups, with both male and female hunting for food until the pups disperse. The male

FIGURE 14.1 A vixen and kit in Mount Pleasant Cemetery, 1998. (Peter L.E. Goering)

red fox is responsible for territory defence, and the territorial boundaries are marked with scent from the subcaudal gland on the upper portion of the tail. Under natural conditions, red foxes rarely live longer than 4 years, but two foxes recently trapped in the wild were aged 8 years 7 months, and 10 years 8 months (Chubbs and Phillips 1996). The main cause of death in red foxes is human action, either hunting, trapping, or vehicle accidents.

In urban or protected areas such as conservation areas, fox populations may reach high densities, making it necessary to control their populations. The main justification for control is that the foxes are predators of many different animals. Using poison in fox control usually is inappropriate because of its toxic effect on other animals. Biologists have assessed ways to control breeding and pregnancy in red fox populations. For more information on red fox population control and the effects of this voracious little predator, see Aanes and Andersen (1996), Lampman, Taylor, and Blokpoel (1996), and Marks et al. (1996).

The incidence of rabies in red foxes also could be used to justify their control. In Ontario, however, the use of vaccines distributed in baits attractive to red foxes dramatically reduced the incidence of rabies in these carnivores (Daoust, Wandeler, and Casey 1996).

The Coyote *(Canis latrans)*

The coyote is a medium-sized carnivore, weighing up to 18 kg, with its head, body, and tail measuring between 1 and 1.5 m. The males are usually larger than the females. The coat is greyish, with a grizzled appearance due to the black tips of the hair shafts. The fur usually has a reddish tint, but overall the colour can vary from pure grey to reddish brown, while the underparts are usually pale or white. Although coyotes resemble wolves and some domestic dogs, they are recognizable by several skull features and by their behaviour.

Coyotes eat a surprising array of food, such as rabbits, rodents, birds, reptiles, amphibians, fish, insects, and fruit. The pups often eat insects, which helps them to practise hunting. Although coyotes are often thought of as active predators, most of the livestock and wild ungulates they eat are scavenged rather than killed by the coyotes. Urban coyotes often kill cats and small dogs.

Foxes and coyotes often occur in the same areas, and both may eat the same food. Coyotes, which are bigger, often tolerate foxes, but at other times are aggressive and may even kill the same foxes they had tolerated previously. The biggest factor in the harmonious coexistence of coyotes and foxes is the presence of food. When food is scarce coyotes usually tolerate foxes, but at a localized food source, such as a carcass or a garbage can, coyotes threaten and attack foxes. To read more about what and how coyotes eat, and what eats them, refer to Gese, Stotts, and Grothe (1996).

The Striped Skunk *(Mephitis mephitis)*

A well-known mammal in the Toronto region, about the size of a small cat, the striped skunk has a distinctive white striped pattern down its black back. The body is stout, the head small, and the tail bushy. The striped skunk has two oval scent glands, one on either side of its anus. These glands are well embedded in the sphincter muscles, allowing the glandular secretion to be forcefully ejected for up to several metres. Skunks are omnivorous, eating small mammals, bird and turtle eggs, plant material, and insects. Their natural predators are primarily large owls, but most other predators avoid them. Skunks, however, are often hit by vehicles.

Skunks are most often encountered in wooded ravines, fields, or rocky outcrops, but they also can live under porches, in vacant buildings, in culverts, and near dumps. They may dig their own burrows, but they often use natural cavities. In the wild in summer, skunks use from 2 to 22 shallow dens scattered over their home range, allowing them a variety of escape routes. Skunks are mainly terrestrial, but when threatened may climb trees. Although they are usually active at dusk and at night, they are sometimes seen ambling about by day.

Skunks are actually docile, but when provoked will respond with a variety of warning behaviours, culminating in spraying the secretion from the anal glands. Skunks' warning behaviours can be grouped into several categories. When a skunk is disturbed, it usually advertises by giving either a 'tail-up' or a 'stomp' warning signal. The 'stomp' behaviour was most common in dense habitat, where visual signals had limited usefulness. In one study, most of the warning signals were given to scientists tracking the skunks and the remainder to both predators and non-predators who disturbed the skunks in any way. Interestingly, in almost half of the observed disturbances, the interactions between the skunk and the disturber ended with the retreat of the skunk.

The Mink *(Mustela vison)*
Mink are medium-sized mammals that have a long, slender body with a long neck and a small, flat head. The legs are short and the tail is bushy and about half the body length. Mink are widely distributed across Canada and North America. Full-grown males may reach up to 60 cm in length, and they are markedly larger than females. The fur of the mink is soft and shiny, and is a valuable commodity. It has thick greyish brown underfur and longer lustrous guard hairs that vary from brown to black and that are usually lighter on the belly. In the Toronto region, mink are often seen along rivers such as the Humber.

Mink are excellent swimmers and are usually associated with water, typically inhabiting stream banks, forest edges, swamps, and lakeshores. Mink are solitary animals, and are usually active at night. During the winter, water levels often drop below the level of ice and the thin layer of air allows the mink to move freely under the ice. They live in dens under the roots of trees; they may build their own or take over the dens of beavers and muskrats. Females may use two dens to raise their pups and have relatively small home ranges. Males usually roam more widely, in territories up to almost 9 km^2 and take shelter in a larger number of dens.

Mink are active carnivores, preying on small mammals such as meadow voles, muskrats, shrews, and cottontails, as well as fish, frogs, and salamanders. Mink are also known to prey on beaver kits. Mink themselves are preyed on by owls, red foxes, coyotes, and wolves.

The Bats

The Big Brown Bat *(Eptesicus fuscus)*
Big brown bats (Figure 14.2) are the only bats that occur year round in the Toronto region. Weighing 15 to 20 g and with a wingspan of about 35 cm, these bats often can be seen flying at dusk as they search for prey. The wing membranes are blackish, the fur on the back a rich golden brown. Known for their voracious appetites, big brown bats, like other insectivorous species, regularly consume half of their body weight in insects every night. Nursing females have even more impressive appetites, often eating more than their own body weight daily.

From spring until well into autumn, these animals typically spend their days in roosts, often situated in the attics or eaves of buildings, venturing out each night to hunt insects. During periods of inclement weather, they may remain in their day roosts; over more prolonged periods of cold, they hibernate in buildings. In June, female big brown bats give birth to their young. Typically, each female in this region bears twins, but farther west in North America, a female bears only a single young.

Despite their name, their small size makes it relatively easy for these bats to get into buildings, often through small cracks and spaces, usually where the roof meets the walls. Big brown bats are easy to overlook in a building (near Ottawa over 80 percent of people with bats living in their attics were unaware of their presence). The small teeth and claws of bats mean that they do no structural damage to buildings, but because of their droppings most people would prefer not to share their homes with them.

> *In the wild in summer, skunks use from 2 to 22 shallow dens scattered over their home range, allowing them a variety of escape routes.*

Female big brown bats often roost in attics and eaves because the warmth of these locations promotes more rapid growth of the young. Bats more often roost in older buildings than newer ones, and in some areas taller buildings with galvanized steel roofs are more apt to harbour bats than neighbouring lower buildings with asphalt shingle roofs. Higher buildings may provide better temperature conditions, because of the exposure of the roof to the sun, and clearer approach paths for flying bats. To read more about what makes a building a good roost for big brown bats, see Williams and Brittingham (1997). Other studies of big brown bats include Brigham (1991) and Brigham and Fenton (1989).

The Hoary Bat *(Lasiurus cinereus)*

The hoary bat, Canada's largest, weighs about 30 g and has a wingspan of about 40 cm. Like eastern red bats and silver-haired bats, hoary bats migrate south for the winter, although we do not know how far they actually travel. In the early autumn, hoary bats often are found on Toronto Island by people looking for migrating birds. Hoary bats appear to use the islands as a stopover before heading south across Lake Ontario, and they may be found there by day, roosting among the foliage of trees and shrubs.

Hoary bats have distinctive echolocation calls, making them conspicuous to a trained observer using a bat detector, an instrument sensitive to the ultrasonic echolocation calls of bats. Surveys with bat detectors rarely reveal hoary bats in the greater Toronto region. Elsewhere in southern Ontario, the greater the level of urbanization, the less common the hoary bats and eastern red bats. These more urban areas are often used by big brown bats (Furlonger, Dewar, and Fenton 1987).

Unlike big brown bats, hoary bats are not known to roost in buildings. They roost among leaves, and the combination of their small size and quiet behaviour makes them almost impossible to locate when they are roosting. Although it was easy for us to determine the tree in which hoary bats with radio transmitters were roosting, we never actually saw one, in spite of several years of effort. Hoary bats feed heavily on moths, typically taking the ones attracted to and disoriented by streetlights. Many of the moths they hunt have ears that

FIGURE 14.2
A flying big brown bat is frozen by flashes as it approaches a camera set up in an abandoned mine. In the background, a little brown bat turns away from the scene. Big brown bats are common in Toronto, where they often roost in buildings. (M. Brock Fenton)

provide an early warning about an attacking bat; this is one reason that the bats succeed on only about half of their attacks. Foraging hoary bats fly at an average speed of 14 to 43 km per hour. To learn more about hoary bats and their behaviour, see de la Cueva Salcedo et al. (1995) and Hickey and Fenton (1996).

The Rabbits

The eastern cottontail (*Sylvilagus floridanus*) is an attractive rabbit with a wide distribution in the eastern United States and Canada. It has a maximum total length of about 44 cm and a maximum weight of about 1.5 kg. Cottontail rabbits are brown and may have black-tipped hairs on the back. The underside is pure white, and the tail is dark brown on top and white below. Eastern cottontails do not turn white in the winter, unlike most jackrabbits and hares.

Favourite habitats of the eastern cottontail are fields, forest edges, thickets, and hedges, where food is abundant, and where they can safely make a nest for their young. Brush piles and herbaceous, shrubby vegetation are among the favourite nest sites. These rabbits are herbivorous: among their favourite foods are grasses, clover, dandelion, and ragweed. During the winter, woody species, including the bark of willow, alder, and aspen, make up most of the diet.

The mating season of cottontails begins as the number of daylight hours increases and the temperature climbs. The female secretes a pheromone to attract a mate, and an elaborate mating sequence follows. Female cottontails, true to their reputation, can produce an average of 35 young per year. Females may bear 3 to 8 litters per year, each with 1 to 12 young. They are born blind and hairless, but they develop rapidly. Cottontail rabbits are not generally long lived, with an average lifespan of 15 months, but occasionally living to 5 years in the wild.

The Rodents

The Woodchuck *(Marmota monax)*

Large (40-80 cm long) stout rodents, woodchucks – also known as groundhogs – (Figure 14.3) usually weigh about 3 kg, but can weigh up to 6 kg before entering hibernation. Their upper parts are brown and may appear grizzled, the belly is a reddish brown, and the feet and tail may be dark brown or even black. Woodchucks are widely distributed across North America, living in forests, along edges of woods and meadows, and even in drainage ditches beside roads. This species is solitary and very territorial, and males and females usually are together only to mate. Mating occurs when the animals emerge from hibernation, usually resulting in the birth of four pups after one month of gestation. Young are usually weaned by six weeks and full adult size is reached in two years, but the normal lifespan in the wild is only two to three years.

Woodchucks are grazers; they often come into conflict with farmers and gardeners because of the damage they cause to crops. They grow extremely fat on grass, clover, and crop plants, and in the spring they eat bark and buds of shrubs and trees. The woodchuck is active mainly in the morning and early afternoon, only rarely emerging at night. As winter approaches, woodchucks gain weight, attaining up to 20 percent of their body weight in fat by the time they are ready to hibernate. Unlike most other species of marmot, woodchucks hibernate alone, making it more difficult for them to cope with cold winter conditions because they do not have the body heat of other animals to help conserve energy. A woodchuck's burrow usually is constructed on a well-drained slope and extends 5 to 7 m underground. The sleeping chamber is lined with grass, helping the woodchuck to survive the winter. Woodchucks may emerge if the weather is warm, but return quickly underground in inclement conditions. This habit is the basis for Groundhog Day, 2 February. If the groundhog exiting its burrow sees its own shadow, it will return to the burrow to sleep, and winter will last for another six

FIGURE 14.3 A groundhog or woodchuck that went up a fence post to avoid a charging dog. Normally one would have expected this animal to seek shelter underground. (M. Brock Fenton)

weeks – or so legend has it. If the groundhog doesn't see its shadow, however, then there will be an early spring. For more information on the hibernation of woodchucks, see Ferron (1996).

Woodchucks are important prey for foxes, coyotes, wolves, hawks, and, when young, even for snakes. They are often hunted for game, destroyed as a pest, or trapped for their fur. The use by woodchucks of roadside embankments and river and stream dikes often allows road kills and flooding to maintain population control.

The Eastern Chipmunk (*Tamias striatus*)

This chipmunk can be very common in urban settings and is easily identified by the bright reddish fur and the series of stripes on its back extending to the base of the tail. The eastern chipmunk has large internal cheek pouches, so that it can collect and store food in its mouth while it forages. Chipmunks are most active in forests and sheltered areas, using brush piles, rock piles, and old buildings as dens. Chipmunks usually nest underground and, although they can climb, they generally stay on the ground.

Eastern chipmunks eat nuts, seeds, berries, and insects as well as other invertebrates and occasionally small birds and birds' eggs. The lifespan in the wild averages approximately three years, although captive chipmunks have lived up to eight years. This small mammal makes a distinctive call, which sounds like a 'chip.'

Eastern chipmunks are territorial, defending small core areas within their home ranges, usually the one in the immediate vicinity of the burrow. The size of chipmunk home ranges varies; there is a general relationship between home range size and the availability of food. These animals have larger home ranges when food is scarce (Lacher and Mares 1996).

Each fall there is a frenzied period of activity while the chipmunks hoard food in larders for their use in winter. The chipmunk uses its sense of smell and its memory to find the food stores. It does not develop as heavy a layer of fat as many animals that hibernate, so it is necessary for the chipmunk to arouse itself and eat from the larder every few days. Hibernating eastern chipmunks curl up into balls, tucking their heads into their bellies and wrapping their tails around their heads and shoulders.

The Grey Squirrel (*Sciurus carolinensis*)

Grey squirrels are common in urban areas. There are two distinct colour phases, black and grey, but in each the underparts tend to be lighter. There is considerable variation in the fur colour, even within the same population. The tail is large and fluffy, often held in a characteristic S-shaped arch over the back while the animal sits. The tail is used for balance, particularly when the squirrels are running and jumping amid the treetops. Grey squirrels moult most of their fur twice each year, once in the spring and again in the fall. The tail, however, moults only in the late summer.

Eastern grey squirrels typically have two litters per year and, compared to other rodents, the development of the young is quite slow. The average litter size is three pups, and they are born blind and hairless. Baby squirrels usually are weaned by 9 weeks, and are independent by 12 weeks. The young disperse from the nest by 6 months of age.

Grey squirrels live in forested and well-treed areas and eat a variety of foods, depending on the season. In the spring and summer they consume buds, fruits, and seeds, as well as birds' eggs and nestlings. In the fall, they eat the ripe nuts of many trees, including acorns and hickory and beech nuts. These nuts are rich in fat, so the squirrels get a layer of fat to help them through the winter. The nuts are also cached for later consumption. When a squirrel finds a nut on the ground, it picks it up, digs a small hole, places the nut into the hole, and pats the earth down on top. Usually the cache is found again by smell or by memory, but when it is forgotten, seedlings often sprout. In a study of squirrel feeding behaviour, Steele, Hadj-Chikh, and Hazeltine (1996) presented grey squirrels with healthy acorns, acorns infested with weevil larvae, and healthy acorns with the caps missing. The squirrels preferred to cache whole, uninfested acorns, and they dispersed them farther than the infested acorns or those with the caps removed. However, they were observed eating the weevil larvae from a significant number of the infested acorns. It appears that grey squirrels play an important role in the dispersal of oak trees, because they choose healthy seed to cache, increasing the likelihood that a seedling will take root if the cache is forgotten. Furthermore, grey squirrels make up protein deficiencies by eating weevil larvae when they can.

Everyone who has maintained a bird feeder will appreciate the problem-solving abilities of grey squirrels. These animals have an uncanny ability to obtain food from the most squirrel-proof of bird feeders. It is no wonder that these mammals thrive in the Toronto region.

The White-Footed Mouse *(Peromyscus leucopus)* and the Deer Mouse *(Peromyscus maniculatus)*

The white-footed mouse and the deer mouse are two of the most common small mammals in North America. They are virtually indistinguishable, and identification is usually possible only for experts. They are small, with a total length of 175 mm, and a mass of 22 g. The adult coat is brown on the upper parts, and the tail is brown above and white underneath. These little mice have large grey ears. Both species of mouse live in many different habitats, particularly dry deciduous forests in an enormous range that includes the eastern half of the United States and much of southeastern Canada.

These mice are prolific breeders, usually producing three or four litters a year, each with up to seven babies. The young are born after three weeks gestation, and grow hair, open their eyes, and are weaned by three weeks. Females can begin breeding at seven weeks of age. The mice rarely live longer than one year, although four-year-olds have been recorded.

Population fluctuations often occur in these species, as in other small mammals. Wolff, Schauber, and Edge (1996) associated population fluctuations with the production of acorns, noting that in years when acorn production was high, food caches lasted for the entire winter, and the mice bred throughout this time. In years when acorn production was lower, caches of acorns lasted only till January, and breeding was suspended until food sources became available again in the spring.

The House Mouse *(Mus musculus)*

The house mouse is a small, 'commensal' rodent. Commensal literally means 'sharing the table,' and indicates the close affinity of house mice with humans. The house mouse was first introduced to North America from Europe by explorers and colonists and is present wherever there are humans. House mice are small, weighing up to 30 g, with a total head and body length of up to 90 mm. The fur is light brown or grey, and the tail is dark.

These mice do not see well: they are colour blind and respond better to motion than to other visual cues. House mice have a very acute sense of smell, which they use to find food and to recognize other mice. They can find salts and sugars with concentrations as low as one part per million. Mice mark their trails with urine and faeces. Their hearing is acute, and much of the communication, particularly by the young, takes place in the ultrasonic range, beyond that of human hearing.

House mice breed continually through the year, and a female produces, on average, eight litters per year. Each litter has four to seven young, which reach sexual maturity and start to breed at one and a half to two months. Although there is the potential for very high population growth, this is curbed by very high infant mortality. The young are blind and hairless at birth. The eyes open after two weeks, and soon after the young follow their mother out of the nest and learn to find food on their own.

House mice will eat almost anything but prefer grains. They are delicate feeders, nibbling rather than gorging like commensal rodents such as rats. Mice will eat at any time, as long as there is no disturbance, and they have no preferred time of activity. The house mouse needs about 3 g of food a day to survive, and if conditions are unstable, or if the population is dense, they will store food in caches, usually in their nests or along their burrow systems.

The house mouse was first introduced to North America from Europe by explorers and colonists.

The Meadow Vole *(Microtus pennsylvanicus)*

The meadow vole is a small, widely distributed rodent, weighing up to 75 g, with a total length, including the tail, of up to 20 cm. Voles are stout animals, with short legs and a short tail, small rounded ears, which are barely visible, and small eyes. The colour of the coat ranges from chestnut to brownish black, the underparts are silvery grey, and the feet are grey. The meadow vole is usually found in grassy areas, especially in meadows and old fields.

Meadow voles eat primarily grasses and sedges, but seeds, carrion, and even insects can be taken. Some meadow voles store food in caches in their tunnels. Voles may strip trees and shrubs of bark during the winter months, causing permanent damage at these sites. They also eat grains that are stored for use by domesticated animals, making them a pest to humans in many areas of their range. Meadow voles make trails through the vegetation, leaving piles of grass stems at areas where they stop to eat. There are several 'toilets' along these paths, which the voles use rather than leaving droppings on the path.

Meadow voles are prolific breeders, with a gestation period of 21 days and the potential for a female to start breeding at age 3 to 4 weeks, before she reaches full adulthood. Litters average 4 to 6 young, which are weaned at 14 days and disperse from the nest at 4 weeks. In the winter, when the snow is deep, meadow voles build a network of tunnels under the snow at ground level. They also make occasional ventilation tunnels, through which they will sometimes leave their winter tunnels and move about on the snow's surface.

Populations of meadow voles are cyclic, but when they are abundant, they represent an important food source for many predators including wolf, badger, ermine, weasels, short-tailed shrew, and birds.

The Beaver *(Castor canadensis)*

As our national emblem, the beaver, with its large incisors, sleek rich fur, and paddle-shaped tail, is familiar to most Canadians. Beaver have many fascinating habits, and they are particularly well known for their ability to alter the environment by cutting trees and building dams. The dam is built from sticks, stones, and mud that are deposited at a narrow point in the river or stream. The dam, which may reach 50 m in length, 3 m in width, and 2 m in height, often continues up onto the banks of the river or stream, and is sealed into place with mud. Once the dam is complete, the lodge is constructed. This is the main retreat of the beaver, and is usually constructed from poles, sticks, and mud laid down in a cone. There are several chambers in the lodge, with a central highly placed sleeping pad, an eating platform, and a plunge hole that the beaver use to enter and exit the lodge. Beaver do not always build lodges and dams, and may live entirely in burrows dug into the banks of rivers and ponds.

Beaver have a complex social system. The basic unit is a monogamous pair and their kits. There is a division of labour among beavers. During the summer, both sexes spend equal amounts of time feeding, travelling, and resting in the lodge, but by late summer and early fall, the females spend most of their time eating, apparently to provide energy for future reproduction. During this same period, males spend most of their time constructing and fixing the lodges and amassing caches of food for the winter (Buesh 1995).

Beaver are herbivorous and prefer foods such as tree bark, leaves, and twigs, as well as submerged plant material such as roots. They prefer alder trees for building their lodges (Barnes and Mallik 1996), and deciduous rather than coniferous trees, particularly trembling aspen, for eating. Beaver cut and cache trees and wood for winter eating. Busher (1996) found that witch hazel, if available, was cached for winter consumption more often than red maple, which was cut and consumed immediately.

The Muskrat *(Ondatra zibethicus)*

The muskrat is a medium-sized North American rodent, weighing 2 kg and measuring 70 cm in length. This aquatic animal shows several adaptations for its life style, including a long, laterally compressed tail, large partially webbed hind feet, and lips that close behind the incisors and allow it to gnaw at submerged vegetation without getting water in its mouth. It has shiny brown fur, with its sides and belly slightly lighter in colour.

The muskrat has lips that close behind the incisors, allowing it to gnaw at submerged vegetation without getting water in its mouth.

Muskrats live in family groups occupying a house, usually in a cattail marsh, river, lake, or stream. The area immediately around the house is actively defended against other muskrats. Population densities, and hence home range size, may vary according to resource availability. During the breeding season, muskrats appear to be monogamous.

Muskrats build houses and dens in much the same way as do beavers. They also build 'push-ups,' or pathways under the ice connected by frozen domes of vegetation. As the ice begins to form, the muskrats will break through and pull up submerged weeds and vegetation. This plant material freezes, and makes an insulated dome (a push-up) over the hole in the ice. While under the ice, the muskrats can then go from push-up to push-up for air while feeding on submerged vegetation.

The Norway Rat *(Rattus norvegicus)*

The Norway rat is a common commensal rodent that is fairly large and stocky with smallish ears and a scaly tail that is not quite as long as its body. It is usually brown, although white or black individuals are not uncommon. These rats live in organized colonies where large males are usually dominant, maintaining and defending territories and harems. If the density of rat populations gets too high, the smallest and weakest individuals are often starved or forced out of the colony. Norway rats can have quite small home ranges and their population densities can be very high when there is adequate food. Rats can be aggressive, and there can be many territorial fights within colonies. Using a spool-and-line technique, Key and Woods (1996) studied the habits of Norway rats and roof rats (*Rattus rattus*). By attaching a spool of thread to an animal and following the thread, these researchers learned that Norway rats spend most of their time on the ground while roof rats were more arboreal.

Norway rats are omnivores, and will eat grains, meats, and carrion. In urban settings, they often thrive on garbage. As with many mammals, rats transport food, and this behaviour is influenced by many factors. There is usually a lot of aggression around food sources, and as smaller rats always carry food away while larger rats carry food less often, it appears that the smaller rats are moving themselves out of harm's way by removing food from the source before they eat it. Rats also actively steal food from one another, so the action of transporting food serves to redistribute food through the colony and also informs other rats about food availability.

The Norway rat is a serious pest because of food and crop destruction and its impact on public health. Because rat populations are often determined by the amount of available food, the first step to controlling Norway rats is cleaning up an area. If garbage is properly stored and disposed of, and if there are minimal places for rats to live, it is possible to keep population numbers down.

MAMMAL WATCHING IN THE GREATER TORONTO REGION

There is no doubt that watching the behaviour of animals can be very entertaining, and at some time most of us have used this as a way of taking ourselves into a different world. In the Toronto region, the most conspicuous mammals to watch are people, followed closely by their pets (cats and dogs), and grey squirrels. Sometimes it's the interaction among these four common mammals that is most entertaining.

Bird watchers often depend on songs to detect and then find their quarry. Sounds are not nearly so useful for detecting mammals, although the conspicuous 'chip' calls of eastern chipmunks provide an obvious exception. Would-be mammal watchers usually find the animals by looking for their tracks and 'sign' (a polite word for droppings). Typically, actual sightings of other mammals in and around the Toronto region happen by chance. For example, on an early afternoon in mid-May 1998, a white-tailed deer was spotted in the hillside backyards between South Kingsway and Riverside Drive south of Bloor Street, while just after dark a coyote was seen exploring a patio and backyard near Royal York and Dundas.

For those who would like to see beavers and muskrats, the best strategy is a quiet evening walk along the banks of almost any of the streams and rivers flowing through the Toronto region to Lake Ontario. If you have a canoe, even better. From a canoe in the Humber River just below the Old Mill, you are able to watch mink, beavers, and muskrats.

Next to squirrels, big brown bats are the easiest to see of the Toronto region's mammals. Walking in a park around dusk is usually the best approach. These bats tend to be quite predictable, leaving their roosts just after dusk and setting off in search of flying insects. Unlike many other mammals, big brown bats are very 'noisy,' using echolocation to detect their prey. But most of what these bats say is beyond the range of human hearing, so you need a bat detector to tune in on these mammals.

THE FUTURE OF THE MAMMALS OF THE GREATER TORONTO REGION

The Toronto region has a rich mammal fauna, but the future of these animals will depend largely on the degree of habitat destruction. In many ways, large cities are the ultimate in human-dominated ecosystems (Vitousek et al. 1997). With expanding urbanization, more and more sensitive species are marginalized into tiny fragments of habitat or disappear altogether, as seen from the changes of the last 200 years (Table 14.1). The native mammals that are common today in the Toronto region are, for the most part, the opportunists that exploit the urban environment. While mammals like the beaver are distinctive and the signs of their presence conspicuous, mammals such as bats are more easily overlooked, even though they may be abundant.

Species composition of the mammal fauna in the Toronto region has changed over time, but diversity remains high, attesting to the ability of mammals to adapt to changes in the surrounding habitat. The challenge for us is to find ways to ensure that native mammals continue to live in and around the Toronto region. As we become more informed about the steps that can be taken to restore ecosystems (Dobson, Bradshaw, and Baker 1997), it should become routine to ensure the continuity of the biota of the area, including the mammals. One factor that will influence our success in this endeavour will be the extent to which humans can tolerate other mammals. Like the mammals themselves, we can expect that people will show different levels of tolerance for different species. So, while one neighbour will find raccoons cute and squirrels delightful, another may hold a completely different view. The central challenge is to develop greater tolerance among people for their fellow mammals.

15

Birds

Madeline A. Kalbach

Many decades ago, birding in the greater Toronto region was generally limited to within a 48 km radius of the Royal Ontario Museum, an area that includes part of the lake with its associated water birds. Bird watchers took public transit to High Park, Cedarvale Ravine, and Rattray Marsh. All that changed when the automobile made it possible for birders to chase rarities at a moment's notice in places like Point Pelee and Long Point. Bird watching and compiling a list of all the birds seen in one's lifetime have since become a competitive and challenging activity. Although it has been popular for decades, fostered by both professional and amateur enthusiasts as well as local ornithological societies, bird watching is currently one of the fastest-growing recreational activities in Canada (Foot and Stoffman 1996).

This chapter reports on bird diversity and bird watching in the greater Toronto area. It centres on the distribution of Toronto's bird populations in the past, present, and future. In so doing, the effects of habitat, seasonal changes, and human intervention are discussed.

HABITAT

The greater Toronto area has a wide variety of habitat providing for a diversity of bird life. The area specifically referred to in this chapter stretches from Ajax in the east to Burlington in the west and from Lake Ontario in the south to the rural fringe north of the urban corridor. This area comprises wetlands, lakeshore, ravines, river valleys, and open country. Table 15.1 shows some specific places in the Toronto region to visit for bird study and observation.

Open country includes such spaces as uncultivated farmland, hydro corridors, railway and subway right of ways, industrial areas, parks, and cemeteries. While many of these habitats are found along the urban corridor, others are found in the rural fringes adjacent to the urban area. Birds of the open spaces include such species as the eastern meadowlark, savannah sparrow, song sparrow, and eastern kingbird. Birds of prey, such as owls and the red-tailed hawk, are also found in the open spaces of the greater Toronto region, searching for mice in grassy fields.

Toronto's parks, cemeteries, neighbourhood gardens, and industrial areas also provide good birding. The University of Toronto's St. George Campus, with its trees, shrubs, and flowers, is host to nesting species such as the house finch, red-eyed vireo, and robin. Mount Pleasant Cemetery is one of the best places in Toronto to see migrant warblers in the spring. The Toronto Islands can often bring a fine day of in-city birding. The peregrine falcon nests on the ledges of tall downtown buildings. Flat-roofed buildings in the city and suburbs make perfect nesting spots for the common nighthawk, and the brick chimneys of the city's older houses provide nesting places for chimney swifts. Some birds, such as the house sparrow, rock dove, and starling, have adapted particularly well to the city: they have become permanent residents and nest there.

> *Bird watching is currently one of the fastest-growing recreational activities in Canada.*

R.M. Saunders

Richard M. Saunders, c. 1980. (Courtesy Sarah Saunders)

Dr. Richard (Dick) Merrill Saunders was an exuberant bird watcher, one of the best-known field naturalists of his day, and a pillar of the Toronto Ornithological Club and the Toronto Field Naturalists from the 1940s to the 1960s.

Born in Gloucester, Massachusetts, in 1904, he received his early education at Clark University and then taught at the American University in Beirut, where he met his wife, Anne. Subsequently he received his PhD at Cornell. From 1931 to 1971, Saunders was professor in the Department of History at the University of Toronto, specializing in modern European history. During his career at the university, he was known for his dramatic lectures and seminars. In natural history circles, too, Saunders was a dramatic speaker, eloquently giving testimony to his love of nature.

Saunders wrote beautifully and informatively. For 25 years he wrote the newsletter of the Toronto Field Naturalists' Club. His first book, *Flashing Wings* (1947), provides a vivid portrayal of bird watching in Toronto at that time. In a chapter in a later volume, *Carolina Quest* (1971), he describes an expedition culminating in a view of the now extinct ivory-billed woodpecker. In his 60s, he turned his attention to wildflower photography. He collaborated with Mary Ferguson on two books on Canadian wildflowers.

Dick Saunders died on 25 June 1998 after a short illness. His presence and enthusiasm made the natural world a richer place.

TABLE 15.1

Selected special places for bird study and observation, Toronto region, 1998

Location	Period
Rouge valley	Spring and fall migration
Don valley	Spring and fall migration
Moore Park ravine	Spring and fall migration
Leslie Street Spit	All year round and especially during migration
Toronto Islands	Spring and fall migration
Mount Pleasant Cemetery	Spring and fall migration
High Park, Grenadier Pond	Winter: water birds
Humber valley	Spring and fall migration
Humber Bay Park	Winter: water birds, winter finches, sparrows, and other winter residents
Marie Curtis Park	Winter: water birds
Credit River harbour	Winter: water birds
Rattray Marsh	Spring and fall migration, nesting season
Lakefront	Fall and winter: water birds

Wetlands are found at the mouths of rivers and creeks in places like Pickering, Scarborough, Mississauga, Oakville, and Bronte. Rattray Marsh, Mississauga, and Humber Valley Park are two examples. It is here that you can find ducks and geese. Mute swans nest in the Rattray Marsh. During spring migration, American coot, common moorhen, sora rail, and various shorebirds can also be seen in the marsh.

The Lake Ontario shoreline from east of Toronto to Oakville is the place to visit if you want to see wintering ducks such as redhead, old squaw, and white-winged scoters. Ring-billed gulls and herring gulls can be seen year-round; but the great black-backed gull is only seen in winter. Occasionally a glaucous gull can also be seen on a winter's day. Large numbers of Canada geese can be seen in any season along the waterfront.

In recent years the waterfront has undergone development in order to provide green-belt area and parkland for urban residents. The Leslie Street Spit, for example, was constructed from landfill. Over the years, shrubs, trees, and other flora found their niche there. Ring-billed gulls, one of the first bird species to nest on the spit in great numbers, were soon joined by common terns and other species. The Leslie Street Spit is now considered a premier birding area during any season of the year. However, it is particularly good during spring and fall migrations.

The greater Toronto region also has an abundance of river valleys and ravines. These wooded areas generally consist of coniferous trees such as white pine and hemlock and deciduous trees like beech, oak, and sugar maple. Food and shelter for the birds are provided by the many varieties of shrubs. The network of river valleys and ravines that snakes northward from the waterfront affords both protection and safety for many migrant songbirds on their northward spring journey. Typical nesting species of these ravines and river valleys include the great horned owl, pileated woodpecker, northern oriole, veery, and red-eyed vireo. Winter residents of the valleys include black-capped chickadees, golden-crowned kinglets, white-breasted nuthatches, and owls.

SEASONAL CHANGE

In Ontario, birds are on the move in every month of the year, and migration is heavy for some six months (Goodwin 1995). June, the main month for breeding, is the one time during the year when migration virtually comes to a standstill, because most birds are busy raising their young.

Winter

During the winter season, birds are at their lowest numbers in the region, but their distribution is far from static (Goodwin 1995). In the greater Toronto region, species such as the red-tailed hawk and blue jay are permanent winter residents, but the individual birds you see in winter may not be the same ones you see during the breeding season. Other common permanent residents seen in winter in the Toronto region include Canada goose, ruffed grouse, rock dove, great horned owl, American crow, black-capped chickadee, white-breasted nuthatch, starling, house sparrow, and the downy, hairy, and pileated woodpeckers (Goodwin 1995).

Some birds that winter in Toronto, the dark-eyed junco and the American tree sparrow, for example, generally migrate to the greater Toronto region from northern Ontario. The numbers of these two species have tended to be relatively stable over the last 35 years, whereas others, like the common redpoll, have erratic annual numbers. In some years, large populations of common redpolls invade the region, but in others its numbers will be low or the bird may not be seen at all. Other winter residents in the Toronto region that tend to be erratic include rough-legged hawk, snowy owl, pine

FIGURE 15.1 A typical winter habitat. (Madeline A. Kalbach)

grosbeak, purple finch, red- and white-winged crossbills, and evening grosbeak.

Annual Christmas bird counts, generally held during a two-week period from before Christmas to after New Year's Day, provide an excellent picture of the numbers of wintering birds in the Toronto area, including water birds. The most common ducks found along the lakeshore in winter are the black, mallard, gadwall, American widgeon, greater scaup, oldsquaw, common goldeneye, bufflehead, and common merganser. Less common species include the redhead and the red-breasted merganser.

Small numbers of Canada geese began to appear on the waterfront in winter in the late 1960s. Their numbers have now mushroomed into the thousands and they are found virtually everywhere along the lakefront from Ajax to Hamilton. The mute swan is also a bird of the waterfront in winter, as is the double-crested cormorant.

In general, winter birds concentrate where there is food and shelter (Figure 15.1). Mixed forests with some

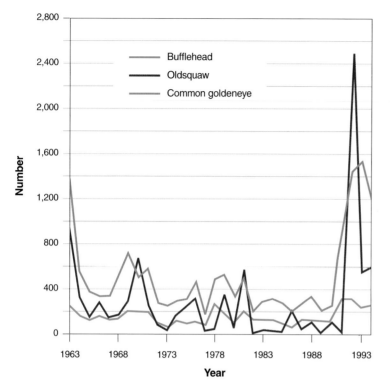

FIGURE 15.2 Numerical distribution of selected duck populations, South Peel Naturalists' Club Christmas counts, 1963-94. (South Peel Naturalists' Club Christmas count data, 1963-94, compiled by Lorelie Mitchell)

conifers make ideal locations, especially if there is a stream or other water source close at hand. Cornfields, weedy fields, feedlots, and home feeders attract many birds. Because of their output of warm water, power plants on the waterfront (Pickering and Lakeview) attract large concentrations of water birds. Food sources – such as the zebra mussel, which first appeared in Lake Ontario around 1990 – can have a major impact on the numbers and location of wintering duck populations. The zebra mussel has accounted for the huge rafts of white-winged scoters, greater scaup, oldsquaw, common goldeneye, and bufflehead often seen in winter at the western end of Lake Ontario off Oakville and Bronte. Figure 15.2 shows the dramatic increase in oldsquaw, common goldeneye, and bufflehead on Lake Ontario from Port Credit to the Oakville-Bronte area from 1990 to 1994. A higher than usual red-breasted merganser count has also been attributed to the mussel.

Spring

In the greater Toronto region, spring migration begins in mid-February with the arrival of horned larks. Snow buntings and winter finches are also on the move in February. Crows can be seen in large flocks near the end of the day as they fly to a communal roost for the night. Movement of waterfowl takes place in March and April. Robins return in March, along with the blackbirds, grackles, and cowbirds. The bulk of spring migration occurs in April and May. It begins slowly, but gradually picks up momentum until the warblers push through in May. Migration of land birds is nocturnal and generally occurs during warm weather. Cold fronts put a halt to spring migration until the next warm spell occurs.

The most productive birding areas in general tend to be in shrubby or treed areas along the waterfront – the Leslie Street Spit, Humber Bay Park, and Rattray Marsh – and in the river valleys and ravines of the region. Warblers and vireos seem to prefer forest edges, and there tends to be a greater diversity of species and larger numbers in these areas compared to open spaces or the forest proper (see Figure 15.3). During the spring migration, the most productive areas in the city are the Mount Pleasant Cemetery, the Toronto Islands, and the Leslie Street Spit.

Summer

By mid-June, the breeding season is well under way, and by the end of July most young have fledged. Overall there is very little movement of most species at this time of the year. One exception is the shorebirds that begin to move from their breeding grounds in July – by August their numbers have increased astronomically.

Migration of land birds is nocturnal and generally occurs during warm weather.

Autumn

Hawks begin to migrate in large numbers in September and can be seen from several vantage points in the greater Toronto region. One of the most dramatic of all bird

FIGURE 15.3 A typical spring habitat. (Madeline A. Kalbach)

HUMAN INTERVENTION

Some birds, such as the mallard, have a wide range throughout the world. Others, such as Kirtland's warbler and the red-cockaded woodpecker are confined to a very small breeding range. In order to become established in an area, birds must be able to cope with the environment and food supply and to adapt to changes when they occur. Those species most successful in this regard will be the survivors in the long run. The red-winged blackbird is an example of a species that has successfully adapted to a changing environment. Earlier in its history, the red-winged blackbird nested mainly in cattail marshes, but land development and wetland drainage have caused this habitat to shrink. The redwing's response was to adapt by nesting on roadsides and in pastures; it has even been known to nest in a crevice of a Toronto subway station.

This section examines the effects of factors such as urbanization, buildings, feeding, and feeders on bird distribution in the Toronto region. In essence these effects could be classified as human intervention. It should be noted that not all of human interventions have deleterious effects on birds. Clearly, as will be discussed in this section, some birds are able to take advantage of human habitation.

Urbanization

Loss of habitat is the most important single factor for a reduction in the numbers of wildlife. As other chapters in this book have shown, the process of urbanization has been responsible for the loss of a great deal of wildlife habitat, including that of birds (McKeating and Creighton 1974). In addition, toxic chemicals and the fragmentation and isolation of habitat are known to be affecting populations of rare and common species alike (Cadman, Eagles, and Helleiner 1994). The fragmentation of the forests, for example, means more forest edges in our suburbs. These edges provide perfect places for nest predators such as blue jays, grackles, and crows (Austen, Cadman, and James 1994). Pesticides such as DDT affected birds like the peregrine falcon (Figure 15.4) and the double-crested cormorant. The latter experienced DDT-induced eggshell thinning that resulted in egg breakage and reproductive failure. Eventually, DDT and some other toxic pesticides were banned, and by the late 1970s and early 1980s the bird populations that had been affected began to recover (ibid.).

migrations is that of the broad-winged hawks. In the early 1960s, John Dales suggested that Toronto and its western outskirts were probably the second-best broad-wing area in North America (Dales 1961). One of the best vantage points for watching the broad-winged hawk migration during the 1950s was a garden near the junction of Highway 10 and the Queen Elizabeth Way. Dales reported that in 1950-8 the largest broad-wing count recorded from this location was 20,800 in 1955; the largest single day's count was 10,300 on 12 September 1955 (ibid.). This area has continued to be a good vantage point for hawk watching. Although the specific garden has been lost to urban development, nearby Erindale College is an excellent place for hawk watching in general. Days when the winds are moderate and from the north and northwest are usually the best time for this activity.

> *The red-winged blackbird has been known to nest in a crevice of a Toronto subway station.*

FIGURE 15.4 Peregrine falcon. (Ted Muir)

The population pressures within the greater Toronto region have increased the demand for housing, shopping malls, and office towers. Such development often occurs on lands that had been wildlife habitat, resulting in a loss of the associated species of birds and other animals. At the turn of the century and for many decades thereafter, clearing the land was a slow, time-consuming process, but technology has improved to a point where it is now possible to clear the land and construct a building or subdivision in a matter of weeks. Such an abrupt loss of woodlots, fields, and hedgerows makes the associated wildlife, including birds, disappear very quickly. Yet, all need not be lost if homeowners, developers, and urban politicians plan carefully when landscaping. Growing native plant species, for example, can provide important habitat for a variety of birds. Other plants, such as sunflowers, provide a good source of food (Figure 15.5).

Feeders

The practice of setting up backyard bird feeders began in earnest in the 1950s (Ehrlich, Dobkin, and Wheye 1988). In general, there is little evidence to indicate that providing them with food has major effects on the survival, population stability, and migration patterns of our birds. However, there is some evidence that range expansion may be correlated with feeders (ibid.). In the Toronto region, beneficiaries of feeding appear to include the northern cardinal, house finch, tufted titmouse, and mourning dove. Also, in the United States, some species, such as the mourning dove, no longer migrate in areas where supplemental food is plentiful (Ehrlich, Dobkin, and Wheye 1988). The same seems to be true in the Toronto region. Figure 15.6 shows that over the 32-year period of the South Peel Naturalists' Club Christmas count in the counties of Peel and Halton, the number of mourning doves increased from 14 in 1963 to a high of 804 in 1985 and was above 300 for each of the next nine years.

Weak individuals and smaller species such as the black-capped chickadee also benefit from feeders (Ehrlich, Dobkin, and Wheye 1988). Anyone who has put up feeding stations has probably experienced increases in starlings and rock doves and noticed that they can be very aggressive when it comes to getting a place at the trough. The more desirable birds must wait their turn, and some may in fact never get a turn. In addition to attracting aggressive bird species, feeders have other disadvantages. Birds who feed on the seed that has fallen to the ground risk attack by cats and other predators. The presence of so many birds also tends to attract hawks that prey on smaller birds (ibid.).

Providing water is another important factor in feeding winter birds. Technology now makes it possible to keep water in birdbaths throughout the winter months. In the short run, providing feeders and water for winter birds may save many from starvation if the natural food supply is frozen over during a major ice storm or

FIGURE 15.5 A planting of sunflowers provides a natural bird feeder. (Madeline A. Kalbach)

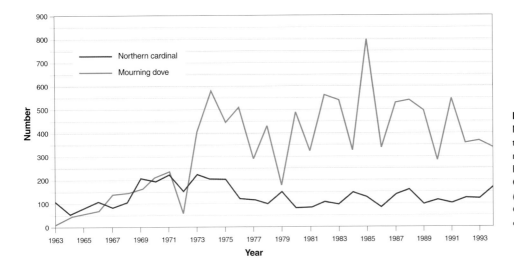

FIGURE 15.6 Numerical distribution of the mourning dove and northern cardinal, South Peel Naturalists' Club Christmas counts, 1963-94. (South Peel Naturalists' Club Christmas count data, 1963-94, compiled by Lorelie Mitchell)

covered for most of the winter season in a year of an unusually large quantity of snow.

In the Toronto area, one of the main beneficiaries of winter feeding is the Canada goose. One of the joys of a winter visit to the waterfront areas of the Toronto region for many people, especially children, is a chance to feed the ducks and geese. Thousands of Canada geese seem to have taken full advantage of this free meal service and no longer migrate in fall. Figure 15.7 illustrates the exponential increase in the number of wintering Canada geese counted in south Peel and Halton regions since 1967. Similar patterns have occurred in the rest of the area, especially along the waterfront. In essence the Canada goose has adapted well to our urban environment.

Canada geese have increased to the point where they have taken over many waterfront parks and other open areas in the Toronto area. Some municipalities are proposing to take action because the geese are perceived as a nuisance, fouling recreational areas with their droppings and being aggressive toward people who come too close during the breeding season. In addition, they often interfere with aircraft at the Toronto Islands airport. Mississauga has examined the possibilities of culling some of the geese by having them captured, either to be sent away for processing at meat plants or to be relocated. Any program that involves killing the Canada geese cannot be undertaken without a special permit because the species is protected under the Migratory Birds Convention Act.

Feeding stations are only one source of food for birds. Species congregate in areas where they can find suitable food, and, consequently, these various feeding grounds make excellent vantage points for birders. One of the best places to see gulls, for example, is at the local garbage dump. These scavengers can be found feasting there at virtually all times of the year. Cornfields provide another food source for specific bird populations: ducks and geese can often be seen foraging for after-harvest scraps in the fields of the urban and rural fringes of the Toronto region where such crops are grown.

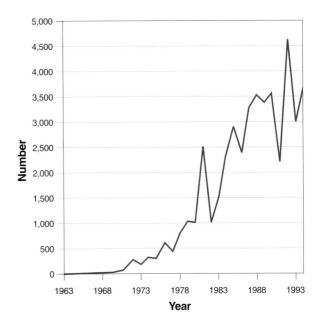

FIGURE 15.7 Numerical distribution of Canada geese, South Peel Naturalists' Club Christmas counts, 1963-93. (South Peel Naturalists' Club Christmas count data, 1963-94, compiled by Lorelie Mitchell)

Lighted Structures and Windows

Only in this century have migrating birds faced the danger of massive artificial obstacles along their flight path. These obstacles are generally tall buildings covered with glass and lit at night. Migrating birds seem to be drawn to them as moths are to lights, especially in foggy or rainy conditions. Huge numbers of birds have been injured or killed by flying into the glass.

In the early 1960s, concerned individuals began to collect birds from the base of some of Toronto's tallest skyscrapers in an effort to save those that had flown into the buildings and fallen to the concrete below but survived. The Fatal Light Awareness Program (FLAP) was formed in 1993, in partnership with the World Wildlife Fund of Canada, to rescue such birds trapped in the downtown core. About 20 volunteers scour plazas and sidewalks beneath skyscrapers in the wee hours of the morning for dead, injured, or disoriented birds. FLAP members collect about 3,000 live birds annually in Toronto, about half of which are able to return to the wild. Unfortunately, each year another 10,000 birds do not survive their encounters with the city's skyscrapers (Abraham 1997). In addition, it is alleged that maintenance staff at some Toronto buildings clean up the carcasses before the arrival of the FLAP volunteers, so these numbers may be low. The range of birds affected is remarkable – over 100 different species en route to and from South America, Mexico, and the United States. On a positive note, groups like FLAP appear to be raising awareness of the severity of the problem. Managers of many buildings now turn off or reduce the intensity of their lights during migrating seasons.

Each year 10,000 birds die from colliding with the city's skyscrapers.

The lighted office towers that dominate the Toronto skyscape are not the only buildings hazardous to birds. Virtually any building with large glass windows, including houses and apartment buildings, can be a threat. Most of us have experienced birds hitting a window of our home. Some break their necks and die, others are just stunned, eventually rallying and flying off. The number of such injuries and fatalities may not, in and of themselves, be a serious threat to the region's bird population. Yet when combined with the hazards of skyscrapers, the reductions of habitat, and a variety of other human impacts, the problem becomes more serious.

FUTURE TRENDS

In his chapter on birds in *The Natural History of the Toronto Region,* James H. Fleming reported 292 species of birds for the Toronto area (Fleming 1913). Eighty-three years later, the number reported by the Toronto Ornithological Club in their field checklist was 374. Of those, 183 species – compared with only 72 in 1913 – have been known to breed in the area. The increase in the popularity of bird watching over the last few decades and a 1981-5 project to compile an atlas of Ontario breeding birds most likely account for the increase in recorded species. In addition, at the turn of the century formal common names were given to all recognized subspecies so that many of the species reported by Fleming are now classified as one species. Holboell's grebe and horned grebe, for example, are now considered to be the same species, namely the horned grebe. Technically, then, the number of species Fleming reported was actually lower than it appears.

TABLE 15.2

Recommended status designations of selected breeding birds of Ontario, 1994

Extinct	Endangered	Threatened	Rare
Passenger pigeon	Peregrine falcon	Least bittern	Red-throated loon
	Acadian flycatcher	Black tern	Black-crowned
	Loggerhead shrike	Short-eared owl	night heron
	Prairie warbler		

Several of the birds listed by Fleming were not as abundant in 1913 as they were earlier and are still relatively uncommon today. Examples include the American widgeon, formerly called the baldpate, and the ruddy duck. Some of the breeding birds noted in 1913, including the piping plover, the lark sparrow, and the golden plover, no longer breed in the Toronto area. Austen, Cadman, and James (1994) indicate the status of some of Ontario's birds as a result of a study of Ontario rare breeding birds. Table 15.2 shows their recommended status for several species found in the Toronto area.

It is clear that the populations of Toronto's birds have diminished as the amount and nature of habitat in the region has changed. Although the overall number of species has increased, some species have become uncommon and others are rare. If forests, wetlands, and fields continue to be destroyed, many more of the Toronto region's bird populations will be affected. Those species that cannot adapt to the changes made by humans will ultimately disappear. We will never again see endless flocks of passenger pigeons darkening the sky, but by managing and preserving some of the habitat that exists today we can be reasonably sure that we will continue to be able to enjoy the birds in the special places of the Toronto region. The future will no doubt show gains and losses, but overall the birdlife of the region has persisted remarkably well, wherever suitable habitat remains.

Part 4
The Special Places

16

From Acquisition to Restoration: A History of Protecting Toronto's Natural Places

Wayne Reeves

Protecting lands for their ecological value is a recent phenomenon in Toronto. The natural features that define Toronto have endured despite a legacy of ongoing alteration for urban ends (Grady 1995), beginning with permanent settlement in 1793, accelerating after 1880, and peaking by 1970 with the end of rapid postwar growth. Traditionally, municipal governments in the Toronto area encouraged private development through limited planning controls and the provision of infrastructure. The latter had an especially pronounced impact on Toronto's natural environment. The valleys, ravines, and lakefront were filled in and reshaped to accommodate roads, subways, and expressways, sewer mains and treatment plants, solid waste disposal sites, flood and erosion control structures, and port and industrial areas (Anderson 1993). Parks were typically small in size, marked by recreational facilities or highly ornamental treatment, and of little ecological consequence. This pattern held until after the Second World War.

Support for Toronto's natural structure comes chiefly from a regional system of protected places that arose as much through fortuitous events (like private donations and natural calamities) as from conscious public policy. The interplay between preservation, development, and naturalization is explored below in three broad phases of activity: early acquisitions and advocacy (1873-1942); the establishment and rapid expansion of the regional green space system (1943-75); and recent naturalization and restoration efforts (1976-present).

Protecting lands for their ecological value is a recent phenomenon in Toronto.

EARLY ACQUISITIONS AND ADVOCACY, 1873-1942

The City of Toronto's first permanent public parks – Riverdale, Stanley, and Queen's – were established in the 1850s (Wright 1983). Located in the Don valley, Riverdale Park was the earliest (1856) and the largest (covering 44 ha by 1883), hosting Toronto's main zoo and playing fields by 1900. Ecologically, Toronto's greatest nineteenth-century achievement was the creation of High Park. City council accepted John G. Howard's gift of 67 ha in 1873 with some reluctance, as the property lay outside the city's official boundaries and was inaccessible to most Torontonians. Perhaps wishing to emulate New York's Central Park, the city purchased another 71 ha for High Park in 1876, and made its last major acquisition of 29 ha in 1930. The site is now 162 ha.

Howard envisioned High Park primarily as a natural area accommodating passive recreational uses (Figure 16.1). For many years, the site's heavily forested character owed more to municipal parsimony than to Howard's wishes or the public's love of untouched nature (Wright 1984). The city focused on underbrushing picnic sites, hauling manure in to prevent blowing sand, planting fast-growing alien tree species, and laying out roads. Apart from a small zoo and some commercial concessions, meagre civic budgets restricted development of the park until the mid-1950s.

A fragment of Toronto Island, granted to the city by the

FIGURE 16.1 Gathering wild lupines in High Park, 1918. Lupines have returned to the park with the introduction of controlled burns in the 1990s. (John Boyd, photographer. National Archives of Canada, PA071052)

Dominion government in 1867, was set aside as a public park in 1880. Island Park, which opened in 1888, was an essay in land reclamation. Marshes, lagoons, and sandy areas were filled with garbage, street-sweepings, and dredged material to make way for lawns, flower beds, and new trees. The bulk of the island, effectively privatized for commercial and residential use, experienced similar alterations. Though the parks commissioner described the island in 1887 as Toronto's 'most beautiful natural feature' quasi-urban development had produced a highly artificial landscape by the Second World War (Gibson 1984). At that time, Island Park amounted to 90 ha of dry land, little more than a quarter of Toronto Island.

Although calls to establish a park system in Toronto date back to 1882 (McFarland 1970), the first systematic proposals were made only in the first decade of the twentieth century. The main advocate was the Toronto Guild of Civic Art. This citizens' group, founded in 1897 with an interest in public art and city planning, published a plan in 1909 outlining various civic improvements for Toronto. The guild was the first body to link an understanding of Toronto's most distinctive assets with a plea for their protection:

> Toronto has a character which in spite of the absence of imposing natural features is definite and pleasing. It is a place that has many attractions and that can be made very delightful to live in, if its attractive points are recognized and developed. The lake shore, the Island, the rivers Don and Humber, the [Lake Iroquois] ridge, the numerous ravines, the facilities for outings both by land and water which make the city such a pleasant abode in summer, all these features can easily be spoiled for want of timely recognition and incorporation in an inviolable plan. There is no beautiful feature in Toronto so salient but that it could be obliterated by speculative building or by such undue commercialism as the establishment of a smelter where there is no real occasion to establish it.
>
> We have kept in view as far as possible, in drawing this plan, the characteristic features of the place, and have endeavoured to incorporate them in the plan, so as to preserve and indeed develop them, and thus to

develop the natural character we have and make of Toronto not just a beautiful city, beautiful in a conventional way, after the model of some other city, but to bring out its own beauty. It is character in a town that makes the dwellers in it love it. Toronto should bring to the minds of those who live in it something which is lovely and pleasant in its own way; so that, when we have been away and are returning homewards, we may feel that, though it is good to see other cities, we are glad to get back to Toronto (Toronto Guild of Civic Art 1909).

For the guild, Toronto's major green spaces could provide visual amenity, shape and guide the location of urban development, and act as recreational corridors. The guild's call for more parks and connecting driveways recognized that the region's intrinsic character would be lost without deliberate planning. The identity of 'Toronto' was derived in large measure from features outside the city limits (like the Scarborough Bluffs and the Humber valley), yet city council had few powers to regulate development in these areas. Convincing Toronto politicians to acquire extraterritorial lands when few parks existed inside the city would be a challenging task.

No comprehensive program of action to maintain Toronto's natural character emerged in response to the guild's plan. Initial planning efforts by the City of Toronto in 1910 and 1912 echoed the driveway concept of 1909, but emphasized the need for small neighbourhood parks and playgrounds (Wilson 1910; Chambers 1912). Acquiring and developing such spaces became the focus of municipal activity (*Park System of Toronto* 1941). Some of these parks included ravines feeding into the Don valley, notably the Rosedale Ravines (15 ha), Reservoir Park (18 ha acquired mostly in the 1870s for the Rosehill Reservoir), and Sherwood and Lawrence Parks (21 ha).

Progress on securing large natural areas came in fits and starts, again relying on private initiative and lands outside the city. Mixing self-interest with civic spirit, real estate baron and Guild of Civic Art member Robert Home Smith struck a deal with the city in 1912 to protect the lower Humber valley. Home Smith turned over 52 ha of valley-bottom land and dedicated a boulevard allowance; in return, the city expended $125,000 on building Riverside Drive and Humber Boulevard and on making park improvements. The roadway serviced Home Smith's five new subdivisions, which straddled the Humber River (Reeves 1992). Important dedications also occurred in the Don watershed. In 1923, Susan Marie Denton donated most of the lands making up Dentonia Park, a 30 ha parcel straddling Massey Creek. Five years later, Alice Kilgour donated Sunnybrook Park to the city. In 1930, the Toronto Field Naturalists' Club opened Canada's first urban nature trail through this 74 ha North York property (Saunders 1965).

The period 1873-1942 saw the protection of some significant green spaces in the Toronto area and the beginnings of a conservation philosophy set in a regional context. The idea of protecting natural areas for nature's sake remained latent, however. For many advocates, recreational development provided the major reason for preservation – as in using the ravines and river valleys for an encircling network of parkways and boulevards. Another rationale was maintaining Toronto's visual identity, which rested on natural rather than architectural factors. But this argument had a dark side. Only those elements of Toronto's natural heritage that satisfied certain aesthetic norms (and thus contributed positively to a sense of the region) were slated for preservation. Those elements that were neither beautiful nor picturesque could be turned to productive ends. As a result, Ashbridge's Bay – some 520 ha of public marsh, sand, and shallow water – was reclaimed for port and industrial use, virtually eliminating the largest wetland in eastern Canada by the Second World War (Desfor et al. 1990).

The reclamation of Ashbridge's Bay virtually eliminated the largest wetland in eastern Canada by the Second World War.

A REGIONAL GREEN SPACE SYSTEM TAKES SHAPE, 1943-75

As the Second World War entered its third year, Canadian politicians turned their minds to the role of conservation and regional planning in postwar reconstruction. The green space ideas sketched out for the Toronto region in the 1940s were enduring and ultimately fruitful. They provided the conceptual basis for a regional parks system that developed rapidly – thanks to legislative and fiscal support from the Ontario government – after 1953.

The Special Places

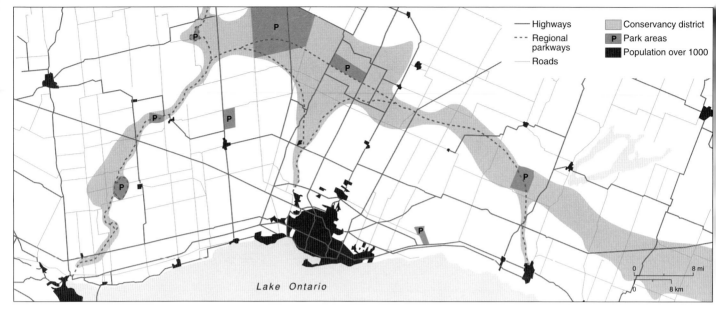

FIGURE 16.2 D.F. Putnam's scheme for a regional green belt in the Toronto region, 1943. The 'conservancy district,' or outer green belt, takes in the Niagara Escarpment and the Oak Ridges Moraine. The inner green belt lies within the Toronto metropolitan area – at lower centre (Toronto City Planning Board 1943b). (Courtesy of the Toronto City Planning Division)

D.F. Putnam

Donald F. Putnam, 1959. (Conrad E. Heidenreich)

Donald Fulton Putnam was born in Lower Onslow (near Truro), Nova Scotia, on 15 August 1903. He graduated from the Nova Scotia Agricultural College in 1924, later obtaining a BSA at the Ontario Agricultural College (Guelph) in 1931 and a PhD in botany from the University of Toronto in 1935. From 1924 to 1930, he worked on a wide variety of agricultural projects, chiefly in Nova Scotia and British Columbia. For most of the 1930s, he was a demonstrator in the University of Toronto botany department, but in 1938 he joined the university's newly established geography department. There he taught agricultural geography, the geography of Canada, soils, and other aspects of physical geography. In 1953, Putnam became chair, a position he held for a decade. He received many honours, including, in 1969, the Massey medal for a distinguished lifetime contribution to his profession.

Putnam is best-known for *The Physiography of Southern Ontario* (1951), a book he researched between 1934 and 1950 with Lyman Chapman of the Ontario Research Foundation. Few people knew Ontario as well as he did. No student ever forgot Putnam's tours of gravel pits and roadcuts through eskers, drumlins, and moraines to furnish the bits of evidence with which to reconstruct the glacial history of an area. He taught his students to be curious about their surroundings and to look for interrelationships to explain their findings. His overriding concern as a professor, researcher, and citizen was for an understanding of the physical environment and its fragile relationship as a resource for human exploitation. Putnam died in Toronto on 23 February 1977.

The fundamentals were laid out in the Toronto City Planning Board's (TCPB) *Master Plan for the City of Toronto and Environs* (1943a). An inner green belt, shaped like an inverted-U, protected the Don and Humber valleys and their tributary ravines from 'encroachment and vandalism.' A low-speed recreational driveway linked these 'metropolitan park reserves.' As distinctive elements in the urban structure, the reserves formed 'barriers between residential and industrial districts, ... break[ing] up residential parts of the City into well-defined separated neighbourhoods, arresting the spread of continuous bricks and mortar to uncontrolled limits' (TCPB 1943b, 17).

In an appendix to the *Master Plan,* one of Canada's leading regional geographers contemplated planning on a scale far exceeding that of the proposed Metropolitan area. D.F. Putnam drew attention to conserving the Niagara Escarpment and the Oak Ridges Moraine, and to reforesting marginal agricultural lands. His proposed 'conservancy district' included parks at Mount Nemo, the Credit Forks, and other sites, and a 'peripheral driveway' linking Oshawa and Hamilton (Figure 16.2). Parks at Heart Lake and in the Rouge valley were among those suggested within the agricultural zone, which would enjoy 'protective planning' that stabilized stream flows and soil erosion. Reserving the lakeshore for public use was also a top priority (TCPB 1943b).

The problem of specifying an appropriate planning and administrative unit for the Toronto region was much discussed in the 1940s. For Putnam, the political had to be reconciled with the biophysical:

> From a geographical viewpoint, the so-called Toronto region is only a small part of a much larger region; its claim to consideration as a regional unit rests largely in its relationship to the Toronto Metropolitan Area. The creation of a rural green belt and the undertaking of any other measures of a conservational nature should be planned in harmony with both these relationships. It is not out of place here to urge the setting up of machinery to study the whole question of conservation needs in Ontario, and particularly in the older settled region (TCPB 1943b).

The 'machinery' that addressed conservation needs became the Conservation Authorities Act of 1946. The act permitted municipalities in a watershed or group of watersheds to develop policies and programs to conserve the natural resources under their jurisdiction. Each conservation authority was to be established on local initiative: municipal councils petitioned the province to hold an organizational meeting; a resolution calling for the creation of an authority was passed; and an order-in-council established the authority.

Amid pressing needs for flood control and reforestation, four such bodies were created in the Toronto region. In 1946, Ontario's first conservation authority was organized in the Etobicoke valley. Authorities for the Don and Humber valleys (1948) and the Rouge-Duffins-Highland-Petticoat watersheds (1956) followed. Comprehensive reports were prepared for each authority by the province's Department of Planning and Development. The reports covered such topics as water, land use, forestry, wildlife, recreation, and history, and laid out proposals for flood control and water conservation, reforestation, and multiple-use conservation areas (Richardson 1974).

Passage of the Planning Act and the appointment of the Toronto and Suburban Planning Board (TSPB) in 1946 set the framework for regional planning in Ontario and the Toronto region. The board considered a variety of intermunicipal matters, including parks and land use, for the 630 km^2 bounded by Steeles Avenue, Lake Ontario, the Rouge River, and Etobicoke Creek. Support was thrown behind the 1943 master plan, and the inner green belt concept was dealt with in detail. The board urged a 'joint acquirement' program on the part of the city and York County. Of the 1,820 ha desired, some 1,100 were owned privately and thus were 'liable to development in a manner not in conformity with the green belt program' (TSPB 1947).

In 1948, the Toronto and Suburban Planning Board was reconstituted over a larger area as the Toronto and York Planning Board (TYPB). The new board also supported the conservation of Toronto's major valleys and ravines, increasing the size of the proposed inner green belt by nearly 50 percent (TYPB 1949). Acquiring lands proved much more difficult, though some progress was made in the Don valley. The board nonetheless remained committed to the principle of planning well in advance of current needs to avoid later regrets.

Creating a system of large urban green spaces became possible with the formation of the Municipality of Metropolitan Toronto in 1953. Metro chair Frederick G. Gardiner provided leadership for the new government, and continuity with previous regional initiatives. While

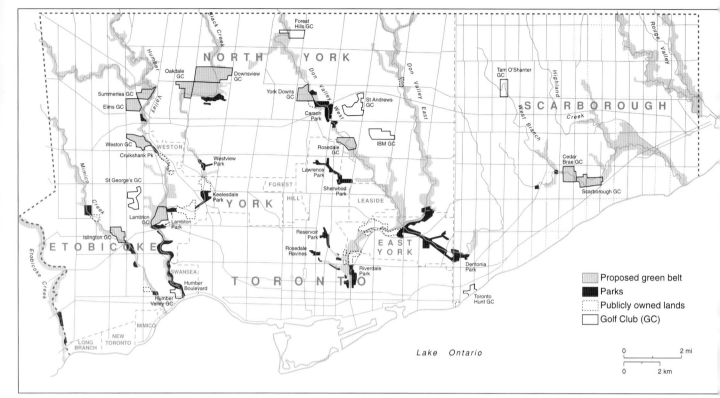

FIGURE 16.3 Metro Toronto's proposed inner green belt system, 1954. The system, focused on the major valleys and ravines, comprises existing large public parks and landholdings, private golf courses, and (for the most part) private valley properties awaiting municipal acquisition (Metropolitan Toronto Planning Board 1954). (Courtesy of the Toronto City Planning Division)

Gardiner's role in promoting urban growth though the provision of hard services has long been recognized (Kaplan 1967; Colton 1980), his support for conservation and regional parks has been overlooked (Gardiner 1948; Reeves 1993).

Led by Gardiner, Metro council adopted recommendations made by the Metropolitan Toronto Planning Board (MTPB) in 1954 to establish a 2,700 ha park system based on the major river valleys (Figure 16.3). With over 40 percent of the system vulnerable to urban development, council authorized the creation of a green belt acquisition fund and directed the Metro Planning Board to prepare an official plan that addressed the river valleys, Toronto Island, and 'such other parks of a regional nature as may be deemed of Metropolitan significance and recommended from time to time' (Metropolitan Toronto Council 1954).

The Metropolitan Toronto Parks Department was formed in 1955. A year later, Parks Commissioner Tommy Thompson outlined the basic philosophy and future scope of the regional parks system. Unlike neighbourhood parks administered by Metro's 13 area municipalities,

> Metropolitan parks ... should be regional in appeal, serving large communities. They should have enough area to accommodate widely diversified interests and activities. Their development should be extensive, rather than intensive, and because they will involve, in total, a very large land area they should be designed to be maintained effectively at minimum cost. Our regional parks should take advantage of the available valley land and be no less than 250 acres [100 ha] in area.
>
> Recreationally they will provide those things that the neighbourhood park seldom offers, but which people increasingly demand. The tempo of modern living and the density of our population makes it essential that nature be preserved in those areas where it still exists. Metropolitan parks should offer

opportunities for an outdoor experience – a basic need of people – in a manner which they can enjoy. But in addition to the day camps, council rings, extensive picnic facilities, bridle paths, nature trails and wilderness areas, they will serve as the laboratories for outdoor education and conservation. Indeed, the whole concept of Metropolitan parks should be consistent with the highest ideals of conservation itself.

By taking the greatest advantage of the natural aspects of these properties, it is hoped that development and maintenance costs can be minimized (Metropolitan Toronto Council 1956).

Thompson based his system on those 'distinctive topographical or other physical features which may merit exploitation for the benefit of the entire region rather than simply for local purposes.' This meant establishing major parks in the lower and middle Humber, Don, and Highland Creek valleys, on Toronto Island, in Vaughan Township, and ultimately in the upper reaches of the Humber and the Rouge.

The fact that the natural features of the Toronto region were largely unprotected meant that acquisition, not development, became Metro's highest priority in the short term. In the 1950s and 1960s, major deviations from this policy occurred only at Toronto Island, the Humber and Don golf courses, and the two 'estate parks,' James Gardens and Edwards Gardens. A unique venture began in 1965 when Metro and the Federation of Ontario Naturalists established Canada's first urban wildflower reserve in James Gardens (Metro Parks 1973).

In Metro's other parks, extensive development typically involved access roads, parking lots, and paved trails, though some areas also featured picnic areas, fire pits, drinking fountains, water taps, and washrooms. Natural conditions prevailed. But while the parks 'retain[ed] as

Tommy Thompson

Thomas (Tommy) W. Thompson (1913-85), a Toronto native and graduate of the Ontario Agricultural College, became Metropolitan Toronto's first parks commissioner in 1955. He worked closely with the Metro Toronto and Region Conservation Authority to develop one of Canada's finest regional parks systems.

In 1978, Metro Council paid Thompson this tribute when he resigned to become general director of the Metro Zoo:

> To the general public, Mr. Thompson is probably the best-known Department Head that Metropolitan Toronto has ever employed – his name over the years became synonymous in the minds of the citizens with their major parklands, and he succeeded beyond all expectations in his efforts to encourage the citizens of the community not only to stand and admire the beautiful areas of countryside that became part of the parks system and

Tommy Thompson (centre, with walking stick) at Toronto Island Park, c. 1970. (City of Toronto Archives series 47, file 1, box 92294-1, with permission: The Toronto Star Syndicate).

> the Metropolitan heritage, but to also become involved actively by using the many and varied facilities provided for their benefit – his famous 'Please Walk on the Grass' philosophy and his personal love of walking (which he developed into frequent personally-led rambles) are prime examples of his own profound love of nature, and his unabashed affection for the community (Metropolitan Toronto Council 1978).

Thompson retired from the zoo in 1981 to become executive director of the Metro Toronto Civic Garden Centre. After his death, Metro council honoured this outstanding parks administrator by renaming the Leslie Street Spit 'Tommy Thompson Park.'

much as possible their primitive character' (MTRCA 1965), turf was maintained in the valley bottoms and some low-lying areas were filled with industrial waste. Large, marshy bays along the lower Humber were subjected to both measures (Reeves 1993). Gardiner applauded these sorts of changes in 1960; he recalled that in 1953 Metro's valleys had been 'sitting idly by as an undeveloped heritage with no successful effort to change them into green belts and recreational areas' (Metropolitan Toronto Council 1960).

Metro quickly assembled an impressive amount of green space in the central and west Don, the Rouge, Highland Creek, the south and north Humber, Black Creek, and on the waterfront (Metro Parks 1977). In 1954, Metro's holdings consisted of one 67 ha parcel; in 1959, the count reached 611 ha; by 1962, the 1,635 ha Metro system was larger than the 13 area municipal systems combined. Rapid growth continued through 1965, when the regional parks amounted to 2,080 ha; in 1974, newly filled waterfront land pushed the total to 3,161 ha (MTPB 1963; Metro Planning 1976). By 1976, major new or expanded parks were in place along the lakeshore and in all valleys except Etobicoke Creek and Mimico Creek. The largest named areas were the Metro Zoo (282 ha), Toronto Island Park (223 ha), G. Ross Lord Park (138 ha), Rowntree Mills Park (104 ha), Ernest Thompson Seton Park (96 ha), Morningside Park (95 ha), and Colonel Danforth Park (81 ha).

Two interrelated factors were crucial to this land assembly: Hurricane Hazel's destructive visit in 1954, and the creation of the Metropolitan Toronto and Region Conservation Authority (MTRCA) in 1957. After taking 81 lives and causing $25 million in property damage in the Toronto region, Hazel convinced Gardiner that parkland serving regional needs was the proper function of low-lying areas like Toronto Island (*Globe and Mail*, 30 October 1969). Hazel was also directly responsible for Metro's creation of Marie Curtis Park in flood-ravaged Long Branch, and for changing the tenor of watershed planning. Whereas the province's conservation reports of the 1940s had taken a rather complacent view of settlement on floodplains, the acquisition of such lands in the name of public safety was pursued vigorously after 1954. The idea of providing for public recreation on floodplains and other conservation lands also gained favour.

The Metro Toronto and Region Conservation Authority, taking in 23 municipalities and over 2,400 km^2, was formed in 1957 through the amalgamation of the Etobicoke-Mimico, Humber, Don, and Rouge-Duffins-Highland-Petticoat authorities. The effectiveness of these four bodies had long been undermined by the representation of some municipalities on more than one authority. Metro's formation, Gardiner's striking of a joint committee to consider greenbelt and conservation matters in 1953, and Hazel's visit prompted organizational change.

A close working relationship developed between Metro and the conservation authority. The authority benefited from the strong financial base provided by Metro; thanks to Gardiner, Metro council gained a powerful voice on the authority's board, exerting political influence far beyond the municipality's borders (Richardson 1974). And, while Metro obtained parkland through direct purchase, by assumption from or arrangement with the area municipalities, by private donation, and by arrangement with the province, especially with respect to properties damaged by Hurricane Hazel, most of the system resulted from two agreements with the authority (Metro Parks 1977).

The first agreement, made in 1961, focused on Metro's valleylands. The authority's *Plan for Flood Control and Water Conservation* (1959) had sought, through floodplain acquisitions and the construction of dams and channel works, to minimize damage caused by flooding. Implementation was funded primarily by Queen's Park and Ottawa. Floodplain lands in Metro acquired by the authority for conservation purposes were turned over to the municipality for development and operation as regional parks. This policy was intended to 'bring about maximum use of those lands which may have a dual park and conservation potential' (Metropolitan Toronto Council 1957, 1961). The central Don, for example, benefited from Water Conservation Scheme 'W.C. 14.' About 287 ha were assembled, giving rise to Ernest Thompson Seton Park, Taylor Creek Park, and the Charles Sauriol Conservation Reserve (Metro Parks 1973). A similar arrangement was reached for waterfront properties in 1972.

By 1966, when about 70 percent of Metro's 2,145 ha park system had been secured through the authority, a measure of strain had emerged between the two institutions. This stemmed from their differing goals and policies. The authority's emphasis on water conservation and flood protection had resulted in some unspecified 'key' park acquisitions 'not being completed.' (As the authority put it, 'Conservation may include parks, but parks are not necessarily conservation' [MTRCA 1967].)

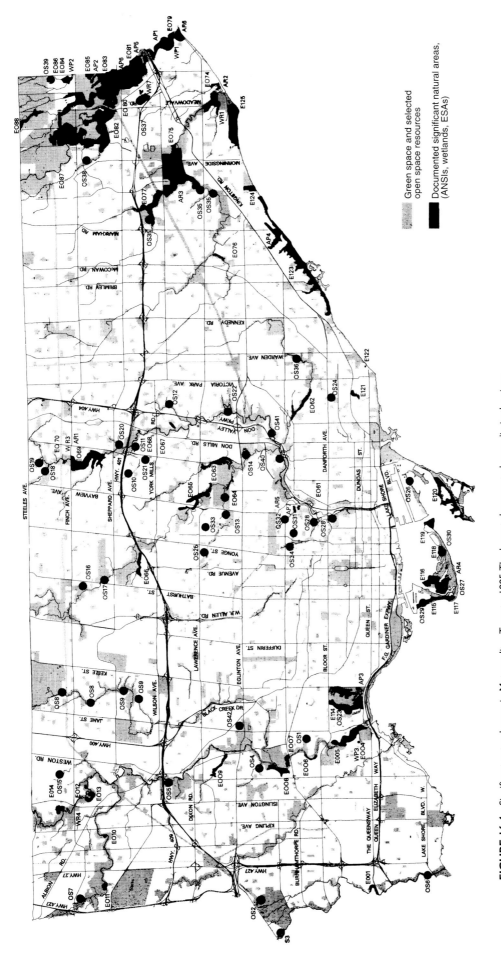

FIGURE 16.4 Significant natural areas in Metropolitan Toronto, 1995. The largest documented areas lie in the river valleys (such as the Highland and the Rouge, upper right) and on the waterfront (such as Toronto Island and Tommy Thompson Park, bottom centre, and the Scarborough Bluffs and East Point Park, centre right). The lower Humber marshes and High Park are at bottom left. Ravines, woodlots, natural areas, and regional lands constituted 71 percent of Toronto's 8,023 ha park area in 1997 (Metropolitan Toronto Planning Department 1995; Toronto Parks, Recreation and Culture Commissioners 1997). (Courtesy of the Toronto City Planning Division)

Metro was also concerned about the amount of 'undeveloped' land in the system. Despite these frictions, the pursuit of mutual objectives had brought much land into the regional parks system, and a 'happy inter-relationship' was said to exist (Metropolitan Toronto Council 1966; MTRCA 1967).

From the regional perspective, the period 1943-75 was marked by great advances in visionary planning, institutional development, and public green space acquisition. The new regional parks system focused as much on corridor continuity (initially along the valleys, later along the waterfront) as on individual units. Nevertheless, the protection of significant natural areas was a by-product, not a primary objective, of flood control and water conservation efforts and a land-extensive approach to meeting regional recreation needs. The heavy engineering used to manage floods and erosion and provide recreational space in the valleys was, by the mid-1970s, modest compared with the lakefilling taking place on the lakeshore.

NATURALIZATION AND RESTORATION, 1976 TO THE PRESENT

Having emphasized rapid system expansion and limited park development for two decades, Metro concluded in the mid-1970s that its valleyland acquisition program was nearly at an end and that the provision of major recreation facilities should be the new regional priority (Metro Planning 1976). This view was not shared by all. In 1976, the Toronto Field Naturalists issued *Toronto the Green*, which included recommendations for the conservation and management of natural areas in and around Metro Toronto. The Field Naturalists demanded that all municipal official plans 'explicitly recognize the conservation of natural areas as an important aspect of planning,' requested that all municipalities adopt ravine protection by-laws, identified eight natural areas that should receive maximum protection, called on the province to initiate a watershed planning program for the Rouge River, and proposed guidelines for managing natural areas:

a) except where an area such as a playing field is specifically required to have mown grass, natural vegetation should be left to grow;

b) dead trees (unless hazardous) and underbrush should be left for the animals and to return to the soil naturally;

c) native shrubs should be planted on slopes to prevent erosion, to control access by people, and to provide food and cover for animals;

d) streams should be left in as natural a state as possible; that is, natural vegetation should be allowed to grow along streams and channelling should be avoided. (Toronto Field Naturalists' Club 1976)

Toronto the Green marked the beginning of four key shifts in green space thinking that, in application, were often intertwined. The importance of functioning natural systems in an urban context was reaffirmed, while recognizing that some natural areas were of greater significance than others. Community-based activism, usually focused on a discrete geographic unit, became an insistent factor in environmental management. Resource managers and decision makers were pushed to develop new management goals and methods highlighting the regeneration of natural features and systems. And watershed-based, conservation-oriented planning came into vogue.

The issue of assigning significance to natural areas was taken up at various levels. Most of the core survey work began in the 1970s. In 1982, the Field Naturalists issued a list of 28 significant natural areas in Toronto, and the Metro Toronto and Region Conservation Authority released its *Environmentally Significant Areas Study* (Varga 1982; MTRCA 1982). Almost all of the environmentally significant areas in Metro Toronto lay within the Metro park system. The authority later brought forward revised criteria to update the status of its existing environmentally significant areas and to evaluate new candidate areas (MTRCA 1993b); the political process of approving the latter began in 1996.

Analysis also proceeded at the provincial and local scales. In 1983, the Ministry of Natural Resources issued the *Maple District Land Use Guidelines* as 'a design for integrated resource development' in the Toronto region (Ontario Ministry of Natural Resources 1983). Within Metro Toronto, the province identified 10 important wildlife resource areas and 9 areas of natural and scientific interest (Hanna 1983). By 1995, a total of 13 areas of natural and scientific interest, 7 provincially significant wetlands, 48 environmentally significant areas, and 42 other significant natural areas had been identified in Metro Toronto (Metro Planning 1995; see Figure 16.4).

Complementing the Field Naturalists' regional perspective, much community effort has centred on specific natural areas. For example, the Friends of the Spit formed

FIGURE 16.5
A community tree-planting event on the central Don River, 1995. The lower Don has been the focus of ecological restoration projects since the mid-1980s. (Stephanie Lake, photographer. Courtesy of the Toronto Parks and Recreation Division)

in 1977 to advocate on behalf of the natural features and processes associated with the Outer Harbour Eastern Headland. The 5 km long Headland – better known as the Leslie Street Spit or Tommy Thompson Park – arose through lakefilling that began in 1959 in an abortive attempt to create new port facilities. When the spit was labelled 'Aquatic Park' in 1972, debate erupted over what level of recreational development should occur (Gemmil 1978). The Friends of the Spit lobbied for ongoing naturalization, passive human use, minimal facilities, and a car-free environment in what was one of the authority's largest environmentally significant areas (Friends of the Spit 1987; Courval 1990).

The approved master plan for Tommy Thompson Park reflects the influence of the Friends of the Spit and, more generally, a new management ethos for Toronto's natural areas. Ensuring that the spit remained an urban wilderness required 'the adoption of the natural succession or ecological approach which relies on natural processes, augmented by minimal intervention and management of the park to achieve over time the diversity of community types outlined in the Master Plan' (MTRCA 1989, 1992). Unassisted natural regeneration will continue in some areas; in others, natural succession will be guided by park managers, who will create the base conditions and initial species mix.

The exploration of alternative naturalization techniques – what Michael Hough (1990) calls 'restoring identity to the regional landscape' – began in Metro Toronto's local park systems. Prompted partly by budget pressures, the City of North York began experimenting with park naturalization in 1982. Mowing was abandoned in selected areas, allowing the slow process of community succession to begin. The City of Scarborough followed suit. Elsewhere, restoration plans and projects emphasizing a higher degree of human intervention were also pursued. Concern over decline in High Park's rare natural communities helped push the City of Toronto to develop a master plan with the local community and to begin restoration work, including controlled burns (Toronto Parks and Recreation Department 1992; Suhanic 1997). In the Don valley, Toronto's Task Force to Bring Back the Don has organized many volunteer planting events and the construction of Chester Springs Marsh (Task Force to Bring Back the Don 1989; Chester Springs Marsh 1997).

These local efforts were soon joined by work at the regional level. In 1990, Metro began a naturalization program with the establishment of a wildflower meadow adjacent to a new paved path in the west Humber. Creating more natural landscapes was intended to help increase the diversity of wildlife habitats, improve the aesthetic quality of the region, and provide community members and park users with an opportunity to develop a sense of stewardship for these areas. This ongoing program has relied heavily on community volunteers, and has seen more than 400,000 native trees, shrubs, and wildflowers planted on sites throughout Metro Toronto (Figure 16.5) (Metro Parks and Culture 1996a).

As for watershed-based conservation planning, the Field Naturalists' call for the province to establish a program for the Rouge valley was ultimately successful. The significance of the Rouge was not in question. In 1956, the Ontario Department of Planning and Development described the Rouge as the 'choicest block of natural unspoiled wilderness' in Metro; in 1987, the valley became one of 36 sites in the Carolinian Canada Land Protection and Stewardship Program. Community groups like Save the Rouge Valley System Inc., founded in 1975, pressed for stronger protection. In 1990, the province announced that the Rouge would become the largest urban wilderness park in North America, spanning about 4,250 ha from Lake Ontario to the Oak Ridges Moraine. Under the *Rouge Park Management Plan* (Ontario Ministry of Natural Resources 1994), a number of 'restoration zones' have been identified. Undertaking restoration projects and implementing the rest of the plan falls to the Rouge Park Alliance, a multi-jurisdictional body with community input. Other Toronto watersheds have also benefited from regeneration-oriented task forces in the 1990s.

New large-scale planning frameworks for the Toronto region accompanied the rising interest in watershed consciousness, significant natural areas, and restoration ecology. The federal and provincial governments co-operated on two fronts in the 1980s, establishing the Royal Commission on the Future of the Toronto Waterfront (1992) and the Metro Toronto and Region Remedial Action Plan (1994).

Municipally, progress was slow on regulating private development in and adjacent to significant natural areas. Metro's green space acquisition program prompted the Toronto City Planning Board to consider the state of Toronto's ravines in 1960. 'There never has been and there is not at present a City policy on the preservation and use of natural parkland,' lamented the Toronto City Planning Board, noting that 340 of the original 770 ha of ravine land had been built over (TCPB 1960). City council quickly adopted – in principle only – the board's recommendations to tighten development controls. Specific legal measures followed at a more leisurely pace. The first zoning by-law amendment came in 1967; in 1969, ravine policies appeared in the city's new official plan. While the Conservation Council of Ontario found in 1971 that little had been achieved, the province did pass enabling legislation that year – though Toronto (to the Field Naturalists' chagrin) did not designate a single ravine under this act until 1976, and did not adopt the implementing ravine control by-law until 1981. By the latter date, Metro's official plan had rigorous conservation policies in place for the major river valleys and the waterfront (Metro Council 1980). The province's two waves of planning reform in the 1990s brought questionable results. At first, provincial policies respecting municipal planning decisions required strong environmental protection, but this approach was reversed after a change in government (Ontario Ministry of Municipal Affairs 1995, 1997).

In terms of conceptual outlook and actual achievement, many advances in protecting and restoring Toronto's special natural places have occurred since the mid-1970s. Metro's park system, Metro Toronto and Region Conservation Authority programs, and area municipal land use policies have generally protected the valleys from incompatible development and preserved their functional capabilities with respect to flood control and the conservation of the natural environment (Metro Planning 1988). The regional park system alone grew from 3,194 ha in 1976 to 4,680 ha in 1996. Though large gains were made in the west Humber and east Don, most (58 percent) of the growth took place along the lakeshore. There, an extra 869 ha nearly tripled the area of public waterfront green space (Metro Parks 1977; Metro Parks and Culture 1996b; Maciejewski and Lebrecht 1997).

However, a closer look at the waterfront reveals that success was mixed. About 800 ha of the parkland was literally new, most having been created since 1972 by the conservation authority. Yet, despite providing new recreational opportunities, the lakefill parks at Bluffers, Ashbridge's Bay, and Humber Bay offered little in terms of new habitat. The authority's extensive erosion-control

The Metro naturalization program has seen more than 400,000 native trees, shrubs, and wildflowers planted on sites throughout the area.

works at the Scarborough Bluffs were similarly found wanting, and also attracted the wrath of earth scientists (MTRCA 1991; Ohlendorf-Moffat 1991). To address these issues, the conservation authority embarked on a series of waterfront habitat-rehabilitation projects in 1992, 'wilding' the shoreline to restore or create wetlands and fish habitat (MTRCA 1993a, 1993c).

This remedial work – taking place on sites not two decades old – indicates how public agencies have fundamentally altered their way of doing business. As late as 1976, the regional planning authority admitted that the conservation of significant natural areas was not a primary driving force for land acquisition and management by the conservation authority or Metro (Metro Planning 1976). This was the same planning authority that, in 1967, had championed massive lakefilling to create recreational land – and the one that promoted green management over engineering marvels in Metro's 1994 Waterfront Plan (MTPB 1967; Metro Planning 1994).

Such philosophical shifts seem to augur well for the future. But deregulation, downsizing, and devolved responsibilities have severely constrained the ability of governments to deal effectively with natural areas in the Toronto region. The challenge will be finding the means to sustain the vision that gave rise to – and that has nurtured – special places like High Park, Ernest Thompson Seton Park, and Rouge Park.

17

Special Places

Betty I. Roots

In this chapter attention is focused on some special places in the greater Toronto region. Their locations are shown on Figure 17.1. The places described here are considered special because they still have ecological integrity and relatively few exotic (introduced) species (see Chapter 6). Some of them are also, as John Westgate wrote in Chapter 1, windows that afford glimpses into the past. Environmental change beginning with the first continent-sized ice sheets over the greater Toronto region is recorded in the Scarborough Bluffs and along the walls of the Rouge River valley and of West Duffins Creek. They are places where vestiges of earlier ecosystems and evidence of their gradual transformation may be seen and studied.

The sections that follow are not intended to be exhaustive accounts. Whole books could be, and indeed have been, written about any one of these 'special places.' Rather, they represent personalized vignettes, reflecting why the places are special to the authors. A result of this approach is a wide variety of styles of writing, mirroring the diversity of the authors. It is hoped that these introductions to the special and unique aspects of each place will encourage readers to explore the special places for themselves and to develop their own appreciation of them.

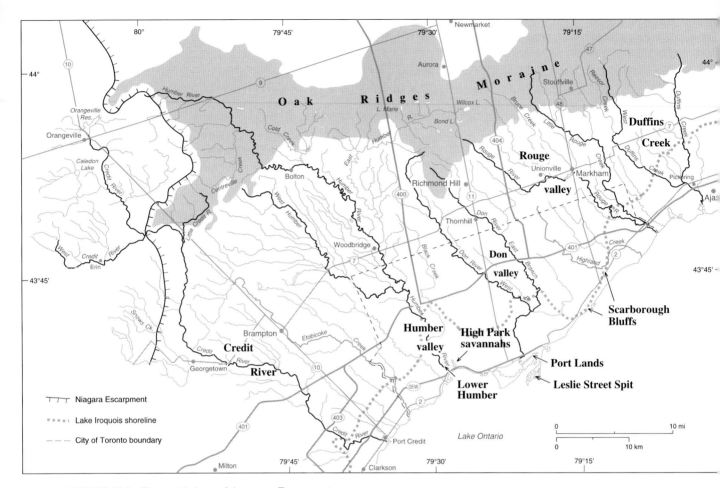

FIGURE 17.1 The special places of the greater Toronto region.

Waterfront Ecosystems: Restoring Is Remembering

Tom Whillans

THE LEGACY

Toronto's shoreline is an illusion – an impression of permanence nurtured in visitors by the apparent immutability of land and concrete, the short-term nature of most visits, and the common window of pleasant recreational weather. Yet shorelines are ephemeral features. The erosional Scarborough Bluffs and the depositional Toronto Islands – highly reshaped in most stretches by people – offer an alternative perspective. It is a perspective that history and biological pattern verify.

Perhaps nothing captures the spirit of the natural Toronto waterfront as effectively as the image portrayed by early writers, such as Kane (1859), Magrath ([1832] 1953), and Scadding (1873), of autumn spawning Atlantic salmon. In the early 1800s the jacklights of some 100 spearers could be seen at night on Toronto Bay, yielding hourly 20 to 50 fish per canoe. The endemic Lake Ontario population was greatly reduced by 1850 and eventually eliminated, probably by habitat change and fishing pressure. The last recorded endemic salmon were caught from the bay in 1874 (Kerr and Kerr, 20 June 1874) and from Lake Ontario, off Scarborough, in 1898 (MacCrimmon 1977).

Of course the bay was very different in the late 1700s and early 1800s. In 1834, the shallow north shore supported extensive emergent vegetation in front of 6-7 m high bluffs (Bonnycastle 1841). Emergent and submersed plants were so abundant in 1793 that they impeded navigation and required a mechanical weed cutter (Simcoe 1934). Inflowing streams were numerous, clear, and cool (Steedman 1986). Sand, gravel, and stone covered much of the bottom of the bay (Aitken 1793; Bouchette 1792; Vidal 1815). Excellent water clarity in 1793 allowed people to spear salmon (Simcoe 1934).

Like many animals that use shoreline waters, Atlantic salmon were present only seasonally during reproductive migration. Lake sturgeon, suckers, and redhorses also swam through the bay during spring spawning (Ure 1858; Nash 1913). Lake herring and lake whitefish migrated in huge numbers during the fall along the south shore of the Toronto Islands and Ashbridge's Bay Beach (Doyle 1893; Nicolls 1833). Lake trout spawned at about the same time in cobble-bedded shallows offshore from Scarborough, south of the islands, and off Port Credit (Kerr and Kerr, 7 March 1881). This seasonal biological glut, like waterfowl migrations, symbolizes the dynamic nature of nearshore ecosystems. The spawning fish sought warmer, well-aerated waters with bottom materials and food for their young. The shore zone functioned like a single-sided river. Heat, nutrients, and suspended sediments that had ingressed from rivers and the open lake were transported by currents along the shore. The resultant high potential biological productivity was attained where the long-shore flow eddied behind protective physical features.

> *In the early 1800s the jacklights of some 100 spearers could be seen at night on Toronto Bay, yielding hourly 20 to 50 fish per canoe.*

It is not now possible to find the same composition of fish species or environmental conditions in these locations. Ashbridge's Bay and Beach are under landfill; the mouth of the Don River is an armoured channel laden with contaminated sediment; Toronto Bay is much deeper and smaller in area. The cobble sought by lake trout between Whitby and Hamilton is presumed to have been removed between 1840 and 1900 by a large fleet of stone hookers, schooners specially modified for the purpose (Whillans 1979). Only the south shore of the Toronto Islands resembles years past, due in part to the artificial headland to the east, which has inadvertently protected the islands. Fish can still be found during spawning runs, but the suckers are the only historical species to persist. Patient observers will find spring congregations in gravelly areas below the lowest dams in the Don and Humber Rivers. The resident fish most likely encountered is the wide-ranging longnose dace. Exotic Pacific salmon – rainbow trout (actually salmon), coho, chinook, and pink salmon – run through the bay to spawn, apparently unsuccessfully, in the local rivers. Nonetheless, the exotics do impart a sense of the former immense salmon runs.

FIGURE 17.2
The 1912 construction of the eastern harbour facilities at Toronto, obliterating the Ashbridge's Bay marshes.
(The Toronto Harbour Commission Archives PC1/1/3041A)

Exotic rainbow smelt also run along these shores and up creeks, offering somewhat of a replacement for the herring and whitefish spawning phenomenon, but now in small numbers compared with 20 to 30 years ago.

Except the central part of Toronto Bay, which was naturally some 9 m deep, the Toronto waterfront from the west end of the Scarborough Bluffs to Humber Bay was essentially a large wetland complex. The rushes that lined the north shore of Toronto Bay followed the former shoreline approximately where Front Street now sits and extended around and into the Don River for about 1.3 km (Collins 1788). In spite of the early clearing efforts, rushes persisted at least through the 1850s (Hind 1852-3). Both Ashbridge's Bay and the Toronto Islands were known to be thick with emergent vegetation and also contained significant areas of wooded swamp. The Mead Hotel on the islands was described in 1887, for example, as surrounded by an 'island of sedge' (*Toronto Mail*, 20 October 1887). Conditions for submersed aquatic plants in Toronto Bay remained good in the 1870s – abundant stonewort, waterweed, and wild celery indicated reasonably clear water. Faunal surveys in the sediment of 1861 and 1872 revealed some six types of mollusc, five of which were widespread in the bay: *Amnicola, Limnaea, Physa, Planorbula,* and *Valvata* (Nicholson 1872; Williamson 1861). This record contrasts tellingly with the three species of sludgeworm (Tubificidae) that dominated exclusively by the 1970s (Brinkhurst 1970).

TRANSFORMATION

In nineteenth-century southern Ontario, wetlands pastured livestock. Hay was in short supply, particularly before the land was cleared and machines replaced beasts of burden. As occurred in the Maritimes and still does on the prairies, the naturally productive marsh grasses and sedges provided common and private fodder. Ironically, many of the marshes that sustained the livestock of pioneers are among the approximately 80 percent of original southern Ontario wetlands that Snell (1987) determined have been converted to other land use – mostly agricultural.

One of the largest marshes in the greater Toronto region was Ashbridge's Bay at the urbanized mouth of the Don River. It represented a twist on the pattern of abuse. In the 1880s its 526 ha received the nutrient-rich runoff of cattle byres associated with the Gooderham and Worts distillery and was by numerous accounts highly polluted, resulting in large fish kills (Kerr and Kerr, 25 July 1881; 21 May 1883). Some 10,000 cattle were processed by that facility in 1883 alone (Kerr and Kerr, 14 August 1883). The putrid condition of the area was used to justify the decision to fill in the marsh in order to construct port facilities in the years around the First World War (Figure 17.2).

Many wetlands along the Toronto regional waterfront suffered a similar fate. Approximately 91 percent of the original wetland between Mimico Creek and Frenchman's Bay was lost, mainly to filling (Whillans 1982). Wetlands of Lake Ontario's northern shoreline must endure harsh forces. The lake levels fluctuate by several metres over decade-scale cycles. Cold waters upwell regularly into the otherwise warm summer shallows. Strong onshore winds, seiches, currents, ice action, and storms buffet the coast. Wetlands have persisted only behind protective bay-mouth bars or in deeply incised bays. Most often these conditions are afforded in the mouths of streams and rivers that occupy the ancient over-sized valleys of periglacial rivers. Even here, the conditions have by no means been constant: river floods and the storm-whipped lake have contributed historically to the breaching of most natural protective structures (Whillans 1980).

The exceptional sand spit that formed the Toronto Islands sheltered the wetlands of Toronto and Ashbridge's Bays. It was human-fortified after storms in 1854 and 1858 created the eastern gap, reportedly because the rate of longshore and offshore transport of sand from the spit began to exceed the supply of sand imported by currents from the Scarborough Bluffs (Fleming 1853-4). Along the north shore of Toronto Bay and around the islands, much of the natural wetland was filled by the late 1880s (*Toronto Mail*, 20 October 1887). Landfilling technology has evolved to produce, over the past 30 years, major artificial headlands at Sam Smith Park, Humber Bay Park, Bluffers Park, and the Leslie Street Spit (Tommy Thompson Park). The latter, in particular, has imparted the stability to the Toronto Islands that they never enjoyed naturally.

THE ENDURING ECOSYSTEM

Recent conditions along the Toronto waterfront are revealing. On the open, exposed shorelines, whether natural or constructed, the aquatic flora consists primarily of attached filamentous algae, sometimes growing more than a metre long. Bottom-living fauna are scarce. The scud is one small (1 mm long) shrimp-like invertebrate that abounds. It can reliably be found in clumps of filamentous algae. Lately, exotic zebra mussels have become predominant on hard surfaces. A few fast resident fish are also found, such as emerald shiners, spottail shiners, and the exotic alewife.

Along protected Lake Ontario shores the picture is very different. Even though plant growth is minimal because of poor water clarity, the diversity and abundance of bottom-living invertebrates and fish is considerable. In addition to the scuds, there are aquatic sow bugs, dragonfly larvae, backswimmers, fly larvae, snails, and many others. Typically in a shoreline wetland eight to ten fish species can be found. The Humber Marshes are a case in point. The marshes are located just upstream of the mouth of the Humber River. They consist of cattails on levee banks that are virtually all elevated above the normal river water level. This and the highly turbid water have resulted in little rooted plant growth in the water itself. In spite of this and the outfall of a sewage treatment plant near the river mouth, the diversity of fish is remarkable. Species include: gizzard shad, spottail shiner, longnose dace, creek chub, white perch, pumpkinseed, largemouth bass, and Johnny darter.

Approximately 91 percent of the original wetland between Mimico Creek and Frenchman's Bay was lost, mainly to filling.

Admittedly, this variety pales in comparison with the natural wetlands farther east along Lake Ontario – Rouge Marsh, Frenchman's Bay Marsh, Duffins Creek Marsh, and Carruthers Marsh – where upward of 25 species of fish might be expected, dominated by esocids (pike), centrarchids (sunfishes), and cyprinids (minnows) (Stephenson 1990). Nevertheless, it illustrates the potential offered by protection alone. The potential is not simply in terms of species presence but also in terms of ecological function such as spawning, nursery, and foraging.

THE RECLAMATION

Most of the protection from external forces that was once afforded by creek mouths and natural sand spits along the Toronto-area waterfront is now gone. Either the inlets have been filled or the built environment is so oppressive that natural ecological integrity has been compromised. The nature of the human alterations is such that their reversal or elimination would be impractical, both physically and economically. Yet one of the lessons of the historic north shore ecosystem is that natural changes in shoreline were frequent. Many, related to water levels and climate, were also cyclical. Historical records describe how in certain years emergent vegetation in marshes was eliminated by high water levels, fish were prevented from spawning by shifts in baymouth bars, or the littoral zone was exposed by low water levels.

Normally, when the causal factors subsided, the natural successional process of recovery followed. Great Lakes littoral ecosystems are typified by a remarkable ecological memory. In the case of emergent vegetation communities, this memory is afforded in part by seed banks that are dormant during poor environmental conditions, but reactivated by suitable conditions. The various mechanisms of homing constitute one form of ecological memory in fish communities. Small microhabitat refuges represent ecological memory in patchy benthic (bottom-living) invertebrate communities. Ecological memory of these types explains the direct recovery of former communities in disturbed locations. Where seed banks have been destroyed or migratory fish runs have been extirpated or benthic sediments have been rendered inhospitable over the complete area of distribution, then recovery depends upon a different sort of memory.

Ecological memory also exists at the ecosystemic level. The artificial spits that were constructed along the Toronto-area waterfront provide examples of this. In protected locations of the artificial headlands, flora and fauna whose habitats were eliminated over the past 200 years have established themselves. Thus prairie plants now exist where no local seed source occurred. Fish spawn in newly created locations where their homing could not have told them suitable conditions can be found. Aquatic invertebrates abound where deep water would have previously precluded a local refuge from existing. The variability of natural ecosystems has enabled physiologically suited and behaviourally adaptive species to seek and/or colonize created embayments and shores.

The opportunity and challenge for restoration ecologists who embrace the Toronto-area shorezone ecosystem is both to monitor the natural colonization and to 'reconstruct' the historical understanding of the fundamental regional habitats as a basis for habitat enhancement and strategic 'naturalization' (that is, the reestablishment of endemics). Ecological development would thus be based on natural inclinations and historical precedents. Ecological functions may not be restored exactly where they were historically, but the artificial headlands offer the fortuitous possibility that they can at least be restored regionally. Emerging from ecological and human memory is a vision for the future.

The Port Lands: The Significance of the Ordinary

Michael Hough

This section is about the Port Industrial District on the Toronto waterfront, known as the Toronto Port Lands. This area is part of the larger Lower Don Lands that stretch from Lake Shore Boulevard south to the North Shore Park, and from Leslie Street west to the inner harbour (Figure 17.3). It is where you will find Metro's hazardous waste depot for dumping old paint cans and used batteries. It is where, at one time, you could get cheap used carpet, old doors, lumber, or scrap wrought iron. Yet for some it's a good place to pick dandelion leaves and local herbs for the kitchen. For others it is for cycling along the Martin Goodman Trail (Figure 17.4), bird watching en route to the Leslie Street Spit, looking out over the lake, or watching the sail boarders negotiating wind and waves in the outer harbour.

But for many others, the Port is, to say the least, run down. The place is a visual mess. Many of its old industrial buildings are falling apart. There is litter everywhere: hunks of concrete and building materials lie scattered about, like gigantic toys left after the kids have gone to bed (Figure 17.5). Outdoor storage areas, transformer stations, rusting ships (Figure 17.6), unmaintained road verges, and overgrown vacant lots, all give clear signals that this is a derelict and, some would say, just plain

FIGURE 17.3 Aerial view of the Toronto Port Lands looking west. (M. Hough and Waterfront Regeneration Trust 1997)

The Special Places

FIGURE 17.4 Cyclists, roller bladers, walkers on their way to the Spit. (M. Hough)

FIGURE 17.5 Scrapyards on the Port Lands. (M. Hough)

FIGURE 17.6 A rusting abandoned ship in the Port Lands. (M. Hough)

ugly part of the waterfront. There is, arguably, nothing 'special' about it – hardly a place to bring dignitaries visiting Toronto.

THE SPECIAL AND THE ORDINARY

Each of these observations tell us something different about the Port Lands. But if this place *is* just plain ugly, why is it appearing in this book? There is, after all, so much of this kind of despoiled urban environment around us, and so little that remains pristine, biologically significant, beautiful – those natural places far from the city, in fact, about which other contributors to this book have so eloquently written. Why do so many people still think of cities as places where nature has been abandoned? And what lies behind this idea of 'specialness'? Answers to these questions lie in challenging perceptions about what is significant, ordinary, beautiful, or just plain ugly. There's a need to look beneath the façade of abandoned urban areas – 'peeling back the rubber mask of the present' as the distinguished stream hydrologist Robert Newbury once said of the Don River – to gain insights into their past and into the environmental, social, and economic forces that made them what they are today (Newbury, personal communication).

I therefore argue three things. First, true observation of the places around us requires that they be seen from the perspective of both human and natural history. Second, ordinary places are often those that we pass by without noticing. Once they are observed and understood, the chances of their becoming special, in the context of the forces that shaped them, are immeasurably heightened. This is so of any urban or non-urban landscape, if we define *landscape* as natural history modified and reshaped by human culture over time. Third, urban landscapes are in constant transition and change, responding to cultural, economic, and natural forces. Nothing stays still. The Port Lands provide an appropriate setting for examining these issues. So this section should be read as a tour guide to the experienced, observed landscape. It is about understanding and reflecting on what is there, rather than a scientific treatise on its flora and fauna.

Looking More Closely

One's first impression of the Port Lands is their flatness. What made them this way? As anyone who has studied the history of Toronto's waterfront will know, this was, prior to 1912, the Ashbridge's Bay marsh, one of the largest and most productive coastal wetlands in the Great Lakes system. As the city grew, sewage and wastes from the Gooderham and Worts' cattle byres, combined with a breakwater built to stop siltation of Toronto Bay in the last century, created stagnant water and pollution in the marsh. This prompted a local newspaper of the day to comment that it was 'a malarial swamp ... teeming with pestilence and disease' (Royal Commission 1990). In 1912 the Toronto Harbour Commission unveiled its plan for a new port to be constructed by landfilling the marsh, a plan that was implemented over several decades, and by 1936 the framework for the port had been completed. Thus the overall character of this new industrial landscape has come to reflect the flat topography of the original marshes. Ironically, while development forces have traditionally flattened natural landforms elsewhere, here there was no need to do so (see Figure 17.2, p. 246).

Other impressions settle on the observer of the Port Lands – its regenerating vegetation and the birds and animals it supports. It is clear that this is a landscape in ecological transition. There is nothing left of the original marshes that once existed, yet productive terrestrial and aquatic habitats abound, created by its low-lying landscape, the lake sands that were used as fill, the fluctuating lake levels, the results of human intervention, and the forces of nature. Seasonally flooded wet meadows and marshy areas can be found at the base of the Leslie Street Spit, along ditches, and in vacant lots that have, over time, come to support breeding geese, ducks, and songbirds. The outer harbour provides spawning habitat for a wide variety of forage and predatory fish. The open fields and shrubby areas provide feeding and breeding habitat for birds such as savannah sparrows, horned larks, eastern meadowlarks, killdeer, and bobolinks (Royal Commission 1990). Overall, some 300 species of plants and over 260 species of birds have been identified in the Port (Kalff, McPherson, and Miller 1991). Snakes find a home in the massive and curiously sculptured concrete structures that have been dumped on the site from demolished buildings and public works projects, and the coyote, a prime carnivore, has moved through the Port Lands to take up residence on the Leslie Street Spit.

> *True observation of the places around us requires that they be seen from the perspective of both human and natural history.*

Woven into this tapestry of re-generating life systems is the continuing human history of the Port and the Don River. Development and industrial activities in these areas once destroyed the original natural environment and left a legacy of contaminated soils, groundwater, and sediments. The Keating channel, for instance, contains an estimated 95 percent of the contaminants flowing from the entire watershed (Task Force, 1991). Yet this legacy of destructive intervention and transformation is also linked to, and shaped by, the industrial heritage of the Port Lands, from which fortuitous ecological renewal has emerged (Figures 17.7-17.12).

RELICS FROM AN INDUSTRIAL PAST

The Port Lands are rich in industrial artefacts that speak to the area's unique history. The magnificent lakers – a few still operational (compared to the rusting hulls formerly seen, Figure 17.6) – docked along the ship channel and docks are a mute testimony to Toronto's historic

FIGURES 17.7-17.12
The Port area: evolution and change in an abandoned oil storage site, as observed and recorded by the author over 20-5 years. (M. Hough)

FIGURE 17.7 Spring, late 1970s, yearly flooding from seasonal rains.

FIGURE 17.8 Spring, 1983, water collects in the depressions around oil tank bases, designed to collect oil in the event of a spill. A wetland community begins to form.

FIGURE 17.9 Several years later a diverse 'natural' community emerges, including wetland vegetation, wet and mesic meadow. A copse begins to establish itself on the bases of the former oil storage tanks.

FIGURE 17.10 Within less than a hectare, these habitats support nesting geese, ducks, muskrat, and songbirds.

FIGURE 17.11 Summer, 1987, the site is bulldozed. Landfill replaces habitat and fills in the depressions.

FIGURE 17.12 1992 to the present, regeneration of the disturbed site takes place. Wetland plants and a meadow community re-emerge on the original wetlands. A new forest of poplar and aspen has begun to dominate drier areas.

FIGURE 17.13 The bascule lift bridge crossing the ship channel. (M. Hough)

(29 species were recorded in 1983, compared to 15 collected in the summer of 1990 along the north shore of the outer harbour) (Royal Commission 1991). It was also a favoured location for wintering birds – ducks, geese, gulls, and unusual predators such as barred owl, northern goshawk, and gyrfalcon (ibid.).

Magnificent views of the inner harbour and city skyline from Polson Quay speak to the extraordinary waterfront location of the city. The first working port of Toronto is today Harbourfront, the place where the city and the lake meet. The contrasting pastoral qualities of the Toronto Islands, the inner harbour that both separates and unites them, and the diverse flora and fauna of the Leslie Street Spit have immeasurably contributed to the city's rich diversity of human and non-human habitats and visual drama (Hough Stansbury et al. 1987). During the summer the Port Lands are alive with people, drawn irresistibly to this unique place that has emerged from an industrial legacy (Figure 17.14).

role in international trade on the Great Lakes. While the Redpath Sugar quay at the foot of Jarvis Street is still operational, international shipping has now dramatically declined. The Cherry Street bascule bridge, which crosses the ship channel and lifts to allow boats to pass into the inner harbour, is a marvel of the industrial technology of its day (Figure 17.13) The silos, cranes, and railway spurs, and the Hearn Generating Station, are reminders of past and present port activities and power generation associated with the port (Royal Commission 1991). Up to 1983, when the Hearn Generating Station was producing electrical power, warm water from its outfall attracted a greater diversity and abundance of fish species than are now present

Spring and fall migrations of birds and butterflies are reminders, too, that the Port forms a crucial wildlife connection between the Leslie Street Spit, the Don Valley System, and regions north and south. As an urban wilderness peninsula projecting into the lake, the spit has become internationally recognized for its extraordinary natural evolution since land filling was first begun by the Toronto Harbourfront Commission in 1959 to create a new outer shipping harbour. In effect, the artificially created spit fulfils migration and habitat functions similar to other natural peninsulas such as Rondeau Provincial Park and Point Pelee National Park, both on

FIGURE 17.14
The Port Lands come alive with people in the summer. (M. Hough)

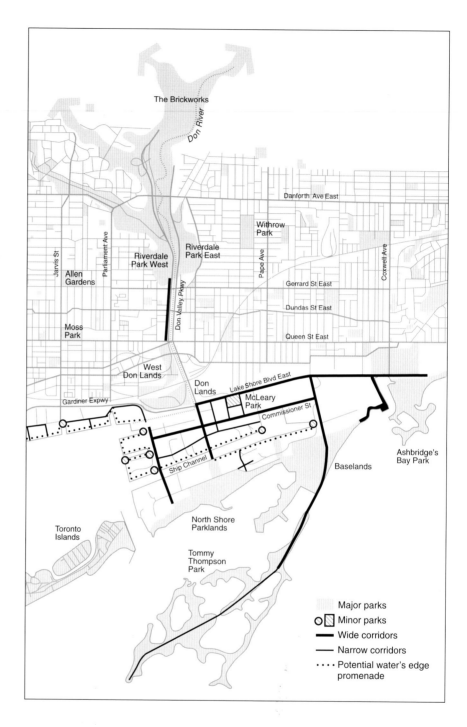

FIGURE 17.15
Long-range planning for the future renewal of the Port Lands: a green infrastructure system will establish habitat links between the Spit and the Don, reinforcing stop-over habitats for migrating birds and butterflies in spring and fall.
(After M. Hough and Waterfront Regeneration Trust 1997)

Lake Erie. The combination of regenerating habitats in the Port's North Shore Park, Cherry Beach, the base of the spit, and in vacant lots, extends these stopover functions. The spit is also where we can observe wave action along the outer shoreline grinding up concrete, brick, and stone into rounded and often beautiful shapes – the product of less than 40 years of lake forces. And it is where wind, sun, the deafening sound of gull and tern colonies in the nesting season, and the limitless vistas of the lake and city backdrop leave you breathless with the experience of urban wildness – a place shaped by people and brought to renewed life by nature.

WHAT IS THE FUTURE OF THE PORT LANDS?

As in many other North American cities, the decline of port-related industries has spurred vigorous debate among local politicians, environmentalists, and labour and community groups about appropriate uses for the Port Lands. In the late 1990s, a consensus has emerged that an industrially zoned area is essential to attract secure, well-paying industrial jobs to the downtown waterfront. At the same time, local residents and environmental groups have insisted that new industrial uses must not impair the Port Lands natural areas. A green infrastructure system for the Port Lands was subsequently proposed by the Waterfront Regeneration Trust (Hough and Waterfront Regeneration Trust 1997) and supported by the Toronto Economic Development Corporation, the city's development agency for the Lower Don Lands (Figure 17.15). Its purpose is to protect the special nature of the area and provide a regional framework for development.

Six basic principles informed the planning strategy. The first principle is to provide a multifunctional environmental framework for development in the Port Lands. Economic, environmental, biological, recreational, and visual values provide the essential framework for the green infrastructure system and for those who work and visit the area. The plan is to establish a network of parks and habitats that will provide an attractive and functionally useful setting for future development and public use (Figure 17.16). The second principle is to protect and restore health and biodiversity. The ecological and human health of the Port Lands – its land, air, and water – must be re-established in the context of the natural systems and human history that have shaped the area. The third principle is to create linkages. Biological and recreational connections between the Lower Don River and the lakefront should be established. Fourth, the waterfront context must be recognized. The hydrological, water quality, habitat, and public access relationships among the Port Lands and the Don River and its watershed should be included when making decisions about the green infrastructure. Fifth, redevelopment should both improve the Port Lands' image and reinforce its sense of place. The redevelopment process should take advantage of, and build on, the special industrial, cultural, and ecological history of this waterfront location. An urban design character that is separate and distinct from other districts in the city should be encouraged. Sixth, the community must be involved. No regeneration process can be successful without the participation of different groups in the decision-making process. They include government agencies, interested citizen organizations, existing businesses, potential development interests, and the general public.

A FINAL WORD

In August 1997 the process of adopting the green infrastructure system as official policy for the future planning of the Port Lands was initiated and supported by the city. The agreement, however, illustrates one of the perennial ironies of such urban landscapes: their natural evolution has been a result of neglect, rather than intent. There is a long-established abhorrence in our

FIGURE 17.16 Corridors and natural parks provide the multifunctional framework for biological continuity and future urban renewal. (Hough et al. and Waterfront Regeneration Trust 1997)

culture of abandoned places. Most transitional landscapes are judged weedy, unkept, or ugly in contrast to the conventional 'beauty' of the civic landscape we find in the city where nature remains firmly under control. Yet, the manifestation of natural processes occurring in urban areas begs the question about the nature of dereliction and what must be done to restore these areas to health. Which *are* the derelict places in the city requiring rehabilitation? Those fortuitous and diverse landscapes representing natural forces at work, or the formal landscapes created by design or by single-minded economic forces?

At least one answer for the Port Lands lies in the way we can reveal the special in ordinary places. It lies in recognizing the interrelationships between the physical and biological processes that made this place what it once was, the human uses that have changed it, and the integration of natural processes and social and economic needs that can ensure its future health. This place is clearly not pristine. To look for this is missing the point. It is a landscape in the process of continual change and renewal, one different from the original marshes that preceded it. The Port Lands are a mix of naturally succeeding and diverse flora and fauna, industrial heritage, and future business; of human use, study, and leisure. This is a special place that reveals itself when we can look and marvel at its complex inner beauty.

Scarborough Bluffs

John Westgate

There is perhaps no other portion of the great inland basin of North America where the strata, showing the different changes which have occurred from the commencement of the Glacial period up to the present, are better displayed than along the shores of Lake Ontario ... On its northern shores ... the banks of the lake are generally low and without features of importance, [but] this general deficiency is more than compensated by the display, perhaps unequalled anywhere round the lake, of glacial strata at the Scarboro' Cliff (Hinde 1878).

These are the opening words of a paper read by George Jennings Hinde before the Canadian Institute on 3 February 1877 and published in its journal in 1878. As a result of Hinde's studies, 'the magnificent shore cliffs at Scarborough ... have the high distinction of being the first place in America where an interglacial formation

FIGURE 17.17 Scarborough Bluffs at the western end of Bluffers Park, Brimley Road. The lower half of the cliff is composed of sands and clays of the Scarborough Formation; the upper half is made up of the Sunnybrook Drift. This site is sometimes referred to as the Dutch Church section, a name coined for its spectacular, although ephemeral, spires. (John Westgate)

Naming the Bluffs

During the French period, the name for the Scarborough Bluffs from Toronto to Port Union was *la grande écorce*, a picturesque comparison of the layered bluffs to the layered bark of a tree. The shore from Port Union to about Moore Point was *la petite écorce*, and from there to the Rivière Saumon (Rouge River) was *presque petite écorce*.

In 1793, the Simcoes arrived at Toronto and began changing Native and French place names to English names. The Don River was given its present name, and le Ruisseau Saint-Jean (earlier Rivière de Toronto) became the Humber. Mrs. Simcoe noted in her diary on 4 August that she and her party 'came in sight of what is named in the map the high lands of Toronto. The Shore is extremely bold & has the appearance of Chalk Cliffs but I believe they are only white Sand. They appeared so well that we talked of building a summer Residence there & calling it Scarborough.' Scarborough is a town in Yorkshire on the northeastern coast of England. The influence of Yorkshire place names was also reflected in the naming of York County and in the decision of Simcoe to change the name of his new settlement from Toronto to York.

was recognized' (Coleman 1933). Meetings in Toronto of the British Association for the Advancement of Science in 1897 and the International Geological Congress in 1913 brought international attention to this site, and interest continues to be strong today. It is evident, therefore, that the Scarborough Bluffs have great stature not only in the physical sense (Figure 17.17) but also in the context of their scientific importance as a detailed repository of evidence for environmental change during the recent geological past.

The Scarborough Bluffs begin at Victoria Park and extend northeastward for 15 km to Highland Creek (see Figure 1.4, p. 18). Over much of their extent, they are capped by the wave-cut Lake Iroquois terrace. However, about 1 km to the northeast of Bluffers Park, Lake Iroquois Bluff intersects the modern shoreline of Lake Ontario, producing a cliff that stands about 107 m above the lake. This marks the highest point along the Scarborough Bluffs and is where the most detailed glacial sedimentary record is exposed – a sequence that can be readily accessed along the ravine behind Fairmont School.

The geological story, deciphered from the sediments exposed at the bluffs, is based on observations made by a number of scientists over more than 100 years of study. Of course, we can never realize the true and complete story because the evidence preserved in the sediments is fragmentary and our ability to determine accurate and precise ages for these deposits is still poor (Figure 17.18). Furthermore, the picture is clouded in places by different interpretations of the same evidence. Nonetheless, these sediments and the fossils they contain have permitted reconstruction of an unusually detailed story of environmental change during the last 100,000 years.

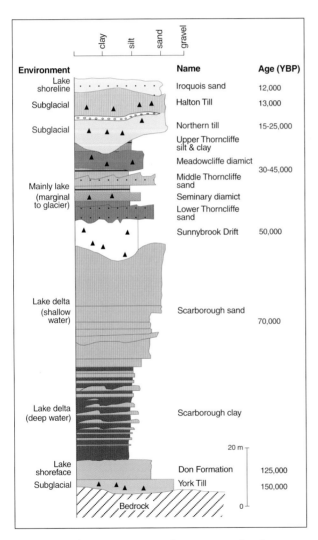

FIGURE 17.18 The composite sedimentary record at the Scarborough Bluffs. The exposed part begins 5 m above the Don Formation. Approximate age is given in years before present (YBP). (Nicholas Eyles)

The thick sedimentary pile at the Scarborough Bluffs can be attributed to the underlying large bedrock valley, known as the Laurentian Channel (Figure 1.6, p. 20). The oldest, exposed sediments are deltaic clays, silts, and sands of the Scarborough Formation that crop out along the entire length of the bluffs. They were deposited by a river flowing south along the Laurentian Channel, eventually debouching into a large lake. The top of these deltaic deposits is about 45 m above the level of Lake Ontario, indicating a relatively high lake level, which was probably caused by a glacier occupying the northeastern part of the basin. As the delta built outward into the lake, the deep-water clays became covered by sands, deposited in shallower waters. Organic material, eroded from the land, was also flushed into the lake and preserved as thin, discontinuous beds or dispersed within the inorganic sediments. Insects, pollen, and plant macrofossils indicate that most of the Scarborough Formation was deposited during a boreal climatic regime, when the mean annual temperature was 2.5°C – about 5°C lower than present. However, fossil beetles in the uppermost sands point to cooler, subarctic conditions, with a mean annual temperature of about 0°C – about 7°C lower than present.

Valleys up to 1 km wide and 70 m deep have been cut into the deltaic Scarborough Formation (Figure 17.19). Some authorities have interpreted them as signifying a drop in lake level – perhaps due to retreat of the glacier out of the Ontario Basin – thereby energizing streams, which then eroded into the deltaic deposits. Others believe these channels may be related to formation of the delta, in which case no change in lake level is required. One of these channels occurs at the Dutch Church section, just to the west of Bluffers Park.

Sunnybrook Drift fills the channels and drapes the delta. It consists mostly of pebbly mud or diamict, but in places (e.g., Dutch Church) the uppermost sediments are thinly bedded silts and clays with dropstones. All experts agree that the latter beds originate in lakes that fronted close to the glacier, the dropstones falling onto the lake floor as they melted out of the floating ice, but the origin of the diamict unit is still debated. One school of thought argues for a subglacial origin, the diamict being plastered onto the ground as the glacier moved across the greater Toronto region. Adherents to this view use the name Sunnybrook Till. The alternate view is that the entire unit is glaciolacustrine, the diamicts being formed by 'rainout' of debris from the floating ice margin onto the lake floor, with the potential for subsequent reworking by gravity processes or bottom

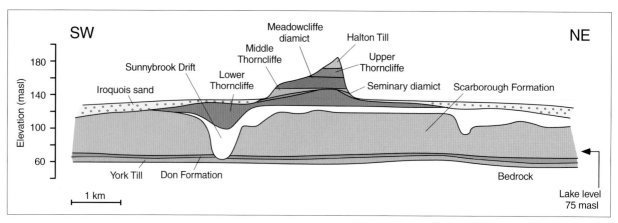

FIGURE 17.19 The distribution of sedimentary beds exposed along the Scarborough Bluffs. The easternmost part of the bluffs, near Highland Creek, is not shown on this figure. (Nicholas Eyles)

currents. Scientists who favour this view use the name Sunnybrook Diamict. In either scenario, climatic conditions are cold and harsh. However, there is evidence that some life existed in these cold, turbid, ice-marginal lakes. The silty beds commonly show signs of having been disturbed by living organisms, pointing to the presence of animal life, and ostracod remains have been recovered from both the stratified and diamict units.

A package of interbedded diamicts and stratified sands, silts, and clays up to 45 m thick covers the Sunnybrook Drift. It includes the Thorncliffe Formation and the Seminary and Meadowcliffe diamicts, and is exposed in the central part of the bluffs. Their planar, conformable, layer-cake geometry, the presence of soft-sediment deformational structures at their bedding contacts, and the lack of any major erosional breaks in the sequence suggest a common origin in a high-level, ice-marginal lake. This interpretation requires the nearby presence of a glacier to maintain the high lake levels throughout this period of sediment accumulation. The major changes in sediment type can be explained by repeated delta growth into the basin and accumulation of sands over diamicts that formed as glaciolacustrine bottom muds with a coarser ice-rafted component.

The Laurentide Ice Sheet moved across the greater Toronto region about 25,000 years ago and deposited the Northern till, which can be seen at the easternmost part of the bluffs, close to the mouth of Highland Creek. This till reaches a thickness of more than 10 m, is grey, sandy, and stony, contains discontinuous sandy beds, and exhibits a highly erosive contact with the underlying Scarborough Formation.

Regionally extensive sands and gravels between the Northern till and the overlying Halton Till – but not seen at the bluffs – show that a brief ice-free episode preceded deposition of the Halton Till by the Ontario glacier lobe, which moved in a northwest to north-northwesterly direction across the Toronto region about 13,000 years ago. This youngest till is exposed at the highest point of the bluffs, just east of Bluffers Park.

Thin, discontinuous glaciolacustrine sands, silts, and clays on the Halton Till bear testimony to the presence of small, ephemeral lakes, ponded against the margin of the Ontario lobe during its final retreat from the area. Twelve thousand years ago this glacier lobe was confined to the Lake Ontario Basin but had stabilized its position for a period long enough for significant wave erosion to occur along the littoral zone of its ice-marginal lake, called Lake Iroquois. The resulting wave-cut terrace, with its cover of beach sands and gravels, is well developed at the Scarborough Bluffs as is its associated bluff (see Figure 1.9, p. 22). Colluvial or slope deposits, in which archaeological materials have been discovered, cover part of the Iroquois Bluff and extend out onto the terrace. They constitute the youngest sediments preserved along the Scarborough Bluffs, excluding the modern beach.

About 1 km to the northeast of Bluffers Park, Lake Iroquois Bluff intersects the modern shoreline of Lake Ontario, producing a cliff that stands about 107 m above the lake.

The steep, dramatic slopes of the Scarborough Bluffs are disappearing. Whereas natural forces were able to maintain them by removal of sediment from the beach zone by wave action, the recent construction of Bluffers Park and of protective berms along the bluffs now prevent wave erosion at the base of the bluffs, leading to more gently inclined slopes, which can be readily colonized by vegetation. As a consequence, the quality and quantity of geological exposures is rapidly diminishing. Presently, berms are being built in front of the famous Dutch Church site, so deterioration of this classic geological exposure is imminent.

The steep, dramatic slopes of the Scarborough Bluffs are disappearing.

Some argue for the bluffs to be preserved as a wilderness area, thereby ensuring continuance of their rugged topography and protection of their associated fauna and flora, but others counter with the view that developments like Bluffers Park allow more people to see and enjoy this scenic treasure. Perhaps the concept of a wilderness area within the confines of the ever growing, populous Toronto region, where Nature can do its work unfettered by humans, is unrealistic, given the associated hazards – but that's another story.

The Savannahs of High Park

Steve Varga

In 1819, John Goldie, on a botanical exploration of North America, stumbled onto what he called 'as good a Botanical Spot as any that ever I was in' (Goldie 1819). He was describing a vast sandy area known as the Humber Plains, extending from the lower Humber River east to what is now High Park. The sands were laid down over 12,000 years ago as a delta in glacial Lake Iroquois. Following the retreat of this water, several creeks cut deep ravines into the sands around High Park.

Goldie and other early naturalists saw an amazing sight on their visits. Near the Humber, the expansive dry sand plain was kept open by frequent ground fires started largely by lightning strikes. The only trees on the plain were small copses of sassafras, the occasional gnarled and stunted black oak, and white pine. Much more common were low shrubs of dryland blueberry and New Jersey tea intermixed with prairie grasses having such evocative names as Indian grass, big bluestem,

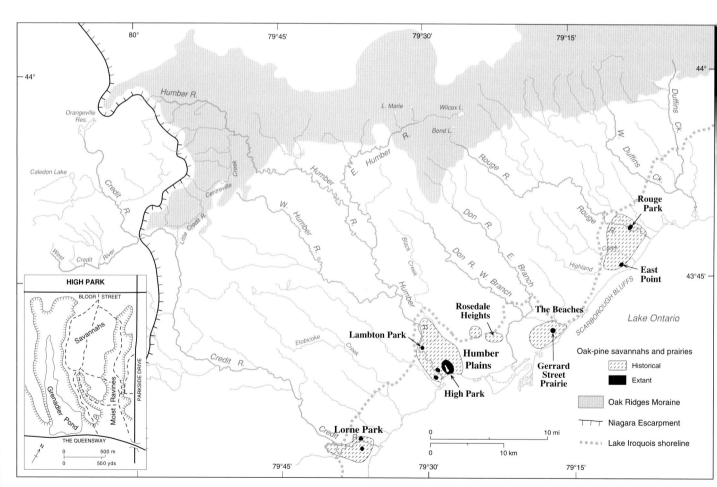

FIGURE 17.20 Historical and present locations of prairies and savannahs in the greater Toronto region. Inset shows High Park's historical savannah/prairie of the Humber Plains and its present extent. Note the location of Grenadier Pond and roadways.

John Howard and High Park

John George Howard, c. 1870.
(City of Toronto Archives SC 268-269)

John George Howard (1803-90) emigrated from England to York, Upper Canada, in 1832. One of the earliest trained architects in Toronto, Howard was also drawing master at Upper Canada College and city surveyor, city architect, and city engineer at various times between 1834 and 1855. He built Toronto's first underground brick sewers (1843) and laid out the grounds at St. James' Cemetery (1844), but Howard's fame comes mostly from designing the Queen Street Mental Asylum (1846) and giving High Park to the people of Toronto.

In 1836, Howard purchased 68 heavily wooded hectares on the lakefront far to the west of Toronto. Within a year he had erected Colborne Lodge, a key example of Regency domestic architecture. Howard advertised villa lots at High Park twice during the 1850s, but these ventures were failures. In 1873, city council agreed to take High Park in exchange for a $1,200 annuity. Howard was appointed park ranger, set about improving the grounds, and, in 1876, persuaded council to buy an adjacent 71 ha estate for $15,000.

Howard was the first great donor of, and advocate for, ecologically significant lands in the Toronto area. High Park remains one of the largest and most important natural parks in the city.

---

and little bluestem. Throughout the growing season, the plain was ablaze with colourful wildflowers. In the early spring there were the yellows of prairie buttercup and star-grass and the blues of arrow-leaved violet, while in June the yellow frostweed, the orange wood lily, and the blue wild lupine were everywhere. Flying in their midst were butterflies such as the karner blue, whose larvae fed on the lupines, and the spicebush swallowtail and scrub-oak hairstreak, which relied on the southern trees of sassafras and black oak.

Around High Park, less frequent ground fires permitted open woodlands to develop, with large spreading black oak trees resembling giant octopi. These trees were set in a naturally park-like landscape of prairie grasses and showy wildflowers. By the time the lupines were setting seed at summer's end, the other wildflowers were putting on their best display. There were the yellows of woodland sunflower, the whites of bush-clovers, the blues of harebell and azure aster, and the pinks of the cylindrical blazing star. Sliding among the grasses was the shy hognose snake, which used its upturned nose to dig its burrows into the soft sandy soils.

> *Savannahs and prairies are now considered one of the most endangered ecosystems on the continent.*

On High Park's moister ravine slopes, there were forests of red oak, red maple, and white pine, and, on cooler north-facing slopes, groves of eastern hemlock harbouring such northern boreal plant species as goldthread, blue-bead lily, and bunchberry. Extensive seepage areas on the ravine bottomlands supported large marshes and speckled alder swamps. Grenadier Pond, in the southern portion of the park, was ringed by sedges, water-willow shrubs, the showy blue flowers of pickerel-weed, and the white flowers of arrow-heads. Grenadier was a lakeshore marsh separated from Lake Ontario by a narrow sand bar. The marsh was home to such birds as the American bittern, least bittern, marsh wren, and Virginia rail as well as many dragonfly species, including such southern ones as vesper bluet and halloween penitent.

These early descriptions of open woodlands at High Park and the plains around the Humber dispel the belief that Ontario was covered in one unbroken stretch of forest. In fact, Ontario once supported over 10,000 ha of prairies and open woodlands known as savannahs, with plants and animals more typical of the tall grass

FIGURE 17.21 Typical plants of the High Park oak savannahs: A – hairy bush-clover (*Lespedeza hirta*), B – pinweed (*Lechea intermedia*), C – big bluestem (*Andropogon gerardii*), D – cylindrical blazing-star (*Liatris cylindracea*), E – harebell (*Campanula rotundifolia*), F – round-headed bush-clover (*Lespedeza capitata*), G – Indian grass (*Sorghastrum nutans*), H – little bluestem (*Schizachyrium scoparium*), I – Canada frostweed (*Helianthemum canadense*), J – wild lupine (*Lupinus perennis*), K – slender-stemmed cyperus (*Cyperus lupulinus*). (Steve Varga)

FIGURE 17.22 Typical plants of the High Park moist ravines: A – rose-twisted stalk (*Streptopus roseus*), B – smooth sedge (*Carex laevivaginata*), C – mountain maple (*Acer spicatum*), D – foamflower (*Tiarella cordifolia*), E – blue-bead lily (*Clintonia borealis*), F – barren strawberry (*Waldsteinia fragarioides*), G – goldthread (*Coptis trifolia*), H – bunchberry (*Cornus canadensis*), I – trailing arbutus (*Epigaea repens*). (Steve Varga)

FIGURE 17.23
J.E.H. Macdonald, *Spring Breezes, High Park,* oil on canvas, 71 x 92 cm (National Gallery of Canada, Ottawa, purchased 1948, #4874)

prairies of the Midwest than of the Ontario forest. They were concentrated on the province's drier sand plains, where periodic natural fires kept back the forests. The largest prairies and savannahs in the greater Toronto region were found on the Humber Plains, including the High Park area (Figure 17.20). Other smaller examples were located on the sand plains around the Beaches, Scarborough Bluffs, and Lorne Park, and on drier river bluffs along the Credit River and the lower Rouge River.

After Goldie's discovery, many of the savannahs and prairies in the greater Toronto region and elsewhere in North America were lost to urbanization and agriculture. Our obsession with eliminating fire, even from these fire-adapted ecosystems, also contributed to the loss. Savannahs and prairies are now considered one of the most endangered ecosystems on the continent. Of those on the Humber, only High Park and a few smaller remnants such as Lambton Park remain today.

Species such as the provincially rare stiff gentian and Virginia yellow flax, which hadn't been seen in the park for decades, are now back.

The inheritance of High Park is due to the foresight of John Howard, who bequeathed the core of the park to the citizens of Toronto in 1873. Until well into the 1950s the park was a wonderful repository of our oak savannah heritage. It supported the region's largest assemblage of rare plant species, over 85 significant ones, with most confined to its prairies and savannahs (Figure 17.21). The remainder were found in the cooler ravine forests and wetlands, many of them rare northern species (Figure 17.22)

A desire for increased recreational development in the park resulted in the loss of some beautiful savannahs and over half of the park's rare plant species, including many of its associated rare insects and its hognose snakes. The park's bottomland wetlands were also degraded and the once wild Grenadier Pond was 'beautified.' Many of the marsh birds present in Howard's day are now gone, and amphibians have been reduced to a few sightings of American toads.

However, the park is now experiencing rejuvenation due to the efforts of the City of Toronto and a large cadre of volunteers. Mowing has been eliminated from most of the park. Small-scale controlled ground fires are being brought back to encourage the growth and spread of the native prairie grasses and wildflowers that are adapted to fire. Exotic plants such as Norway maple and European shrubs like buckthorn and honeysuckle are being removed by hand to allow more space for our native species. Many of the significant species, including those that have disappeared from the park, are being reintroduced. Volunteers collect seeds of such native species as the woodland fern-leaf and shrubby St. John's-wort, the latter recently rediscovered at the doorstep of John Howard's Colborne Lodge home. After collection, the seeds are germinated in the park's greenhouses and planted out in the savannahs. Species such as the provincially rare stiff gentian and Virginia yellow flax, which hadn't been seen in the park for decades, are now back. There is even hope that when wild lupine numbers increase to a healthy size, the karner blue can be restored to its old range. Last seen in the park in the 1940s, this butterfly has been wiped out in the province due to the loss of its savannah habitat. Even Grenadier Pond is being brought back to life by allowing its waters to fluctuate again, cleaning up its inflowing stream, and bringing back some of the marshes that once lined its shores.

To see this rejuvenation first-hand, the best High Park savannahs are found on the uplands just northeast of Grenadier Pond and, in the eastern portion of the park, southwest of the College streetcar loop. Here you can find magnificent black oak trees over 200 years old and many of the rare prairie and savannah shrubs, grasses, and wildflowers that John Howard and John Goldie so loved. A large central upland of oak savannah just north and east of Grenadier Restaurant is being restored to the way it appears in J.E.H. Macdonald's 1912 painting, *Spring Breezes, High Park,* which hangs in Canada's National Gallery (Figure 17.23). The painting depicts a landscape of scattered black oak trees in a blue expanse of wild lupines waving in the spring breezes.

The rejuvenation of High Park sets a wonderful example for restoring and expanding our other remnant prairies and savannahs. There is no reason that the now urbanized sand plains that once supported large prairies and savannahs cannot do so again in vacant lots, neighbourhood parks, along our highways, roads, and railways, and in our front and back yards.

Oak Ridges Moraine

David McQueen

The year was 1978, and we were heading north from Ajax, along Durham Regional Road 23, looking for country properties. At first there were no surprises. The road was straight, climbing gently through conventional, southern Ontario farmland. Then came a railway crossing, and a much steeper hill. On the far side of the hill there unfolded a startlingly different vista. Before us lay a still-higher ridge, thickly forested with a mixture of oak, maple, pine, spruce, poplar, and birch, among other trees. And at the bottom of the intervening valley was the sort of tree-fringed, Group-of-Seven lake that we expect to find displayed on the walls of the McMichael Gallery. Except for the absence of massive granite, it was as though some giant hand had whisked us hundreds of kilometres north to the fringes of Algonquin Park.

We had discovered the Oak Ridges Moraine (Figure 17.24).

The moraine is indeed a 'special place,' differing markedly from the countryside both south and north of it, and not just in appearance. It has, for example, its own microclimate, several degrees cooler than urban Toronto's. People are still skiing on the moraine when the grass is greening in Metro.

Most important of all is the natural water processing that goes on beneath its surface, strongly influencing both its own ecology and that of the larger Toronto bioregion.

Physically, the moraine is a sandy, gravelly ridge extending roughly 200 km from the Niagara Escarpment in the west to the Trent valley in the east. It varies in width from 4 to 24 kilometres, and reaches a maximum thickness of 240 m in Richmond Hill. Its highest point is roughly 400 m above sea level.

It was formed during the last ice age, which ended about 12,000 years ago. Prior to that, two distinct glacial

FIGURE 17.24 Crossing the moraine. (David McQueen)

lobes, pushing from north and south respectively, advanced, melted back, then advanced again, scraping up large amounts of rock, soil, and other debris and depositing it in a trough between them. After the final melt-back, this accumulated material remained as an irregular ridge.

In some places, blocks of ice were trapped beneath the glacial sediments. As they slowly melted, the sediments above them collapsed into the spaces vacated, forming the basins of the numerous 'kettle' lakes that characterize the moraine.

The first humans to visit the moraine were Native peoples. Although they settled parts of the moraine where a reliable water supply was available, the main importance of the moraine to Native peoples was as a hunting and gathering area. The preponderance of oaks then on the moraine meant an abundant supply of acorns in the late fall, when deer congregated in large concentrations for the rutting season. During this time they fattened on acorns for the winter. Because deer are fairly solitary animals at most other times of the year, the late fall was a preferred time for Native peoples to hunt them in well-organized mass drives.

An interesting controversy concerns whether the First Nations periodically burned off parts of the Rice Lake plains and other places in southern Ontario, such as High Park, to encourage new prairie plant growth of a kind attractive to deer. Some writers have asserted this to be the case, while others, agreeing that burning for hunting was a practice south of the Great Lakes, could find no evidence of it farther north (C. Heidenreich, personal communication; Day 1953). Proponents of the latter theory believe that some plains of this 'prairie' type in southern Ontario were maintained on porous sands and gravels, such as those in parts of the Oak Ridges Moraine, by periodic droughts and accidental fires caused by lightning. In some places forest clearance by Native horticulturalists may have played a role in creating a habitat favourable to grasses.

Much more drastic was the impact of the first significant European incursions onto the moraine, beginning in the early 1800s. The lumbermen and settlers vigorously cut over much of the hitherto abundant tree-cover, utilized the cleared land for a style of agriculture often ill-suited to the sandy moraine soils, and erected many water-driven grist- and sawmills. Among the results were the creation of veritable deserts of blowsand and dunes, and the interruption of salmon and other fish spawning runs by mill dams. Stream flows became more turbid and irregular, fluctuating between extremes of flood and drought.

Remedial action began in the 1920s, with the planting of soil-stabilizing 'agreement' forests on abandoned farms and other desertified tracts. A further big step was the decision in the mid-1940s to conduct a major demonstration project to rehabilitate the severely desiccated Ganaraska watershed northwest of Port Hope. The ultimate results of this were the creation of the Ganaraska Forest (now one of the largest in southern Ontario) and the establishment of the first in an evolving series of watershed-specific conservation authorities now covering much of the southern part of the province.

The Oak Ridges Moraine has justly been characterized as a giant rain barrel – a huge natural system of water filtration and storage.

Private landowners also played a role in moraine reforestation. One of the most notable was the late James Walker, who sold his extensive Uxbridge-area tree farm to the Metropolitan Toronto and Region Conservation Authority.

These efforts had a positive impact: much desert was converted back to forest, mill dams were eliminated, and stream flows improved. Rebounding fish populations and increasing supplementation of plantation forests by natural regeneration gave promise of returning ecological health.

Today, however, the moraine, while increasing in significance as a green belt, faces new threats to that very status. On the one hand, the growth of population and built-up area in the greater Toronto region has steadily enhanced the moraine's scarcity value as a multifunctional water resource, green belt, and 'soft' recreational space (for hiking, cross-country skiing, horseback riding, and the like). But that same urban growth has also stepped up pressures for two quite different, competing uses of moraine lands: urban sprawl, and the extraction of aggregates (sand and gravel) for road building and other purposes (Figure 17.25).

Why should these uses concern us? What is the case for making special efforts to preserve the essential, natural-heritage features of this great landform? The answer may conveniently be divided into three parts: water, natural habitat, and landform conservation.

The moraine has justly been characterized as a giant rain barrel – a huge natural system of water filtration and storage. Especially where the sandy soil has been stabilized by trees and other vegetation, an unusually high proportion of rain and snowmelt falling on the moraine penetrates it deeply. The water thereby becomes both clean and cold as it 'recharges' the moraine's underground aquifers. This groundwater is eventually 'discharged,' surfacing again in various ways. Some of it, for example, is drawn up through private and municipal wells. More than 400,000 residents of the greater Toronto region depend for their water supplies on these and other sources traceable back to the moraine.

Even after allowing for the intervention of wells and other human water-taking devices, much of the moraine's groundwater re-emerges at or near its lower flanks in the form of springs, wetlands, and streams, which in turn provide the headwaters and all-season 'base flows' of southward-running rivers such as the Rouge, Don, and Humber, and Duffins Creek flowing into Lake Ontario (Figure 17.26). And, since the moraine is also a watershed, it likewise provides high-quality headwaters for north-flowing rivers such as the Nottawasaga, Holland, and Nonquon, emptying into Georgian Bay, Lake Simcoe, and the Kawartha Lakes.

We cannot 'save' a river such as the Don without saving its headwaters.

The significance of the moraine for the several watersheds discussed elsewhere in this volume will be obvious. As the (Crombie) Royal Commission on the Future of the Toronto Waterfront (1992) pointed out, we cannot 'save' a river such as the Don without saving its headwaters.

A second reason for affording special protection to the moraine is the extraordinary richness and diversity of plants and animals supported by its varied habitats: deep forest, fringe forest of birches and other sun-loving species (Figure 17.27), rolling meadowlands, and kettle lakes. It is a natural 'bank' of biodiversity – a delight today, and an insurance against an always uncertain tomorrow.

Third, in terms of landform conservation, it is important to ensure that 'the form, character and variety of landscapes within the Moraine will be maintained to minimize disruption to natural processes, to maintain visual character and attractiveness and to retain the

James Walker and the Oak Ridges Moraine

James Walker spent more than half of his life turning 400 ha of barren wasteland into beautiful forest, planting more than 2 million trees over 57 years in an area that is three times the size of High Park.

Born in Toronto, Walker graduated from the University of Toronto in 1928. He studied law in England, where he was called to the bar in 1931. In 1932 he was called to the Ontario Bar and began a life-long association with the law firm of McCarthy and McCarthy (currently McCarthy Tetrault).

Walker's interest in reforesting sections of the Oak Ridges Moraine began in 1934. He first saw the land on a cross-country ski trip in the early 1930s, when it had been stripped of its massive white pines in futile efforts by settlers to farm the area. It caught his fancy, and he bought 1.6 ha of dry bare land, ravaged by wind and rain, and a small cabin for $350. In 1947 he bought more land, and more trees. He began planting 100,000 quick-growing poplars annually and, with the help of a small contingent of people, is believed to have ended up with the most land ever reclaimed by an individual in Ontario.

His land is now part of the larger holdings of the Toronto and Region Conservation Authority.

James Walker, mid-1990s. (The Toronto and Region Conservation Authority)

FIGURE 17.25
Large gravel pit, Whitchurch/Stouffville Township, York Region. (David McQueen)

educational and interpretive value of this unique landform' (Oak Ridges Moraine Technical Working Committee 1994).

There is considerable arbitrariness in breaking up the case for the moraine into three parts, as we have just done for convenience of exposition. Fundamental ecology tells us that everything in and on the moraine ultimately connects with everything else. Thus, for example, landform conservation is vital to maintaining the moraine as a natural waterworks, and this in turn is essential to conserving the vegetative cover that stabilizes the soil and so helps to maintain the landform. And the wild creatures that seek food and shelter in moraine woodlands assist the natural regeneration of those same woodlands by the tree and shrub seeds that they bring with them.

Successive Ontario governments, recognizing the impossibility of assigning conventional market values to things like wetlands, natural habitat, and landscape, have moved a certain distance to provide special protection for the moraine. In 1991, interim planning guidelines for the part of the moraine lying within the Greater Toronto Area were established. And in 1991-4, a broadly representative technical working committee devised some, though not all, parts of an implementation strategy for these guidelines (Oak Ridges Moraine Technical Working Committee 1994).

As of late 1997, nothing further had been done to complete this strategy and put it into general practice. What *has* occurred, however, has been a set of very major 'streamlining' changes in legislation and regulations affecting land-use and the natural environment in all parts of Ontario, including the moraine. Space does not permit a full discussion of these here, but they are certainly causing great concern among naturalists and other environmentalists. Among the particular changes that concern them are drastic reductions in the funding and independence of conservation authorities, and the massive downloading of responsibilities for planning and the environment onto municipalities, many of which lack the organization and expertise to carry them out properly.

Meanwhile, it is important that as many people as possible become aware of the moraine, learn something of the issues of competing land uses that will have to be resolved in some way or other, and develop their own informed judgments.

As part of this process, personally visiting the moraine is highly recommended. The Ganaraska Forest and Forest Centre near Kendal, entrances to which are well signed along Durham/Northumberland Route 9, northeast of Bowmanville, make a good starting point. So does the portion of the white-blazed Oak Ridges Trail that begins on Dufferin Street north of Eaton Hall on

▶ FIGURE 17.26
Headwaters of Duffins Creek, Glen Major.
(David McQueen)

▼ FIGURE 17.27
Birches, Glen Major.
(David McQueen)

Seneca College's King campus, then wends its way west by forests, lakes, and monastery.

But there are many, many more moraine locations well worth visiting. To learn how to access them, refer to the publications *Oak Ridges Moraine* (STORM Coalition 1997) and *Oak Ridges Trail Guidebook* (Oak Ridges Trail Association 1997). To those seeking deeper and more detailed insights into what they are visiting, the draft publication of the Oak Ridges Technical Working Committee, entitled 'The Oak Ridges Moraine area strategy for the greater Toronto area' (Oak Ridges Moraine Technical Working Committee 1994) provides a good introduction to the available literature. The background studies listed there, along with their references, will take you deeper still, if that is where you would like to go.

Credit River

Michael J. Puddister

The Credit River is located in some of the most diverse landscape found within the province. The river and its watershed have faced immense change in recent history as a result of the way we have utilized its wealth of natural resources. The region has evolved from a valley of towering white pine to an area of intense agricultural expansion, and then to high-rise office and condominium towers and residential development. Coupled with this evolution has been the development of a new relationship between the watershed residents and visitors and their environment, a relationship that is now based on the conservation of the river's resources, rather than their exploitation.

The lands draining to the Credit represent almost 1,000 km^2 of urban, urbanizing, and rural area (Figure 17.28). Starting at its headwaters near the town of Orangeville, the river gathers tributaries from Erin, Mono, and Caledon, and then tumbles over the Niagara Escarpment at the village of Cataract (Figure 17.29), where, after meeting drainage from Acton, Georgetown, Brampton, and Mississauga, it continues to flow southerly for a total of 90 km to the former village of Port Credit on Lake Ontario.

The Niagara Escarpment is one of the most dominating landforms in southern Ontario, and its influence is felt strongly in the Credit watershed. It contributes to striking differences in climate, vegetation, and land use above and below the scarp face. Above the escarpment, natural areas are common, with cooler temperatures and longer snow cover, water quality is generally good, and the river contains a self-sustaining coldwater fishery characterized by brook and brown trout. Below the escarpment, the fertile soils of the Peel Plain have historically been developed for agricultural purposes and are now being rapidly urbanized, generally from south to north.

It was estimated in 1992 that the watershed of the Credit was 15 percent urbanized, and within 20 years this is expected to increase to approximately 35 to 40 percent (CVCA 1992). Such an increase would be equivalent to a city the size of St. Catharines or Vaughan being added to the watershed by the year 2012.

The name Credit River is based on the trading practices of the Mississauga Indians in the early 1700s. Every spring the Mississauga traded furs with the French at the mouth of the river in exchange for goods, which were provided in advance – on credit. Augustus Jones listed the river in 1796, giving it the Indian name Messinnih, which he translated as 'trusting creek – credit.' Earlier it was given as Rivière au Credit on a French map dating from 1757 (Roulston 1978).

Some of the earliest records are filled with amazing facts. The townships in the area of the Credit River were surveyed between 1806 and 1822. It appears from the surveyors' notes that they were met by a gigantic primeval forest. The uplands were covered with stands of maple and beech as well as some basswood, oak, and elm; on the valley slopes hemlock was common; and near the lake they recorded principally pine and oak (Ontario Department of Planning and Development 1956). The 1806 survey documents timber reserves of pine often 150 feet high, and oak up to 50 feet tall along the banks of the Credit (Corporation of Peel 1967). At the time, the area was especially valued for its timber, which provided the much-needed masts for the British Navy.

Through this period, the settlers continued to arrive, moving upstream, harvesting timber for fuel and building materials. Sawmills were established, and gristmills followed as the early agricultural industry began to take hold. As Alex Raeburn, a long-time resident of the watershed said so clearly in 1997, 'the Credit was their livelihood ... it was their power.' In addition to being able to utilize the river for industry, the settlers found that the land within the watershed also had much to offer. Wild game such as ducks, pheasant, partridge, wild turkey, deer, and rabbit were hunted, and other animals such as beaver, squirrel, wolf, bear, fox, weasel, and skunk were taken for their skins (Zatyko 1979).

One of the early villages to be established along the Credit was what is now known as Terra Cotta. In 1866 it was referred to as Salmonville, as it was said that 'the salmon in the Credit River at that point were so plentiful that they could easily be speared with a pitchfork' (ibid.). Unfortunately, this bounty was not to last. A victim of its own success, the waterpower of the river, which was highly treasured by the early settlers, resulted in the

The Special Places

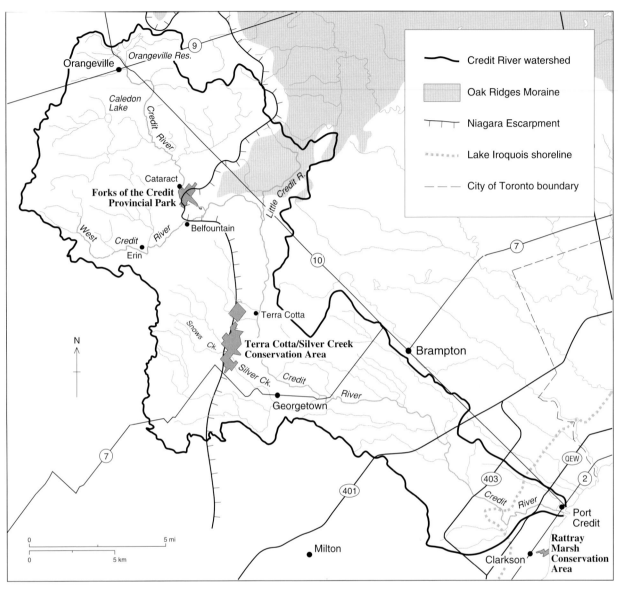

FIGURE 17.28 Credit watershed, illustrating lands owned and/or managed by the Credit Valley Conservation Authority and the location of Forks of the Credit Provincial Park and Niagara Escarpment.

loss of the primary fishery. Smith's *Gazetteer* of 1846 states that 'from the great number of mills erected on the river during the last four years, the fishing is destroyed, the salmon being unable to make their way over the dams' (Clarkson 1977). The habitat of the salmon was also affected by logs floating downstream and by contaminants being discharged from the sawmills.

The upper reaches of the Credit system were first explored by European settlers in 1818 with word that the Caledon Hills held gold. However, there was no gold to be found. The settlers did, however, soon discover the limestone rock on the high cliffs at the Forks of the Credit, and a quarry mining industry began. The Forks, as it is also known, is the area east of the picturesque village of Belfountain, where the main river is met by the west branch, known as the West Credit River. 'By the 1880s the valley had assumed the appearance of a frontier town. Houses to shelter the 400 or more miners and their families formed the hamlets of Brimstone and Forks of the Credit, while stone cutters and skilled tradesmen commuted by wagon or walked from Belfountain' (Trimble 1975). The quarries at the Forks supplied stone for the construction of the provincial Parliament Buildings in 1886 and around the same time for the old Toronto City Hall.

In memory of Roy Trimble, a long-time resident of Belfountain, a new hiking trail was opened in the spring of 1998. The trail takes the explorer downstream from the Belfountain Conservation Area on the West Credit River, alongside the very place where these tradesmen were engaged in their craft so many years ago. If you look closely you can see physical evidence of a past industrial landscape nestled among the trees and shrubs.

The quarries at the Credit Forks supplied stone for the construction of the provincial Parliament Buildings and of old Toronto City Hall.

After the quarry industry waned in the 1930s, the Forks of the Credit soon became a popular recreational area. Nature had begun to repair itself, with trees and grass reclaiming the mills and quarries, and the area reverted to conditions similar to what had existed over 80 years before. Yet in the southern portion of the watershed, urban development pressure was intensifying.

CONSERVATION BY THE COMMUNITY AND FOR THE COMMUNITY

The Credit Valley Conservation Authority (CVCA) was created on 13 May 1954, reflecting the advent of a new concept of personal and community responsibility in conservation (Ontario Department of Planning and Development 1956). The CVCA, as with other conservation authorities, was created through local initiative, with municipal and provincial support, recognizing the importance of managing resources on a watershed basis. The *Credit Valley Conservation Report* (1956) described the natural environment of the watershed in this way: 'Much of the watershed, both land and water, is more spectacular than other areas close to Greater Toronto with its population of more than one million. Hence, the watershed is visited by great numbers of people, including hunters looking for upland game and waterfowl, fishermen seeking fish, and a growing army of naturalists interested in the opportunities to see and enjoy the varied forms of animal life.'

The community leaders of the day recommended the creation of conservation areas, or watershed parks, in the more desirable sections of the valley. There was a vision that this system of protected areas should also be connected with a nature trail and linked to the Caledon Hills and Niagara Escarpment. There was a recognition, even at that time, that pressure from future land uses would intensify, threatening the natural richness of the Credit River – immediate action was required. While 48.5 ha of public parkland existed at the time, the watershed was only 3 percent urbanized

FIGURE 17.29 The Credit River tumbles over the Niagara Escarpment at the village of Cataract. Ruins of one of the first hydroelectric plants (c. 1890) can be seen in the foreground. (Robert Morris)

(Ontario Department of Planning and Development 1956). As development in the watershed continued, the conservation authority became more directly involved in the land-use planning process by providing environmental, technical, and regulatory support to the member municipalities. This role has continued to evolve to the present day: the authority is now recognized as one of the leaders in ecosystem-based watershed management (Royal Commission on the Future of the Toronto Waterfront 1992).

The management of public lands has also evolved through the development of a strategy for the watershed's conservation areas. Its goal is 'to protect the Credit River Watershed's significant and representative ecosystems, and offer sustainable natural heritage appreciation and recreational benefits to its watershed residents' (CVCA 1994). The amount of public open space has grown to 2,990 ha, with the authority owning or managing 47 natural heritage properties, consisting of 2,337 ha, or 78 percent of the total (CVCA 1996).

A CENTRE FOR ENVIRONMENTAL LEARNING

In 1958, the Credit Valley Conservation Authority purchased the first parcel of land that eventually led to the creation of the Terra Cotta Conservation Area in 1960. Along with the adjacent Silver Creek property that the CVCA began to assemble in the early to mid-1970s, this area truly represents some of the most spectacular natural scenery and biodiversity in the watershed. The Terra Cotta Conservation Area was designated by the CVCA board of directors in 1990 as a 'Centre for Environmental Learning,' reflecting a shift from providing traditional recreational facilities to instead providing opportunities for nature appreciation and protection.

As a testament to their ecological importance, both the Terra Cotta and Silver Creek areas contain portions of provincially designated areas of natural and scientific interest (ANSI), and both have been designated as environmentally significant areas by the authority. In addition, the Silver Creek area also contains a Class 1 provincially significant wetland. A total of 67 community types and 403 plant species have been documented within the Silver Creek valley (Varga, Jalava, and Larson 1994). This study identified a small population of American ginseng, which is considered nationally and provincially rare, and Canada milk vetch, spring clearweed, yellow water buttercup, and water pimpernel, all of which are designated as regionally rare.

The areas also contain a wide diversity of animal life. The common white-tailed deer is often seen bounding through the woodlands or nearby fields, and you might catch a glimpse of the not so common porcupine hiding in a hemlock tree, or the pickerel frog resting along the edge of the ponds. Much more significant is the habitat that the areas provide for breeding birds. A total of 63 bird species can be found in the Silver Creek valley, including the nationally vulnerable and provincially rare cerulean warbler, Louisiana waterthrush, and red-shouldered hawk, as well as the locally significant hermit thrush, blue-winged warbler, and golden-winged warbler (ibid.).

The Bruce Trail Association was incorporated in 1963, and a marked trail, 644 km long, was completed in 1967.

THE NIAGARA ESCARPMENT – A PROVINCIAL TREASURE

In 1960, the same year that the Terra Cotta Conservation Area was opened, a citizens' committee was formed to study the feasibility of a hiking trail along the entire Niagara Escarpment. The Bruce Trail Association was incorporated in 1963, and a marked trail, some 644 km long from Niagara Falls to the tip of the Bruce Peninsula, was completed in 1967 (Gertler 1968). Such a mammoth task required the cooperation of hundreds of landowners and agencies who recognized the importance of this ecologically significant landform and the need for controlled public access.

The trail passes through the Terra Cotta and Silver Creek Conservation Areas, linking the Niagara Escarpment and the Caledon Hills to the Credit River system, fulfilling a portion of the earlier vision set out in the *Credit Valley Conservation Report* of 1956. If you walk along the Bruce Trail, near the edge of the Silver Creek valley, and you pass under the majestic white pine, you begin to imagine what it was like along the Credit River when the settlers arrived in the early 1800s.

On 10 March 1967 the Ontario government announced a wide-ranging study of the Niagara Escarpment with a view to preserving its entire length (Gertler 1968). Protection of land was a key theme in the study that became known as the Gertler Report. It clearly recognized that the escarpment was at risk from development pressure and competing demands for land. One of the key components of the report was an ambitious acquisition program aimed at protecting large tracts of this significant landform. It identified additions to the Terra Cotta Conservation Area land holdings, and the purchase of the Credit Forks, as early priorities. The report described the Credit Forks area as 'some of the most spectacular scenery in southern Ontario' (Gertler 1968).

As a result of the Gertler Report and a task force study, the Niagara Escarpment commission was appointed to prepare the Niagara Escarpment Plan. Through these efforts and with strong public support, the plan was finally approved on 12 June 1985. It has no doubt had an impact on the protection and restoration of the landscape. Flynn and Mersey (1997) report that over the last 20 years there has been an increase in total forest cover of 34.6 percent in the Forks area.

The Niagara Escarpment Plan is Canada's first large-scale environmental land-use plan. It balances protection, conservation, and sustainable development to ensure that the escarpment will remain substantially as a natural environment for future generations (Ontario Ministry of Environment 1990). In 1990, the United Nations Educational, Scientific, and Cultural Organization (UNESCO) named Ontario's Niagara Escarpment a World Biosphere Reserve.

THE PEOPLE FINALLY GET THEIR PARK

The Forks of the Credit is one of the outstanding scenic attractions in southern Ontario. Through this reach the Credit falls quickly over the escarpment, creating a picturesque gorge that contains both exposed bedrock cliffs and steep wooded slopes. It is a breath-taking scene in the height of the fall colours. As early as 1936 the *Peel Gazette* had stated that it was 'too bad there isn't a park at the Forks, so thousands could enjoy the beautiful hills' (Greenland 1972). It took almost 50 years, but the 261 ha Forks of the Credit Provincial Park, which contains the Credit Forks Area of Natural and Scientific Interest, became a reality in 1985.

Due in part to the dramatic relief in the Forks area, there are forests on the west side of the Credit River that have retained their primeval character. They are considered by some to be the best example of pre-settlement vegetation along this section of the Niagara Escarpment (Gould 1984; Kaiser 1990). Within the park, Gould reported a total of 408 species of vascular plants, 6 of which are regionally rare (pondweed, sedge, milk vetch, spurred gentian, speedwell, and aster); 55 species of breeding birds, 3 of which are uncommon to the region (mourning warbler, Northern waterthrush, and scarlet tanager); and 15 species of mammals (of which the river otter is considered to be uncommon in the region). Kaiser also documented two provincially rare plants, American ginseng and hart's tongue fern. Moreover, the Credit River through this area is considered to have some of the best spawning habitat for both resident brook and brown trout.

A COMMUNITY SAVES A MARSH

In *The Natural History of the Toronto Region* J.H. Faull reported in 1913 the presence of 'several small ponds near the beach just west of Lorne Park' and their 'offering much to the collector.' In the 1950s the area was explored numerous times by the prominent professor Alan F. Coventry, resulting in several significant discoveries (Hussey 1990). A one-time president of the South Peel Naturalists' Club, Coventry played a prominent role in the eventual protection of the area.

Rattray Marsh, as the area eventually became known, is located on the shore of Lake Ontario, west of the mouth of the Credit River, within the boundaries of Mississauga. The marsh contains a high diversity of plants and animals. Four hundred and fifty plant species have been recorded, of which the bushy, or Lower Great Lakes cinquefoil is provincially rare (Riley 1989). In addition, the area contains 23 regionally rare plants and 50 considered to be regionally scarce; 230 bird species; 22 mammals; 15 fish species; and a variety of reptiles and amphibians (Hussey 1990). The wetland area is an important stopover for migrating waterfowl and a wide variety of other birds. This is due in part to the fact that it represents one of the last remaining Lake Ontario

> *In 1990, UNESCO named Ontario's Niagara Escarpment a World Biosphere Reserve.*

FIGURE 17.30
The Rattray Marsh Conservation Area on the Lake Ontario shoreline – one of the last remaining shoreline wetlands on the lake. (Credit Valley Conservation Authority)

shoreline wetlands. It is a birders' paradise – the place to be at the end of May or August.

The Rattray Marsh Conservation Area (Figure 17.30) was created as a result of a concerted effort by a number of local naturalists. It is named after James Halliday Rattray, who left the property to his estate in 1959. Almost immediately, the fight for its protection began. In 1963 the property was purchased by development interests. Undeterred, the local community – led by Ruth Hussey and members of the South Peel Naturalists' Club, with assistance from the Credit Valley Conservation Authority and the University of Toronto, among others – continued to pressure the provincial and local governments to acquire the lands for conservation purposes. Despite these efforts, the new owners registered a plan of subdivision over a portion of the property in early 1964. The plan called for a number of residential lots – and the marsh was to be dredged to create a marina.

Finally, in 1971 the first parcel of land (10 ha) was acquired by the Credit Valley Conservation Authority, with the remaining 23 ha being purchased in 1973 (CVCA 1997). The local community continues to be involved through the Rattray Marsh Protection Association, assisting the CVCA with the management of a trail system, providing educational tours, and patrolling the area as

FIGURE 17.31 Atlantic salmon reintroduction in the Credit River headwaters – the restoration of the watershed continues. (Robert Morris)

designated conservation officers. The association won the 1997 Clearwater Award of the Washington, DC, based Waterfront Center, for its outstanding efforts. The members' commitment was summed up this way by Mississauga councillor and authority member Pat Mullin: 'Theirs is an effort which stands out as a beacon of success for others. It is an example of how vision and strength of a community can come together to not only protect an ecology, but replenish a community's spirit and sense of natural wonder' (CVCA 1997).

THE SALMON RETURN

The early settlement of the Credit watershed resulted in the loss of the salmon fishery through the construction of numerous dams and the sedimentation of the river channel from massive clearing of trees. Over the years, through the dedicated effort of agencies, interest groups, and willing landowners, most of the dam structures have been removed. The land is also now in a more stable condition, due in part to reforestation and other stewardship practices. Thanks largely to these actions, the Credit River has become the most popular fishing river in Ontario, with brook, brown, and rainbow trout, bass, pike, and other species. However, perhaps the greatest testament to the restoration and health of the Credit fishery is the reintroduction of Atlantic salmon (Figure 17.31).

One of the Credit's tributaries has produced the best survival and growth rates of juvenile Atlantic salmon in North America.

The provincial Ministry of Natural Resources considers the salmon reintroduction program to be of more than just local importance: the Atlantic salmon is a provincial heritage species (Ontario Ministry of Natural Resources 1990). As part of an experimental program, the ministry is working in cooperation with the Credit Valley Conservation Authority and numerous volunteers and fishing clubs. Juvenile salmon have been stocked at the Forks of the Credit on the main river, and within three major tributaries. So far, the experiment has been a success, as one of the Credit's tributaries has produced the best survival and growth rates of juvenile Atlantics in North America.

The Credit River watershed has undergone an incredible evolution – from an environment of primeval forest and abundant wildlife, through a period of exploitation and disturbance, and then to the current period of restoration and protection. This transformation has been driven by our changing relationship with the natural world around us. We will never again enjoy the wilderness that was once the essence of the Credit River. Through the collective actions of the watershed community, however, we can ensure that this incredible river system is protected, enjoyed, and nurtured for the future. As Heather Broadbent, heritage resource officer for the town of Caledon, said in 1997, 'I think it so comforting to look at the valley of the Credit and see how natural systems are being restored, wetlands are coming back, and the whole ecosystem has every indication that it's going to recover.'

Humber Valley

John H. McAndrews

The lower Humber valley from Etienne Brulé Park, just north of Bloor Street, south to Lake Ontario features wild floodplain land within urban Toronto. Twenty-five-metre-high bedrock cliffs of 450-million-year-old Ordovician marine shale define the 300 m wide valley. In postglacial but prehistoric time, the river flowed in a sinuous course to erode meanders in the bedrock cliffs; later the river assumed a straighter course down the valley. On alternate sides of the modern river, marshes grew in these old meander sites (Figures 17.32,

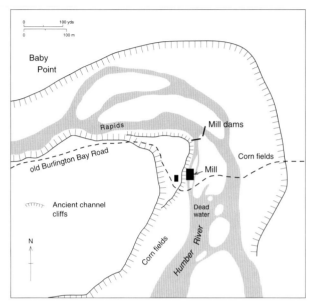

FIGURE 17.33 King's Mill area in the early nineteenth century, showing a group of islands at the head of the still-water estuary. These islands of valley fill are similar to deltas in lakes. Note the river ford crossing of the Burlington Bay Road and corn fields growing on the floodplain of what is now Brulé Park. (After Lisars 1913 and Hawkins 1834)

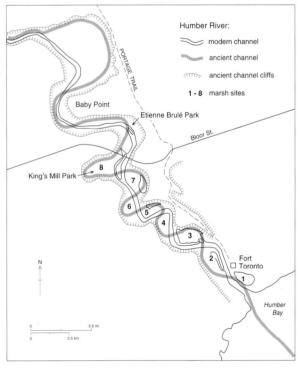

FIGURE 17.32 Lower Humber River valley showing ancient river meander scars and the sites of historic ponds and marshes. Meander sites 2, 3, 4, 5, and 7 are now partly occupied by shallow ponds; sites 6 and 8 have been filled since 1948 to produce parkland (site 6 has a marina); and site 1 has been destroyed by road construction. Sites 7 and 5 support forest. The historic Humber Portage trail began at Fort Toronto – the site is now occupied by a Petro-Canada service station – and follows Riverside Drive north. Baby Point is the site of historic and prehistoric Indian villages. (After Weninger and McAndrews 1989 and Robinson 1965)

17.33), but in the past two centuries these marshes have mostly disappeared under a cover of mineral soil. This soil is either landfill rubble used to create parkland or flood-deposited sediment that now supports forested levees enclosing clay-bottomed ponds (see Figures 17.34 and 17.35).

The most interesting part of this section of the valley is Pond 7, which can easily be reached from the southeast end of the Bloor Street bridge by following the path down the cliff to the floodplain and its pond. Beavers episodically colonize this pond and fell cottonwood and white ash trees. Water birds such as blue heron and duck are also common. Occasionally large introduced fish can be seen – carp in the pond and Pacific salmon in the river. The floodplain forest is dominated by alien Manitoba maple, tree willow, and Norway maple in addition to the native cottonwood, white ash, and white elm. Wild grape vines hang from the trees in this lush forest. Weedy herbs cover the forest floor, including the

FIGURE 17.34
View of Humber River in 1910 to the southeast showing site 6 marsh, now filled to form parkland with a marina. Across the river behind levee with trees is site 5. (Archives of Ontario, Humber River collection C219-0-0-0-3)

FIGURE 17.35
View of the Humber River in 1910 looking northwest from its mouth. The channel in the foreground is bordered by marsh-covered levee. Site 2 is in the centre and site 3 is to the right. Note the drying racks for nets used by commercial fishers. (Archives of Ontario, Humber River collection C219-0-0-0-10)

native giant ragweed, the alien and attractive Himalayan balsam, and the not so attractive garlic mustard.

These aliens and weeds thrive in this seasonally disturbed floodplain environment. In spring, during snowmelt, the river rises, carrying ice blocks downstream. When the blocks run aground at a bend in the river, an ice block dam forms across the river, causing water to rise and spread over the floodplain. Ice blocks floating downstream strike and scar floodplain trees up to 3 m above normal water level. This flood water also carries suspended sediment; sand is deposited on the levees along the channel, and the finer silt and clay in the backswamp and pond where the current is slower. There is also local erosion on the levee surface, especially near tree trunks where there is local turbulence during floods. This deposition and erosion encourages the growth of weedy herbs that, after the flood subsides, develop quickly on the fertile soil beneath the forest canopy. On the other hand, the perennial floodplain trees must tolerate periods of waterlogged soil, which accounts for the absence of upland trees such as oak, sugar maple, and beech.

The lower Humber valley has also been a special place in human history. From Lake Ontario upstream to Bloor Street, where the rapids begin, the river is navigable; it forms the southern end of the early historic Toronto Portage to the Upper Great Lakes. Until the late eighteenth century, canoes from windy Lake Ontario entered

FIGURE 17.36
View of Humber River in 1910, looking south, showing valley fill islands and the ruin of the Old Mill on the site of the historic King's Mill Reserve. In the left foreground is the Beltline Railway; in the middleground Catherine Avenue crosses the river through modern Brulé Park at the site of the early ford of the Burlington Road; and in the background is the site 8 marsh, now filled for parkland. (Archives of Ontario, Humber River collection C219-0-0-0-5)

the relatively calm river mouth to be unloaded for the portage northward over the Oak Ridges Moraine to the Holland River. At this point, canoeing began again down river, across Lake Simcoe and along the Severn River to Georgian Bay. Just above the rapids on Baby Point, there was a seventeenth-century Seneca Indian village, Teiaiagon, as was discussed in Chapter 4. Near this place, the French built a trading post in 1720, the first European settlement in the greater Toronto region (Robinson 1965). This place was also a crossroad, because it was located at the most convenient ford on the Humber River for people walking along the shore of Lake Ontario. For these early people, the floodplain also provided fertile soil for growing corn, and the river itself was a fishery for Atlantic salmon.

In response to a growing interest in the area, in 1750 the French erected Fort Toronto to the east of the mouth of the river. After 1760, the fort was succeeded by a trading post, which persisted for the rest of the century. In 1793, Lieutenant Governor John Graves Simcoe built a sawmill on the abandoned site of a French sawmill just below the rapids on the site of what is now the Old Mill, a ruin that dates from 1850 (Figure 17.36). Here a dam was built and water was diverted to power a sawmill and later, after road building, a gristmill. For these reasons, the first European settlement in the greater To-

From Lake Ontario upstream to Bloor Street, where the rapids begin, the Humber is navigable; it forms the southern end of the early historic Toronto Portage to the Upper Great Lakes.

ronto region was along the lower reach of the Humber River. However, Simcoe rejected the Humber Portage as the route to Georgian Bay and opened Yonge Street to replace it.

What then were the geological events that produced these landforms that made the valley so attractive for human travel and settlement? A good place to begin is around 13,000 BC, when the continental glacier melted out of the Lake Ontario basin but persisted in the St. Lawrence valley. With the valley plugged with ice, glacial Lake Iroquois filled the Ontario basin to an elevation of 130 m above modern sea level at Toronto, well above the present Lake Ontario level of 75 m. This lake, which had its shoreline near Lawrence Avenue, deposited sand over the lower Humber valley region (Sharpe 1980). When the ice melted from the St. Lawrence valley about 12,500 BC, Lake Iroquois drained to the low level of early Lake Ontario. A valley offshore from the Humber River indicates that the prehistoric river eroded the bedrock to 115 m below the modern level of Lake Ontario (Lewis et al. 1995), which was probably the surface of early Lake Ontario. However, the lake still drained to the sea because the sea was 40 to 50 m lower than today. Since then, Lake Ontario has risen to its present level because of postglacial crustal rebound (see Chapter 1). This rebound caused flooding, which formed still-water estuaries and embayments that serve as harbours for towns such as Port Credit and for the former commercial fishing boats and now the modern pleasure craft marina on the lower Humber River.

Crustal rebound was not the only factor in flooding: distant stream capture also contributed to a relatively brief episode of shoreline flooding, which has left its imprint on the Humber valley. Until about 4000 BC, the upper three Great Lakes (Lakes Superior, Michigan, and Huron) discharged to the sea via the Ottawa River. Southward crustal rebound tilted their basins so that there were also outlets at Sarnia and Chicago, and flow through the North Bay outlet to the Ottawa River diminished. By 2000 BC, all of the Great Lakes discharged through Lake Ontario as they do today. Because the Lake Ontario outlet to the St. Lawrence River was not adapted to this larger discharge, Lake Ontario rapidly rose about 15 m to perhaps 2 m above its present level and formed estuaries along the shore. This event, known as the Nipissing Flood, which began 4,000 years ago, helps to explain the valley landforms.

Until about 4000 BC, the upper three Great Lakes (Lakes Superior, Michigan, and Huron) discharged to the sea via the Ottawa River.

The lower Humber valley displays two stages of postglacial development. Before 2000 BC, the river-eroded meander loops into alternate sides of the valley. Since then these loops have been abandoned, and the river now flows in a relatively straight channel. The timing and cause of this channel change has been worked out by studying the sediment beneath the floodplain ponds (Weninger and McAndrews 1989). Sediment cores lifted from beneath meander ponds 3, 5, and 7 (Figure 17.32) penetrate to river channel gravel and contain sediments deposited since 2000 BC.

In the meander pond of site 7, beneath 50 cm of water, we lifted a 590 cm long core of soft sediment before being stopped by hitting the channel gravel (Figure 17.37). Overlying the gravel, which dates to just before 1700 BC, is silt deposited in an estuary formed during the 1,400 years of high water that marks the Nipissing Flood. In this silt, fossil pollen and seeds are both sparse and poorly preserved, indicating seasonal drying. About 100 BC, this mud-flat silt was replaced by organic mud containing well-preserved fossil pollen and

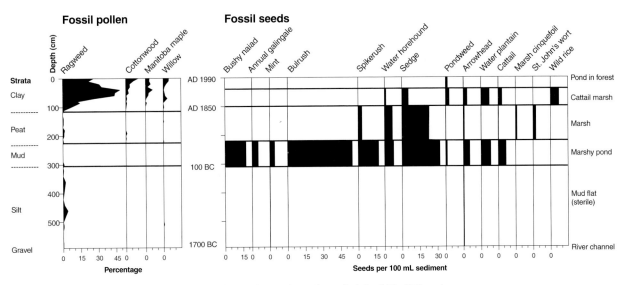

FIGURE 17.37 Fossil pollen and seed diagram from sediments beneath pond of site 7. The 590 cm long core was lifted from beneath 50 cm water. The chronology is based on calibrated radiocarbon dates. Pollen percentages are calculated on counts of 200 tree pollen; only selected pollen types are shown. Note that pollen of floodplain trees is historic. The seeds of selected wetland plants show that a sterile mud flat, deposited during the Nipissing Flood, was succeeded by a marshy pond and marsh after the flood had receded. In the nineteenth century, the succession to the cattail and wild rice marsh was probably due to increased mineral sedimentation.

seeds of pond and marsh plants, indicating that a river levee had formed isolating a pond surrounded by marsh. This new environment was a response to the waning of the Nipissing Flood in Lake Ontario; the river extended south in a relatively straight course through the mud flat. The new floodplain slowly accumulated sediment to keep pace with the renewed rise of Lake Ontario due to crustal rebound.

About AD 800, the pond filled in to become a marsh. This marsh persisted until the nineteenth century, when deforestation of the river catchment caused increased flood frequency and intensity; these flood waters carried soil eroded from newly tilled fields. Over-bank flooding was intensified, especially in spring, when the farm fields were deeply frozen. Sand levees were enhanced and clay began to encroach on the marshes, causing the dominance of cattails. Since the early twentieth century, the marshes have disappeared under a deposit of sand and clay, probably the result of intense soil erosion during road and building construction. Fossil pollen indicates that elm, willow, cottonwood, and Manitoba maple then invaded these newly enhanced levees on the sites of former marshes to form the modern floodplain forest and pond that beaver have come to inhabit.

In October 1954, Hurricane Hazel caused a record flood that peaked 6 m above normal water levels in the Humber.

To a casual visitor the Humber valley appears to be a benign landscape. However, in October 1954, Hurricane Hazel caused a record flood that peaked 6 m above normal water levels. The turbulent water swept away homes and caused loss of life (Kennedy 1979), but there is only a little evidence of landscape change from this tragic event. In the clay-bottomed ponds, this flood deposited a layer of coarse sand; on the levee, white ash tree rings are relatively narrow for the five years after the flood, indicating diminished growth perhaps due to flood erosion that exposed and killed roots.

The future of the Humber marshes is both certain and uncertain. In the longer term, Lake Ontario will certainly continue to rise at a rate of about 20 cm per 100 years, and the lower valley will again become flooded from wall to wall. By the middle of the third millennium, the estuary will broaden and expand upstream, forming new levees and marshes as it drowns them downstream. Further public demand for ball fields, marinas, and rowing courses, however, may entomb the marshes beneath earth fill and cause the river to be channelized further. On the other hand, public appreciation of the surviving marshes may cause them to be preserved or even expanded with their cottonwood trees, cattail, heron, salmon, beaver, and deer.

Don Valley

Louise Herzberg and Helen Juhola

You can perhaps best appreciate the size of the Don valley and how it controlled Toronto's shape by flying into the city in a small plane. As the plane follows the shoreline from the Leslie Street Spit to the Toronto Islands, the most prominent features are the wide green valley and the Prince Edward viaduct.

This was the valley Elizabeth Simcoe described in her diary. This was the valley that artists painted – Seton, Jefferys, Staples, Brigden, Thompson, Coleman. This was the valley of Redruff, the last of the Don valley partridges, and Silverspot, the wise old crow of Sugarloaf Hill – characters that Seton created with his magic pen. North of the viaduct one can see the Don Valley Brickworks quarry. It was from this great hole that Coleman, who was principally a geologist as well as an artist, developed an account of glacial history – an account that placed Toronto and this quarry on the world's geological map.

For those who grew up beside it, the valley was a place for country walks with the family on Sunday afternoons. You could go south to the old Riverdale Zoo, north to the forks of the Don, west to Bloor and Yonge. By the late 1970s, you could find footpaths everywhere – from the lake to Toronto's border at Steeles Avenue and in every ravine. By the 1990s the Metro Parks Department had made many of these paths asphalt trails, accessible even to wheelchairs.

For those who wish to wander in the Don valley there are still wonderful experiences to be had: in April, listening to toads trilling in Park Drive ravine; in May, finding skunk cabbage and marsh marigolds flowering at Todmorden Mills; on warm summer evenings, enjoying the healing scent of balsam poplar; on summer days, tasting woodland strawberries and wild black raspberries; in fall, feeling pine needles and leaf litter underfoot; in winter, watching muskrats swimming under the ice on an oxbow pond.

Industry has been part of the valley almost from first settlement, and industry brings pollution and degraded habitat. By the 1940s the Don River had become very polluted. It could run all the colours of the rainbow, depending on what chemicals the paper plant upstream had released into the river. In the late 1940s people began to condemn the pollution in the valley and tried to do something about it. They believed that the valley's greenness and its wildlife should be preserved as an asset to the city.

As this idea grew, so did the city's population. Residents needed new roads, new sewers. For many practical reasons the valley was the place to put these services, despite their conflict with the ideal of green space. In the late nineteenth century the valley had become a rail corridor. In the 1950s it became an expressway corridor, large parts of it claimed by the Don Valley Parkway and the Bayview Extension. Power, gas, and oil lines were buried in the valley or carried through it. Low areas, especially wetlands, became convenient garbage dumps.

By the 1940s, the Don could run all the colours of the rainbow, depending on what chemicals the paper plant had released into the river.

In 1987, we carried out a study of Todmorden Mills. We found that despite intensive use over the past 200 years, much remained that was natural. More than half of the plants identified were native to the area. In the floodplain forest we found two non-native species, crack willow and Himalayan balsam, plus the natives Manitoba maple, rare yellow jewelweed, and even uncommon turtlehead growing in a ditch. Many of these plants were wetland species, and the wetland habitats they require have disappeared from the tablelands of Toronto.

Amphibians and reptiles have taken advantage of the wetter slopes, as well as of the rubble piles created during the long human history of the area. Toads were found almost throughout the site, and their tadpoles teemed in the old riverbed. The eastern milk snake – a rare species of reptile for Toronto – was seen near the foundations of the buildings, and northern brown snakes were noted emerging from piles of rubble in the woods. Mammals have found places to hide. We saw muskrats and raccoons and someone else observed a beaver trying to cut down a tree in the picnic area. Fish are returning to the less polluted river; for example, we saw white suckers spawning. Many species of birds feed, rest, and nest in the area. Red-tailed hawks nest on hydro poles near the site, and a pair of kestrels nested in the old mill chimney.

A new habitat has been created along the western roadside boundaries of Todmorden Mills. Toronto's extensive use of salt on its roads during winter has resulted in soils adjacent to these roads becoming very salty. Only salt-tolerant species, mostly non-native, can survive there – for example, salt marsh sand spurrey and scarlet pimpernel. The salty roadside borders of the Don Valley Parkway and Bayview Extension play host to such salt-tolerant species.

The Don valley had something that no other valley in Ontario has. It had Charles Sauriol. You couldn't put anything over on Charlie Sauriol when it came to the valley, which many developers learned to their detriment. He walked, skied, planted trees, laid trails, built weirs, fought vandalism, raised bees and collected honey, camped out with Scout troops, organized fabulous steam train trips through the valley, communed with wildlife, studied the valley's history, and wrote books, most of them about the Don.

Flood control adopted after the devastation wrought by Hurricane Hazel in 1954 led to much denaturalization of valleys. It has created urban rivers with an environment of channelled river, mown grass, and decreased wildlife habitat. It took many years for people to become uneasy with the result. Residents wanted a more natural valley, where the needs of wildlife were taken into consideration, where pollution was prevented, and where natural habitats were protected.

One of the first reactions against the urbanization of valleys and ravines occurred in Park Drive ravine. In the early 1980s, Ontario Hydro had just buried a cable in this ravine. To complete the task Hydro wished to tidy up the ravine by improving the road, laying grass verges on both sides of it, and eliminating the little swamp at the lower end. Rosedale residents objected. They did not want their ravine urbanized. They wanted their little swamp. They did not want the road, let alone grass verges along it. They did not want park benches. They were perfectly happy sitting on logs. Ontario Hydro gave them what they wanted. The idea of a natural landscape was precisely what Charlie Sauriol's Don Valley Conservation Association had promoted in the late 1940s and early 1950s.

Since the early 1980s many new groups have formed, all with much the same conservation goals as Charlie Sauriol's group. Friends of the Valley, the Task Force to Bring Back the Don, Friends of the Don East York, the Don Watershed Task Force (part of the Metro Toronto and Region Conservation Authority), Friends of the Don (Headwaters), the North Toronto Green Community, and the Waterfront Regeneration Trust – all are actively involved in restoring the valley to a semblance of its old self.

Other groups, not directly tied to the Don valley, have had lasting impacts on it. The contributions of the Toronto Field Naturalists has been most important. Over many years this group, through its education program of nature walks and relentless communication with city authorities, brought about changes in such matters as ravine bylaws, regulation of the destruction of trees and other flora, regulation of contour changes such as filling in ravines, and the planting of native flora to extend habitat for wildlife. In the latter instance, the planting of native sugar maple instead of European Norway maple was recommended, as the non-native tree multiplies rapidly and casts such dense shade that nothing grows beneath it. The Toronto Field Naturalists have also consistently advocated the purchase of valley lands for use as parkland.

Today conservation groups are actively trying to create and restore habitats. They are scouring the valley for rubbish, cleaning up the river, planting wildflowers, building boardwalks in wet areas, and tracing the routes of long-forgotten Don creeks that have become underground sewers.

At the end of the century, we seem to have come a long way from the attitude that prevailed in the city's parks department in 1914. At that time, the commissioner of parks, speaking of a beautification scheme for Riverdale Park, proudly proclaimed: 'It is sad but true ... the Don River can never be made clean. Everything will be done, though, to dress [the park] up and make it presentable' (*Globe and Mail,* 2 April 1914). Dressing up Riverdale involved hiding the dirty river by raising the banks and dynamiting what was left of the natural contours in order to make terraces. In another 80 years, will our current attempts at renaturalizing the Don valley seem as naïve as the ideas proclaimed in 1914?

Those interested in the Don valley should visit the Charles Sauriol Conservation Reserve in the East Don valley. The reserve was established as a tribute to Sauriol in 1987.

Charles Sauriol

Charles Sauriol, mid-1990s. (The Toronto and Region Conservation Authority)

Charles Sauriol (1904-95) was born in Toronto and lived his 91 years in and around the Don River valley. His father was employed in the dredging and straightening of the Don and, ironically, it was on this foundation that Sauriol's love of nature — and his fierce loyalty to the valley — took root. For years the wilderness surrounding the river provided opportunities for Sauriol to camp and hike.

Urban areas gradually spread northward, however, degrading much of the valley. Sauriol helped organize the Don Valley Conservation Association in 1947 and later became its powerful spokesperson. Through this group he promoted conservation ideas that placed the association and the valley in the vanguard of the conservation movement in Ontario. Many of Sauriol's conservation ideas about the valley are being revived and implemented today.

When Hurricane Hazel struck a devastating blow to the Toronto region in 1954, Sauriol's message was finally driven home: these river and valley systems were natural corridors and could be interfered with only at our peril. In response, the Metropolitan Toronto and Region Conservation Authority, in which Sauriol was to play a tireless and dedicated volunteer leadership role for over a decade, was formed. Sauriol urged that such lands had to be purchased and kept free of development, dedicating himself to this goal and becoming a key figure in the authority's intention to acquire the land. The authority's principal goal was to achieve flood control, thereby preventing loss of life and damage to property. It achieved this control in two ways: by purchasing thousands of hectares of valley land, thus removing them from private hands, and by landscape engineering. The latter involved building dams, regulating valley use in flood-prone areas, and reinforcing riverbanks with gabion baskets.

Sauriol later joined the Nature Conservancy of Canada, in which he set out to accomplish on private land nationwide what he had achieved in the Metro Toronto region. His tireless efforts raised over $20 million, which was used to acquire environmentally significant lands. In Ontario alone, this translated to over 500 natural areas.

Sauriol's inexhaustible energy and tireless work on behalf of the conservation of nature and, specifically, the Don River valley, were recognized throughout his lifetime. Between 1956 and 1995, he received 39 awards, citations, scrolls, and recognitions — an average of one per year — including the Order of Canada. To ensure continuation of the vital work he began, the Conservation Foundation of Greater Toronto established the Charles Sauriol Environmental Trust Fund, dedicated to preserving natural and heritage resources.

Gabion baskets. (Credit Valley Conservation Authority 1988)

Duffins Creek

Harold H. Harvey

Duffins Creek lies in an area formerly occupied by Iroquoian peoples who were ancestral to the Huron. According to Augustus Jones, the Native name for the creek was Sin.qua.trik.de.que.onk, which he translated as 'pine wood on side' (Ontario Department of Planning and Development 1956). The Seneca had built their villages Gandatsekiagon and Teiaiagon near the mouths of the Rouge and Humber Rivers, respectively, the beginning of their trails to Lake Simcoe. Compared to these locations, Duffins Creek was less attractive. The sand bar at its mouth barred boats from entering, and canoes could be paddled upstream only 6 km. The trail upstream led to the Pefferlaw River and thence downstream to Lake Simcoe. Apparently Duffin, the creek's namesake, was a trader, 'a genial Irishman' who set up in business briefly near the present site of Pickering. The name Duffin appears on the map of Augustus Jones in 1791; subsequently, spelling has varied as Duffin, Duffin's and Duffins Creek.

The Duffins Creek watershed is the most intensively studied in the greater Toronto area, in terms of hydrology, water quality, water temperature, and fish and invertebrate fauna. The headwaters of Duffins Creek extend into Uxbridge, the northwest corner of Whitchurch, and 10 km² of Markham Township. The remainder of the watershed is in Pickering Township.

The Duffins Creek watershed is the least urbanized and the least in need of rehabilitation of any of the Toronto area watersheds. In 1975, only 1.36 percent was urban area, increasing to 6.27 percent by 1992. Yet, if continued, this rate of urbanization would lead to 14 percent of the Duffins watershed being urbanized by 2020; elsewhere 14 percent urbanization has resulted in the loss of aquatic species (Weaver and Garman 1994).

There is great uncertainty over land development in the Duffins Creek watershed. Three factors will be of particular significance in determining the course of development. The first consideration is the future of two enormous parcels of land that were acquired by the federal and provincial governments over several years.

The Duffins Creek watershed is the most intensively studied in the greater Toronto area.

The federal acquisition was to be the site of the second international airport serving the greater Toronto area. The provincial acquisition (the 2,850 ha Seaton Lands) was to be the site of a community of 90,000 people, in part relating to employment at the new airport. This land has sat idle for decades, and industrial, commercial, and residential development has thus far bypassed the Duffins watershed to the east of these two parcels. A second factor in future development is that the government currently discourages development on agricultural lands, though this could change. The third factor is Highway 407. Any eastward extension of the highway will largely determine future growth in the adjacent areas in Pickering Township. Clearly the potential for large-scale development within the watershed remains.

As yet, there is no watershed plan for Duffins Creek. Thus large wooded areas along streams may be recognized as having an important hydrologic function, but small woodlots are not. Further loss of forest cover will intensify hydrologic extremes, promote water warming and channel widening, and lead to loss of sensitive species.

Prior to Hurricane Hazel, flooding in the region was reported periodically in spring, usually as a consequence of ice blockage of the stream channel. A serious flood in June 1890 destroyed six dams and eight bridges. Hurricane Hazel, on 15 October 1954, was centred over Etobicoke Creek and the Humber River, yet it did considerable damage in the Duffins watershed (Figure 17.38). Although the kame moraines of the headwaters are quite permeable to precipitation, the soils downstream are relatively impermeable clays. Urban development results in watertight surfaces with 40 to 100 percent runoff, enhancing flood peaks. The upper reaches of Duffins Creek have small kettle lakes and extensive wetlands; in addition there are eight reservoirs, plus community and sport fishing ponds. These help to moderate peak discharge downstream.

The type of forest cover in the Duffins Creek watershed reflects the soil conditions: white pine and oak on

FIGURE 17.38 West Duffins Creek at the 8th Concession Road, a bridge destroyed by Hurricane Hazel. (H.H. Harvey)

light and easily worked soils, maple and beech in heavier but richer soils, and black ash, basswood, and white cedar on wet and heavy soils. Deforestation of the watershed – leading to slight to severe erosion – was driven by timber for export, the obligatory clearing of the front of farm lots before a patent could be obtained, the market for wood ash, and the demand for saw lumber in the Toronto area. Timothy Rogers brought 21 families of Quakers to Pickering in 1809, and built the first sawmill and gristmill in Pickering Township. Lumber exported from Pickering Township reached 3 million feet in 1845, but by 1850 logging was in rapid decline and the industry shifted northward, due to both diminishing supply and the advent of portable and fixed steam mills. The floods in the last three decades of the nineteenth century resulted in dams being lost and mills being abandoned.

Settlement was gradual, and the frontier people who settled the Duffins watershed were experienced backwoods farmers. Their subsistence life style – modest log cabin, small clearing with stumps in place, family garden based on a shovel or harrow, diet supplemented with deer, salmon, and passenger pigeons – had no major impact on the environment. The backwoods era ended by 1850 with the opening of the lattice of roads. Villages grew up where roads met streams. Stouffville, on the northwest branch of West Duffins Creek, came into being with the building of Abraham Stouffer's mills, in 1817-24. A village of 12 households emerged by 1837 where Kingston Road crosses Duffins Creek. The village of Major (Whitevale) started with John Major's sawmill in the 1820s; the mill burned in the 1870s.

The Civil War in the United States increased the demand for cereal grains, especially barley, for export – indeed, the 1860s were known as the 'Barley Years.' Yet at the same time, farmers in the area were complaining that the soil was losing its fertility and crops were therefore lighter. This reduced farm income was soon aggravated by a decline in grain prices, resulting from competition from United States and western Canadian grain. Local farmers responded by increased cutting of woodlots and draining wetlands, abandoning the long-established rule of one-third of the farm in woodlot. By no means was all of the Duffins watershed clearcut; some mature forests were retained, especially in the Glen Major area. The rural population was in decline by the 1870s due to the incorporation of villages and to the movement of people west in search of better farmland.

In the first half of the nineteenth century, Atlantic salmon were still to be found in Duffins Creek. Writing about Wilmot Creek in the 1860s, a fisheries officer commented that the salmon were 'formerly ... so abundant that as many as a thousand and upwards have been taken by torch and spear in a single night.' Another contemporary concurred, noting that the Wilmot Creek salmon 'were so plentiful ... that men killed them with clubs and pitchforks – women seined them with flannel petticoats – and settlers bought and paid for farms and built homes from the sale of salmon.' Yet on Duffins Creek itself, the salmon could not pass upstream beyond the first dam. Gaining any kind of access into the creek was sometimes a problem for the fish. In the 1870s, it was reported that the mouth of Duffins Creek had to be opened eight times during the upstream salmon migration because storms on the lake had blocked it with sand and gravel (Huntsman 1944).

Yet by the end of the nineteenth century, the Atlantic salmon had disappeared from all Lake Ontario tributaries, including Duffins Creek. The many possible causes of the decline and extinction of Atlantic salmon in Lake Ontario were identified by Huntsman (1944) as overfishing (although he

In the 1870s, the mouth of Duffins Creek had to be opened eight times during the upstream salmon migration because storms on the lake had blocked it with sand and gravel.

noted that the decline continued even when the fish were protected from harvest); forest removal; more rapid runoff; clearing of the streambed; pollution with sawdust and farm and factory waste; silting of spawning beds; and blockage of salmon ascent.

Reintroduction of Atlantic salmon into Duffins Creek has been attempted twice in the twentieth century. In 1945-7, over 35,000 fry were released at a total of 27 sites; however, only 3 percent of fry planted survived to smolt migration in their third year (McCrimmon 1954). Mackay (1963) noted that only one adult Atlantic salmon was known with certainty to have returned to Duffins Creek, and the attempt at reintroduction was temporarily abandoned. Currently the Ontario Ministry of Natural Resources is once again attempting to establish Atlantic salmon in selected streams, and in the mid-1990s Duffins Creek was stocked with fry and small fingerlings (J. Bolby, personal communication).

In the eastern headwaters of Duffins Creek the Glen Major Angling Club was established in 1895, with the purchase of an old mill site. The sport fishery for brook trout was enhanced by frequent plantings. Rainbow trout were planted in Duffins in 1961 and 1965 but could already have been established via stocking at Glen Major in the 1930s. There is now a naturalized population of rainbow trout in Duffins Creek.

Exotic fish species – 34 between 1819 and 1974 – have invaded the Great Lakes (Emery 1985). Few of these have entered Duffins Creek. The carp, stocked officially in North America in 1877, is widespread in the Great Lakes. It is relatively common in the lower reaches of the Humber River and Mimico Creek but rarer in Duffins Creek. However, Speirs and Falls (1992) reported carp ranked fifth in abundance at the mouth of the stream. The gizzard shad, a long-time resident in the Great Lakes, albeit rare in Lake Ontario (Scott and Crossman 1973), was ranked sixth in abundance at the mouth of Duffins Creek by Speirs and Falls. The two species of Pacific salmon, chinook and coho, planted extensively in the Great Lakes, have been captured near the mouth and upstream to the beaver dam at Greenwood (D.A. Jackson, personal communication).

The indigenous species of fishes appear to be maintaining their presence in Duffins Creek. The number of species in the cold headwaters is small. In the Glen Major area, for example, there are five indigenous species, dominated by brook trout (Speirs and Falls 1992).

Downstream at the Claremont Conservation Area, Speirs and Falls reported 9 species; at the level of Greenwood Conservation area 20 species; and at the mouth of Duffins Creek 38 species. For the whole of Duffins Creek, the count stands at 58 species.

The Natural History of the Toronto Region (1913) identified the fishes to be found within a radius of about 20 km of Toronto. No mention was made of fish species found in Duffins Creek. At present, despite the episodic trauma of floods, the deforestation of most of the watershed, and the gradual urbanization of the area, Duffins Creek continues to support a diverse community of fishes. The greatest loss has been the extinction of the Atlantic salmon and, in the lower reaches, the replacement of brook trout by brown trout.

FIGURE 17.39 West Duffins Creek along the Seaton Hiking Trail. (H.H. Harvey)

SPECIAL PLACES IN THE DUFFINS CREEK WATERSHED

There are several places of particular interest within the Duffins Creek Watershed. The Seaton Hiking Trail follows West Duffins Creek south from Highway 7 some 10 km into Grand Valley Park, south of the 3rd Concession, and offers hikers a range of habitats with varied flora and fauna (Figure 17.39). The Pickering Museum Village is a remarkable collection of historic buildings, mostly dating back to the mid-1800s. The village mounts a diverse program of heritage activities and skills for children and adults. The Greenwood Conservation Area straddles West Duffins Creek from a mid-point between the 5th and 6th Concessions south 2.5 km to the Canadian Pacific Railway line. The 2 km^2 of parkland include a variety of natural environments. The vicinity of Claremont has sites of interest such as the Claremont Conservation Area; the remains of the bridge across the 8th Concession washed out during Hurricane Hazel; and the birthplace of Tom Thomson, a stone house built by his grandfather, Thomas Thomson, about 1850. Of the many early dams on Duffins Creek and its tributaries, one remains at Whitevale, between Highway 7 and 5th Concession. The Ajax GO Station houses the turbine of the nearby Elmdale Mill, first built in 1867 and twice destroyed by fire. This list is only a brief sample of the rich variety of cultural, natural, and historic special places in the Duffins Creek watershed.

Rouge Valley

John L. Riley

Every youth should be spent in discovery, and part of mine was in discovery of the Rouge valley 25 years ago. Crawling through cattails to get eye-to-eye with a bittern on her young. Dozens of black tern careening and crying over their nests. Wheeling flights of waterfowl lifting out of the marsh and heading south. New botanical discoveries on every visit. It was one of those great summers, which led to other explorations farther upstream and, eventually, to efforts to get the Rouge recognized as a special place and a park.

The Rouge Park is in eastern Toronto and York and western Durham, in and between the valleys of the Rouge River and the Little Rouge Creek south of Steeles Avenue. It is more than 2,860 ha, the largest park in the greater Toronto region and one of the largest urban parks in Canada. It now seems to be just a matter of time until the rest of the Rouge valleys north to the Oak Ridges Moraine are added to the park. Time and nature have treated the Rouge well, and the valley is a fine example of what benign neglect can do for an area after a century and more as a hard-working near-urban landscape (Figure 17.40).

Archaeological sites of former Iroquoian villages dating from 1300 to 1400, some supporting 1,000 to 1,500

FIGURE 17.40 The Rouge Park stretches from its Lake Ontario beaches and marshes northward through the rapidly developing suburbs of Toronto, Durham, and York Regions. The wide, naturally vegetated valley of the Lower Rouge make it a very special place in the future of wildlife in the greater Toronto region. (J.L. Riley)

people, have been found in the area. From about 1667 to 1687, the Senecas settled the north shore of Lake Ontario and built a large village commanding the first rapids above the lake. This was Gandatsekiagon, referred to in French records. Croplands were part of the Native economy, and the first British survey of 1793 still showed 'old Indian fields' close by the river mouth. It is likely that these dated to the Mississauga residency of the area. The same map shows '230 acres of land set apart for His Excellency Major General Simcoe, to complete his military allowance' (Smith 1793). This acreage showed good taste on his part – the land was, and still is, the core natural area of the lower Rouge, the tract including both the Rouge and the Little Rouge Rivers, the tableland between them, and their confluence.

The next land surveys were by Galbraith in 1833. His lot-by-lot descriptions of the marshes and the forests of the area noted, for example, the dry, southfacing slope east of the river and south of what is now highway 401 as having the same vegetation as today: 'Scattering trees of pine and oak with an almost impenetrable thicket of briars, hazel and other sorts of brush' (Galbraith 1833). Throughout the forests were superstorey white pines, with crowns high above the deciduous canopy, and large, sandy plains of red oak, white oak, and white pine.

John Goldie, the Scottish-Canadian botanist, had described the same area in 1819: 'Before mid-day I passed a creek [the Rouge] which lay very low, so that the road is very steep on each side. All the declivity on the east side was completely covered with *Penstemon pubescens* [*P. hirsutus,* hairy beardtongue], such a quantity of which I never expected to see in one place. For a number of miles today I passed through barren sandy pine woods which it is possible will never be cleared' (Goldie 1819). There are records of a fire there in 1802, and the abundance of hairy beardtongue then (and of fire scars in trees from that era) suggests that open woodlands with a history of ground fires were frequent, much like the oak-pine woodlands of the Iroquois Lake plain westward past High Park and the Humber.

Goldie's prediction that the area 'will never be cleared' lasted a few more years. By 1851, 73 percent of Scarborough remained in forest, but by 1850 there were 14 mills in the park, and by 1891 forests remained on only 8 percent of Scarborough. The woods were highgraded first for the large oak and white pine for the square timber trade – this was the era of great ships. The supply of local squared timber was largely exhausted by 1850. The era is remembered in the mast road still evident across Simcoe's land allowance, leading down to the confluence of the two streams.

Tree cores show that many of the park's mature trees and woodlands date from that period of intensive logging up until about 1860. And history shows them lucky to be left standing. Development was often intensive and intrusive in the lower valley, and much of the area is a postindustrial site, a testament to natural healing.

The former shoreline of glacial Lake Iroquois provided ready building materials, and gravel pits opened in 1940 at the current Beare Road landfill site, at the present Toronto Zoo site, and in the rivers themselves, with 45 ha of river bottomlands producing gravel in the 1950s along more than 6 km of river.

Corridors of all kinds crossed the valley. Highway 401 was begun in 1939, its width now approaching 16 lanes of traffic. Major crossings also occurred at Meadowvale Road and the Beare Road landfill. Hydro corridors date from the 1920s, with additional lines in every decade until the 1980s. Cross-valley rail lines and pipelines were added from the 1920s right through to the 1960s. Recreation developments included Rouge rivermouth resorts, boathouses, picnic pavilions, swimming, boat rentals, refreshment stands, picnic grounds, and parking lots. Similar facilities were installed in the valley north at Highway 2, at Twyn Rivers Drive (with a ski hill), and at Steeles. The 1970s saw the construction of the Metro Zoo. All of these activities needed forest cover removed to accommodate them.

Homes were built to overlook the lower marshes, starting in the 1920s. The Depression stopped an ambitious 'Venice of the North' development, but housing spread rapidly after the Second World War. By the 1940s there were dozens of cottages built along the lake and the lower river and, by the 1950s, another 50 cottages were built along the river.

But the valley and the tablelands between the two valleys will probably see no more housing. Hurricane Hazel hit Toronto on 15 October 1954, destroying many bridges in the Rouge, rafting bottomland houses

Rouge Park is the largest park in Toronto and one of the largest urban parks in Canada.

FIGURE 17.41 The Rouge Park includes outstanding exposures of ancient glacial deposits and fossil strata from the last interglacial period. In the vicinity of Twyn Rivers Drive and the Rouge River, the park is a mosaic of relatively pristine upland maple-oak forests, regenerating cedar-ash-poplar bottomlands, and old fields reminding visitors of the farming history of the valley. (J.L. Riley)

downstream, carving back the lakeshore sand bar, and establishing forever, it is hoped, that our valleys are to be respected and conserved. Everyone then took a hard look at land conservation and land-use planning across southern Ontario. The local precursor to the Toronto and Region Conservation Authority summarized its evaluation of the Rouge: 'This area contains the choicest block of natural unspoiled wilderness in the lower reaches of any of the valleys. This conservation area offers the best possibility of any of the valleylands immediately adjacent to Metropolitan Toronto for the development of a large-scale wilderness parkland and nature preserve' (Ontario Department of Planning and Development 1956). Over the next 35 years, the conservation authority acquired lands in the lower valley, and the province acquired other lands in anticipation of airport development farther to the northwest. By 1995, about 80 percent of the valley south of Steeles Avenue was publicly owned.

At the same time, research interest in the Rouge was increasing. Parts of the valley were identified as provincial areas of natural and scientific interest (ANSIs), as environmentally significant areas (ESAs), as provincial wetlands, and as key natural areas within the Carolinian life zone of southernmost Canada. Guides to the plants and wildlife of the area were published, such as Riley (1978; 1980) and Varga et al. (1991). A dedicated organization, Save the Rouge, was formed and set about, with great skill and perseverance, to awaken community interest in a formal park being established in the valley. In March 1990, the province announced the establishment of the Rouge Park south of Steeles, and in April 1999 an additional 660 ha was added northward along the Little Rouge Creek (Figure 17.41).

This is a park with its origins in a hard-working rural landscape. The lessons of time – that the rough terrain was of little use for farming; that hurricanes can be defining; that public lands can be acquired; and that the best management can be passive – have bequeathed us a park of national interest.

The park contains geological sections of great importance for understanding the last two periods of continental glaciation and for studying earth movements in the lower Great Lakes area. Collectively, the natural areas of the lower Rouge valleys, lakeshore marshes, and adjacent tablelands form the most significant system of linked natural areas along any of the lower river valleys draining into northwestern Lake Ontario. This system represents the only remaining large corridor of natural open space extending from the northwest shore of Lake Ontario into the interior of the greater Toronto region. North of Highway 401, the valley corridor averages more than 2 km in width.

Its biological diversity is outstanding. Straddling two life zones, more than 100 native vegetation community types have been identified. The park contains 531 native plant species (28 percent of Ontario's native flora), including several that are nationally and provincially rare (such as ginseng and the shy bulrush) and 92 others that are now rare in the region. Another 231 Eurasian plant species occur in the park, most of them widespread waifs from agriculture and horticulture, and some of them, such as dog strangling vine, real problems for park managers.

At least 225 bird species have been observed inside the park, and 123 species of birds are documented as breeding in the park. National and provincial rarities like Cooper's hawk and the orchard oriole continue to breed in the park. Fifty-five species of fish have been documented in the park, including the rare redside dace and central stoneroller, as have 27 mammal species, including some now very rare in the Toronto area, such as northern flying squirrel, ermine, beaver, coyote, and white-tailed deer. The park's 19 reptile and amphibian species include the regionally rare mudpuppy, blue-spotted salamander, map turtle, and Blanding's turtle.

It is remarkable that these species and their habitats survive in Canada's largest city, and that they survive in a connected natural area of sufficient size and condition to be ecologically viable. Bottomland farm fields have reverted to forested swamps. Worked-out gravel pits on the river are filling with thickets and wetland pockets. The rivers are finding new and natural channels, and the forests are maturing and expanding. The aerial photographs taken since the 1930s tell a great tale of natural healing, which accelerated dramatically after the hurricane of 1954.

The park is not without problems. There are no more black tern in the marsh. Gone are the American bittern, marsh wren, and American coot. The valley faces aggressive non-native species, high densities of predatory raccoons and house pets, episodic forest pathogens, and stabilized Lake Ontario water levels. More and more people want to get across the valley corridor on the limited roadways. More and more people want to visit the park. Trail proliferation and waterfront overuse are obvious, and on-site park management is negligible.

Still, an ambitious park-management plan is now in place, and park staff, advisory committees, and public agencies are working to turn the park into reality. Fully 53 percent of the park will be treated as nature reserve, and another 15 percent is zoned for eventual restoration to native vegetation. These are noteworthy aspirations. Discussions are under way with municipalities in York Region to ensure continuous protected corridors north to the Oak Ridges Moraine.

The Rouge Park hosts 27 mammal species, including some now very rare in the Toronto area, such as northern flying squirrel, ermine, beaver, coyote, and white-tailed deer.

The 1913 *The Natural History of the Toronto Region* made no reference to the Rouge River valley on the eastern boundary of Toronto. Its quiet lakeshore marshes and beaches were for good fishing, good swimming, and good birding, accessible from the railway station just east of the 1856 rail bridge across the river. That quiet rural landscape is no more. The steady roar of Toronto gets louder each day. The old landscape of hard-working, productive farms mixed with woods and steep valleys is gone, replaced by urban development shoved right up to the edge of the valleys.

This is the challenge of the Rouge Park. What will be the judgment of time, when we look back 86 years from now, as we are now looking back over the past 86 years? Will the black tern, American bittern, American coot, and red-shouldered hawk, which lived in the valley not too long ago, come back to the Rouge? Will the otter, last seen in the 1950s, return again? Can we keep the Rouge as a place of discovery, as it has been for so many visitors before?

18

Discussion and Conclusions

D.A. Chant and Henry A. Regier

DISCUSSION

In an essay published in 1998 (*Kitchener Record,* 10 February 1998), David Crombie, a former mayor of Toronto, sketched his vision of the greater Toronto region as a rapidly emerging 'city region.' Crombie referred to the natural systems: forests, wetlands, rivers, and valleys. He also referred to the connectedness of things; the evolving urban ecology; the importance of place; and the mutual interdependence of economic growth, environmental health, and social well-being. These five perspectives form as good a set as any for our purposes here, and we use them as a framework for this discussion.

Forests, Wetlands, Rivers, and Valleys

The greater Toronto region has evolved with a larger legacy of natural things, species, and spaces than most city regions that emerged in the twentieth century in North America. But no part of our city region is now in a pristine state, and none can be restored fully to such a state by direct human actions.

Deforestation in the Toronto region peaked late in the nineteenth century, when the value of forests for lumber and firewood was overtaken by the forests' non-timber values. Deforestation slowed gradually over the twentieth century, with forest stands of various sizes remaining in many stream valleys and in the Oak Ridges Moraine and Niagara Escarpment areas. Many of the woodlots that did persist were grazed by domestic animals, causing severe degradation of the forest soils and the substorey. So much still remains to be done to ameliorate past degradation, especially with respect to linking the surviving forests together with quasi-natural corridors on the tablelands.

As we have seen from some of the earlier chapters, most of the coastal zone of the more urbanized part of the greater Toronto region is artificial. The part of Toronto south of Front Street is built on landfill. Ashbridge's Bay, extending some 2 km east from the mouth of the Don River and protected from the open lake by a sandspit formed by materials eroding from the Scarborough Bluffs, was filled in with excavated materials early in the twentieth century. It was then zoned for industry and transportation, with no effective safeguards against gross pollution or for promotion of aesthetic values. Remediation of some contaminated hotspots was begun late in this century but progress has been slow.

Materials from excavations for buildings and from dredging shipping channels in the harbour were also dumped on the Toronto Islands following construction of retaining walls along their northern shores, which destroyed the original shelving beaches and wetland fringe. A large area of land was constructed with fill at the western end of the islands for the Toronto Island Airport. These changes destroyed natural habitat for inshore aquatic species and for water birds.

Though initially unplanned and unforeseen, other landfilling actually created new aquatic habitat. The large Leslie Street Spit, which 'grew' gradually over the last three decades of this century from a base of the infilled Ashbridge's Bay, was allowed to 'renaturalize' to some degree on both the land and water sides. These peninsulas, with the wetlands that are evolving in their bays, serve to mitigate somewhat the previous permanent losses of much natural shoreline and wetland habitat.

All the smaller streams between the Humber River and Highland Creek were first channelized with hard

A Future Special Place: Taddle Creek

The fact that our creeks are all underground represents how far we are distanced from our environment; the fact that they still flow represents hope.

Anonymous

▲ *University College, Taddle Creek in foreground, c. 1860s. (Archives of Ontario F507 st. 353)*

◀ *Boys bathing in the creek at Wychwood Park, 1916. (detail, John Boyd, photographer, National Archives of Canada PA069862)*

Taddle Creek once tumbled over the ancient shoreline of Lake Iroquois and flowed down through the forest of pine and balsam to Lake Ontario. The old town of York was founded where the creek emptied into the lake. The creek provided a habitat for fish and wildlife along its banks and was a common gathering place for fishing, swimming, skating in season, or just for relaxing.

As the city grew, the stream was contaminated, channelized, and finally buried. Where the Taddle was once a watershed, it is now part of a sewershed. The only observable remnants of the creek are dips in the road or embankments that reflect the original topography. Philosopher's Walk on the St. George campus of the University of Toronto still retains Taddle's natural landscape.

The creek is deeply entwined with the human and natural history of this region. It was of life-enhancing importance to the Aboriginal peoples long before European settlers arrived. Today, Taddle lives in the memories and imaginations of local residents in a rich, informal trove of 'Taddle Tales.'

A grassroots alliance of residents' associations, business and community organizations, institutions, and local environmental groups in the centre of Toronto are working to give new life to the creek. The common vision is to 'bring back' parts of the creek using alternative methods of stormwater management; habitat restoration and naturalization; environmental education; and recognition and respect for human heritage.

We see this process helping to reconnect the Taddle Watershed communities to each other and to the natural world that sustains us. For more information, see the website at http://www.web.net/taddle or contact The Taddle Creek Watershed Initiative, 219 College Street, No. 345, Toronto, ON M5T 1R1.

materials and then buried in underground conduits. These buried streams, still flowing, include the Taddle, which runs under the main campus of the University of Toronto, and the Garrison and Baldwin. As noted in Chapter 3, it is almost impossible to constrain a stream indefinitely with concrete and steel: inevitably it breaks out and creates problems for the constrainers. The main stem of the Don River was not buried, although the Lower Don was channelized and given a right angle turn, where silt still gathers, as it entered Toronto Bay. This interference guaranteed the need for subsidized dredging continuously thereafter. Many of the Don's tributaries in turn were buried. As we enter the twenty-first century, work is beginning toward reversing some of this burial and channelization, and toward revitalizing some reaches of these tributaries and other streams, such as Taddle Creek, that still flow in their subsurface straitjackets. Meanwhile, upstream in the watershed, urban developers are reluctantly giving up channelization and burial as quick, though temporary, technofixes.

By the 1960s, industrial and other pollution had transformed the lower Don valley into a sordid ecological slum through which the fetid river oozed. Irresponsible people dumped trash in the weedy floodplain with impunity. Toronto's Pollution Probe staged a funeral for the Don in the late 1960s. People had turned their backs to the stinking rivers and waterfront, and those who could afford them built backyard pools in suburbia and summer cottages on lakes far away from Toronto (which were also becoming increasingly polluted). Learning did not come easily for otherwise progressive people careless about their waste products.

The Lower Don is being rehabilitated or regenerated in part. Revitalization will presumably continue for decades, but full restoration to a pristine state will be impossible. Exotic weeds are being suppressed in favour of native plants, wetlands are being re-created on the floodplain, the river is being renaturalized – all in part. People stressed by the noise, congestion, cars, concrete, glass, and plastic of the crowded city core can now walk the Lower Don for respite.

The wetlands at the mouths of the Credit, Humber, Highland, Petticoat, Rouge, and Duffins streams have fared rather well by comparison with Ashbridge's Bay and the Lower Don. 'Disturbance-dependent' wetlands like these are pre-adapted to tolerate many of the disturbances caused by humans upstream in the watersheds. The increasing flooding from upstream may have compensated in part for decreased lakeside flooding resulting from controls on the fluctuation of lake levels, but this is conjectural.

Toronto's valleys, especially where the streams have eroded upstream into the glacial Lake Iroquois shoreline, have been subject to benign neglect. Many of the land portions of these valleys have remained as, or have regenerated to, reasonably healthy ecosystems. But the streams flowing through these valleys are usually ecologically less healthy, though not as degraded as they were in the mid-twentieth century.

The forests, wetlands, rivers, and valleys of the greater Toronto region, initially so heavily degraded, are generally in a somewhat better ecological state at the close of the twentieth century than they were half a century earlier. This improvement has been accomplished through the grassroots participation of thousands of people over the decades, and with leadership on some issues by the conservation authorities, by federal and provincial government agencies, by a royal commission chaired by David Crombie, by municipal planning offices, and by the occasional private interest. But there is much left to be done, especially in mobilizing private efforts.

Preserved green spaces and restored rivers do not return things to their original natural state. They simply provide artificial substitutes.

Toronto has done a good job in preserving green spaces as parks and has made progress in 'restoring' degraded features such as the Don River – but such green spaces and restored rivers do not return things to their original natural state. They simply provide artificial substitutes. This is not to say that such activities are not worthwhile – they are. But let's not fool ourselves that we have done anything to turn the clock back to a time when the natural systems of the greater Toronto region were in their original, undisturbed state.

The Connectedness of Things

In 'old-growth' ecosystems there generally are strong and persistent ecological connections among combinations of species, and between particular species and features of the habitat. At the other extreme, species that can survive in severely disrupted and degraded habitats or in hard unnatural environments usually have only weak and transient interconnections with other species and weak linkages to features of the habitat.

Among the animals in old-growth ecosystems, species that become large and old as adults tend to dominate the smaller, short-lived species through predation and competition. These large dominants in turn relate harmoniously to the large plant dominants, notably trees and shrubs. Such dominant plants and animals have evolved to use each other on a sustainable basis.

Dominant mammals, birds, and fish generally migrated seasonally. Birds in particular linked the Toronto area to the south through migration from there in spring, and to the north through migration from there in fall. Some fish species entered the rivers from the lake to spawn in fall and others in spring. Large mammals migrated in and out of the coastal zone. In effect, some of the 'information' necessary for the self-organization of old-growth ecosystems was contained in some of the dominant animal species that moved in and out of these ecosystems in annual cycles. This 'information' includes the genetic code, learning by individuals within a population, and co-evolved behaviour that permits ecological associations to persist. Such information is related to the concept of biodiversity.

If a dominant species that used the Toronto area seasonally was severely harmed here, then there was a carry-over effect into the ecosystem that the species used at other times of the year. Thus, ecological harm done in one locale can be transmitted to other locales through migratory species. Interconnections abound. Generally a migratory species is most vulnerable during the breeding season, which in turn affects the ecosystems in all the various habitats in which it lives. This was understood centuries ago, and regulations have generally outlawed the hunting of animals in a population during their reproductive migrations as well as the taking of excessive numbers of animals during the breeding season. Unfortunately, these regulations did not suffice to protect valued species, such as some fish, for example, that were also caught intensively at other times of the year and were sensitive to pollution and physical alteration of their habitats. The story of Atlantic salmon in Lake Ontario, extirpated from the lake by the end of the nineteenth century, provides a particularly graphic example.

Humans generally value old-growth dominant species more than small opportunistic species. In harvesting these selectively, and inevitably over-harvesting them, some of the information necessary for the old growth association to regenerate is lost. The loss of a dominant species also results in the loss of some of the capability of the ecosystem to recover autonomously. Smaller, less valued species became dominant. Many introduced non-native species of animals and plants alike thrive as 'weeds' in such disturbed situations. Nevertheless, to have such 'weeds' in desolate habitats is often better than to have no organisms at all, so the good services of these species in such situations should not be ignored.

Many small tracts of forests, wetlands, and upstream tributary basins, especially on the Niagara Escarpment and in the Oak Ridges Moraine, were never seriously degraded. None of these is entirely isolated from all the other quasi-natural areas, but the level of exchange of animals and plants between them is at best low. So, if harm befalls one such tract, there is little opportunity for migration of species from a similar association to assist with regeneration. To re-create and improve connections between all natural areas in the Toronto region has become a general goal of naturalist groups as well as all relevant agencies at four levels of government.

A strongly connected web of natural areas may permit the occasional wolf, bear, or moose to find its way into the valleys bordering the cores of the cities of the Toronto region. But the natural areas will never be sufficiently large to enable such species to reoccupy their ancient haunts on a permanent basis. Their occasional reappearance in our urban areas may be taken as evidence that spatial connectedness of natural areas is improving.

Evolving Urban Ecology

Numerous groups of different peoples have also migrated to the Toronto region. In the centuries before the arrival of Europeans, various fishing and hunting groups lived here, followed by Iroquoian agriculturalists and finally by the Seneca, Iroquois, and Mississauga during the historic period. The latter were largely displaced by western Europeans, who were eventually joined by people from many other areas of the planet. Currently people of numerous language groups and of diverse cultural backgrounds interrelate well in the Toronto region. Up to this time, however, it has been mostly people of western European descent who have shown greatest concern about environmental degradation – for much of which this group itself was responsible.

Invariably each human coming to Toronto, whether immigrant, tourist, or returning traveller, brings other species along. Each human is an ecosystem of sorts and may carry viruses, bacteria, protozoa, fungi, and perhaps mites, lice, and worms. Some of the organisms that come with their human hosts transfer to other humans, epidemiologically. The disastrous epidemics of communicable diseases that struck Native peoples as a result of transmission from Europeans may be largely forgotten, but currently new strains of influenza can cause a local epidemic within months of their discovery half a world away. Acquired Immune Deficiency Syndrome (AIDS) presumably will be followed by other surprises in the next century.

Newcomers frequently brought pets, domestic animals, plant seeds, and pests along with them. The beneficial species generally require intensive care to thrive here, but some, like the purple loosestrife, become naturalized and turn into pests.

Among the non-native species that have become naturalized are opportunistic plant species that can thrive in locales that are less hospitable to native plants. Intensely urban areas that contain little natural habitat can be colonized by such opportunistic non-native species. They may be accepted as better than nothing in an otherwise barren cityscape, but if they have objectionable qualities, such as pollen to which people are allergic, they then may be controlled by harsh chemical methods that may also harm native species and even humans.

Any broad policies that foster human migration, tourism, and transportation lead inevitably to an increased risk of introduction of new exotic or non-native species, including human pathogens. The spectrum of introduced non-natives covers the full range of the various kingdoms of species. This problem has never received adequate attention in Toronto, or indeed elsewhere. Policies to prevent such unwanted introductions have been piecemeal and enforced only spottily. Thus there has always been an implicit subsidy for a general policy of 'openness,' in that the costs associated with these risks have been ignored and implicitly externalized.

The issue of non-native species relates to evolutionary information that has accumulated over millions of years. Each organism, species, and ecosystem has its own 'library' of genetic and other information, the bits of which have co-evolved in a particular regional context over aeons. Some isolation between related sets of information, at each of these three scales, appears to be necessary for the survival of each. If we carelessly mix this information through thoughtless introductions, inevitably we sacrifice some native information and undo some of the evolutionary products of countless generations before us. Conservationists who value the indigenous biodiversity that has come down to us in a particular area like the greater Toronto region seek to limit the risks of loss of such evolutionary information.

The Importance of Place

A cluster of cities and towns in a rural fabric make up the greater Toronto region. In the abstract, the region can be thought of as an artificial system of nodes with a network of interconnections. These nodes are the pyramidal city cores, and the connections are the highways, subways, water mains, sewer mains, electrical power mains, trunk telephone lines, and electronic communication means. This complex of cores and corridors was not consciously 'designed' before it came into being: the complexity would have overwhelmed any planning or political process. Rather this 'cultural network' evolved by a process that was partially self-organizing and subject to a set of ultimately weak constraints on what developers were permitted to do.

Again in the abstract, there is a comparable complex of cores and corridors in the quasi-natural parts of the landscape that have not been strongly urbanized. This complex includes parts of the lakeshore, the coastal wetlands, the river valleys, scattered woodlots on the tablelands, nature preserves, conservation areas, and relatively large parts of the Niagara Escarpment and the Oak Ridges Moraine. The river mouths, larger wetlands, larger forests, and persisting wild lands on the escarpment and the moraine may be taken to be the nodes. The water side of the lakeshore, the river valleys, some electric utility corridors on the tablelands, and the more continuous stretches of quasi-natural habitat on the escarpment and the moraine may be taken to be the corridors connecting the core areas of this natural network.

> *In Aboriginal and pioneer communities it tended to be the natural that was destructive of the cultural. Now the reverse is true.*

The cultural and natural networks are offset from each other. Thus the natural network persists or is regenerating in the spaces in which the cultural network has not yet strongly intruded. Natural cores must be buffered from the cultural cores because of the destructive propensities of the latter. In Aboriginal and pioneer communities it tended to be the natural that was destructive of the cultural. Now the reverse is true.

The corridors of the cultural and natural networks must cross in numerous locales. It has become conventional to engineer these crossings so as not to impede completely the role of either the natural or the cultural corridor. That role, put simplistically, is to transmit and transport things that contain mass, energy, and information from core to core, in each of the networks.

In the space between the cultural and natural networks, as defined above, lies an intergradation zone of low-density residential and commercial properties with semi-natural parks sprinkled throughout. Farming continues in some parts of this intergradation zone in the Toronto region. This zone acts to buffer the most urbanized cultural network from the least urbanized natural network.

The 'sense of place' that Torontonians have appears to relate to a commitment to have all three major components of the landscape – the cultural network, the natural network, and the intergrading 'desirable human space' – in vigorous health. Healthy states emerge where humans have strong commitments to appropriate personal, corporate, and civic ethics with respect to nature. Enforcement of firm regulations should be needed only with occasional miscreants, and should not have to be the norm of everyday life. However, too many people and organizations still abuse nature in our region if they are confident that they won't be caught.

The sense of place also relates to an implicit desire that a person's home, say in the zone between the two networks, be within walking distance of one or the other network – say, a subway or a stream valley. Toronto has always struggled to prevent the automobile from becoming the 'dominant species' in all parts of the city region. Recent residential developments in exurbia are troubling in this respect.

Economic Growth, Environmental Health, and Social Well-Being

Late in the twentieth century, Toronto has been recognized repeatedly as one of the world's most desirable places to live. The city is not the centre of a vast economic empire that brings tribute here from elsewhere in the world. The Toronto region does not have breathtaking landscapes, it suffers from quite a long winter, and many areas are severely degraded ecologically. Also, the city has an underclass of poor and homeless people, with street people occasionally dying of exposure. But we seem to have done better than other city regions in striking a tolerable balance between economic growth, environmental health, and social well-being. Of course, it may simply be that the whole world has a long way to go to reach acceptable levels and balances with respect to these goals, and that we look good by comparison.

Perceptive people have long argued that progress toward a good life for humans depends on developing synergistic interactions between the initiatives toward these goals of economic growth, health, and social well-being. The view that these factors are necessarily competitive – that to serve one kind of interest is to disserve the other – is a discredited one. Late in the twentieth century, some governments appear to have slipped back into a view that if laissez-faire economic interests are freed from constraints, adequate benefits will trickle down to the advantage of social and natural interests. (A less tolerant view would be that government simply caved in to raw political pressure from shortsighted economic interests.) Such governments in the Toronto region claimed that the city could become a primary node in global trade if we just voluntarily placed more trust in entrepreneurs to be duly considerate of environmental health and social well-being. From this perspective, relying on a large government apparatus to serve the latter two interests would drive taxes up and lessen the interest of global entrepreneurs: it would be better to use government funds to provide direct subsidies to entrepreneurial initiatives. Some of Toronto's people who would become wealthy as a consequence of this largesse would demonstrate their civic virtues by contributing to social and environmental causes. Decades ago Canadian-born economist and ex-farmboy John Kenneth Galbraith drew an ecological analogy: if a horse is overfed with oats, a sparrow may find some undigested grains among the horse's droppings. One hopes that the recent flip back into a discredited political philosophy will not persist for long.

Aside from party politics, the prospects for achieving a more responsible balance between economic growth, environmental health, and social well-being sketched above are promising. We 'hit the wall' with regard to environmental health with the intense localized degradation of the mid-twentieth century. The ecological slums then so obvious were partially rehabilitated during the last three decades of the century. Valley ecosystems near urban cores have been recovering gradually from the effects of more intense and destructive earlier uses. Urbanization of farmlands to residential and commercial properties, however, still causes serious ecological degradation, in spite of improvements in the practices of some developers.

There is no overall environmental policy at the level of the greater Toronto region. Perhaps the perspective sketched above – of a cultural network offset from a natural network with a hybrid natural/cultural buffer in between – could serve as a schema for a vision for the whole region.

In 1986 the report *Our Common Future* was published by the United Nations Commission on Sustainable Development, chaired by Grö Harlem Brundtland, a former prime minister of Norway. Programs related to all three goals enunciated above were deemed essential to sustainability. This view was reiterated by the UN Conference on the Environment convened at Rio de Janeiro in 1992 and by the UN Conference on Population and Development in Cairo in 1994. The single most important policy worldwide toward this set of goals appears to be the self-empowerment of women, with the second most important being the empowerment of disadvantaged groups generally. Granting more freedoms to dominant entrepreneurial males does not rank as high. The Toronto region has been progressing, but too slowly, toward these interrelated goals.

CONCLUSIONS

This book has presented an overview of the changes in the natural environment, both physical and biological, that have taken place as a consequence of human population growth and urbanization in the greater Toronto region since the Royal Canadian Institute last reviewed these issues in 1913.

We have adopted an ecosystem approach to emphasize that the many changes that have occurred are interrelated: no one organism or physical feature can be viewed in isolation. These features are part of a highly complex natural web, and impacts on individual species and on specific parts of the physical system have a ripple effect that may affect many other parts of this web, sometimes quite far removed from the site of the initial impact.

Change in the greater Toronto region, however measured, is a continuum. What we see around us today is a product of the past and what we will have tomorrow will be a product of today. This change has been the result of a complex of processes in the evolution of a system.

The many individual processes that have been involved in this change are described in the preceding chapters. Some are natural processes; some the consequences of human activity. The natural processes include erosion and siltation, plant succession, some changes in animal populations and species composition, postglacial climatic changes, and so on. Human-directed changes include pollution, urbanization, the destruction of animal and plant species, housing and road construction, population increase and concentration, energy production and consumption. There are overlaps between the two: for example, some erosion is natural and some is caused by the clearing of land and other human activities. But one thing is certain: rates of change have accelerated rapidly over the last 150 years, and the kinds of changes have multiplied dramatically.

There is no overall environmental policy at the level of the greater Toronto region.

Ecosystems evolve whether we like it or not. It is fashionable to decry change, but simply preserving the status quo or going back to a more pastoral, 'natural' past is neither possible nor desirable. We cannot go back, and we would not want to if we could. Back means 'Muddy York' – poor sanitation, poor public health, high infant mortality, low standards of education, even lower standards of public morality, poor food with limited choice, and all the other things that made life less pleasant for the majority not so long ago. When we were children in Toronto, in the 1930s, before the advent of freezers, winter vegetables were limited to potatoes, turnips, carrots and cabbages. Where once it was common, malaria no longer occurs in the greater Toronto region, unless we bring it back from some exotic and expensive holiday venue – elsewhere in Ontario it delayed by years the construction of the Rideau canal system from Kingston to Ottawa in the nineteenth century.

Some of these changes are a result of natural events. As noted, land erosion occurs in even the most pristine environments, though it is greatly accelerated by human activities. Some species of organisms have become less common or, conversely, more abundant, as a consequence of the ebb and flow of ecosystems; some have even disappeared as a result of the evolution of ecosystems. Changes in climate, even without considering the human-induced changes such as the urban heat island in Toronto, have caused changes in the flora and fauna of the greater Toronto region: some species have invaded this area from the south, whereas others have been driven farther north. These itinerant species have had significant impacts on the natural system quite apart from the impacts resulting from human activities.

However, most of the changes that have occurred to the natural environment in the greater Toronto region are a consequence of rapid growth of the human population, and of the urbanization that continually spreads the built environment and the services required to support it over an ever wider area. There also have been major changes in technology that have had severe effects on the natural system: industrialization and the wastes it produces, the automobile and other forms of transportation, pesticides, the building of airports, the destruction of much of the waterfront and the original marshes, and all the rest.

Not all changes are 'bad,' from a natural perspective. New opportunities have been created for new organisms, though many of these are opportunistic, turn into 'pests' from the human perspective, and may themselves have serious impacts on the native fauna and flora, thereby distorting the natural system even more severely.

The impacts of some of these changes have resulted in the complete elimination of some species from the greater Toronto region. Some impacts have simply greatly reduced the abundance of some species – reptiles and amphibians, in particular. Other impacts have created changes that favour some plants and animals to the point where they have become nuisances to humans – raccoons, pigeons, Canada geese, or native weed species, for example. And yet other changes have provided new environments for some species that otherwise would not occur in the greater Toronto region: some years snowy owls are attracted to our airports, which in winter resemble the Arctic tundra, their usual habitat.

Some alien species have been introduced deliberately into the greater Toronto region and some have spread to the region as a consequence of their introduction elsewhere – the English sparrow, for example, or the starling, both deliberately introduced to New York State late last century, and many plant species as well. Some species have invaded the greater Toronto region environment accidentally, but as a direct consequence of human activity. Such species include purple loosestrife, termites, lampreys, alewives, and zebra mussels.

In short, some species have gone from the greater Toronto region, some have come to it. But the greatest changes of all have been in the mix of species that constitute the natural environment of the region: some species have become less abundant while others have thrived and become more abundant. The result is an 'un-natural' natural environment whose composition and dynamics have been determined largely by our human impacts over the last two centuries or more.

Some of the most disturbing features of these changes are that they are cumulative and that they sometimes creep up on us – creeping incrementalism. Each year, changes grow: new roads, new subway lines, new subdivisions, new technologies. And some of them are so gradual that we scarcely notice them from year to year, until finally we look back over a decade or two and suddenly recognize how destructive they have been. There is an old fable that you can place a frog in a pot of cold water, put it on the stove, and gradually heat it. The frog does not take fright at the gradual increase in temperature and eventually it quietly dies when the temperature reaches the lethal point. Too often we humans do not recognize the creeping incrementalism of the environmental changes that directly affect us.

Another disturbing feature is that a sensitive ecosystem can be 'saved' many times but can be lost only once, for all time. Over the last three decades, ordinary citizens have become deeply involved in saving such ecosystems in the greater Toronto region. Citizen protests in the 1970s 'saved' the Pickering area from the ambition of the federal government to build a new international airport there, and the issue seemed permanently settled. However, in 1998 Ottawa announced that it was looking at this possibility once more. A marsh in Mississauga was protected from development ten years ago by the direct intervention of local residents in hearings before the Ontario Municipal Board. But once again developers have turned their rapacious eyes to this marsh, and, presumably, the local residents will once again have to 'take to the streets.' If they lose this time around, the marsh will be gone forever, despite the protection it was given a decade ago.

On balance, it will come as no surprise that most of the human impacts on the natural environment of the greater Toronto region have been harmful to the plants and animals that occur here or formerly occurred here. As these impacts have intensified, regrettably the natural system has too often been ignored. With the benefits of hindsight, it is easy to see that we generally have not done what we could to mitigate these impacts. We seem, even today, to ignore scientific knowledge and fail to use this accumulated knowledge to avoid many of the mistakes of the past.

There seems to be little hope of preventing further growth and development in the greater Toronto region, and most people probably would not want to do so: the 'growth ethic' is still a powerful force in our culture. But facing the inevitability of future change, we must understand and accept the fact that not all change is bad and instead attempt to direct future change so as to optimize human well-being and the well-being of what remains of the original natural system. We must take charge and ensure that future change is for the better, not just for our human systems but for the natural systems as well.

As noted in the Introduction, the Royal Canadian Institute is concerned about the continuing dichotomy between those who know about the natural system and those with planning and development responsibilities. Unfortunately, the will of the latter usually prevails. It is the hope of the editors and authors that this book will accomplish three major goals. First, we hope that the information presented will encourage people to re-evaluate their relationship with the ecosystems of the greater Toronto region. We believe that as a species we are capable of learning from our past mistakes and the heedless way we have treated the plants and animals that share, or used to share, this environment with us. To that end, we have documented the changes, mostly harmful, that have occurred to the natural systems of the region as the human population has invaded this area, and we have tried to reveal the major root causes underlying these changes. Second, we intend that the book establish a base-line of information on the natural environment of the greater Toronto region in the year 1999, as did the 1913 book, against which future changes can be measured and for which future decision makers can be held accountable. Finally, we hope that this book will play a part in bringing together those who have special knowledge and insights into the natural world with those who will be making future decisions on development so that the impacts of the latter can be mitigated and muted by what the former already know.

We are not advocating 'scientific imperialism.' We do not believe that scientists have all the answers – far from it. But we do urge that knowledge of natural systems, of the way they work and their fragility, whether from professional ecologists or natural historians, be given a much more assertive role in the equations that will direct future development and growth in the greater Toronto region.

Understanding the changes of the past can lead to more informed and focused future directions. Such understanding can steer us into the future. We, all of us, must come to understand the processes that have been and continue to be involved in order for us to achieve these objectives in a coherent, rational way. No more unplanned creeping incrementalism. No more ungainly sprawl.

The fact of natural change and the consequences of the human invasion of the greater Toronto region are inescapable and irreversible. So, what direction do we want to go from here, and how best can we achieve the objectives of improving human well-being while at the same time maintaining biodiversity and arriving at sustainability? Finding practical answers to these questions is the challenge that we face as we move into the next millennium.

The History of the Royal Canadian Institute

Conrad E. Heidenreich

The Institute began as an attempt by a group of architects, land surveyors, and civil engineers to incorporate themselves into a provincial professional association. Their first meeting was held in the office of Toronto architect Kivas Tully on 20 June 1849. The Hon. H.H. Killaly, who had been chairman of the Board of Works, Upper and Lower Canada, was elected president, and over subsequent meetings a prospectus for the proposed association was drawn up as well as the draft for a constitution. This constitution did not meet with unanimous approval and was referred back for amendments. When it was brought forward again on 8 February 1850, only two members bothered to show up for the meeting – Sandford Fleming (Sir Sandford in 1897) and Frederick Passmore, both civil engineers and surveyors. These two took it upon themselves to move, second, and pass with marvellous unanimity a series of resolutions that broadened the original aim of the association. Thus was born the Canadian Institute, an organization almost unique in the intellectual wilderness that was then Upper Canada. By the end of the year the Canadian Institute had 64 members, weekly meetings, and Canada's leading scientist William E. Logan (Sir William in 1856), director of the Geological Survey of Canada, as its new president.

On 4 November 1851, the Canadian Institute was granted a Royal Charter of Incorporation by Queen Victoria. This Charter is still the legal basis from which the Institute operates. The mandate of the Institute as set down in the Charter and as it developed in later years was to encourage and actively promote advances in the 'Physical Sciences, the Arts and Manufactures in this part of our Dominions.' More specifically, the Institute was to encourage scientific thinking and original research; to create a museum 'to promote the purposes of Science and the general interests of society'; to raise matters of public concern to which the pure and applied sciences could make a contribution; and to promote the best current scientific work through a lecture and publication program.

It was to be 'a society where men of all shades of religion or politics may meet on the same friendly grounds; nothing more being required of the members of the Canadian Institute than the means, the opportunity, or the disposition, to promote those pursuits which are calculated to refine and exalt people.' (Interestingly, from early on a few women belonged to the Institute, though almost all were associate, rather than full, members.) This was far removed from the original aim to create a professional association for architects, engineers, and surveyors, and many members of these professions resigned from the Institute. Other people, however, flocked to the Institute and in 1852 it had a membership close to 200. This had risen to 500 members by 1855, about 2 percent of Toronto's population.

In August 1852, the Institute launched the first number of its monthly periodical, the *Canadian Journal, a Repertory of Industry, Science and Art*. The *Journal* not only ran abstracts and sometimes the full text of important papers given at the Institute meetings but also reported on scientific and technical advances in other parts of the world. From 1852 to 1855, the editor was Professor Henry Youle Hind, followed by Professor Daniel Wilson (Sir Daniel in 1888), two of the intellectual giants of nineteenth-century Canada. In 1878, the name of the series was changed to the *Proceedings* and in 1889 the *Transactions*. Besides informing its readership of important intellectual advances the publication served to build a reputation and a library for the Institute. The journals were distributed internationally, often reciprocally, so that by 1890, for example, the Institute had exchanges with 485 scientific institutes around the world. The Institute's efforts to build a library were greatly enhanced in 1855, when the Toronto Athenaeum, a prominent literary society with a large library and membership, united with it. By the end of the century the library was judged to be the best scientific library north of Washington.

As the nineteenth century progressed the Canadian Institute had grown in membership and complexity. Lists of members and executives read like a who's who in the intellectual life of Toronto and to a lesser extent the rest of Canada. Presidents before 1900 include three chief justices, nine University of Toronto professors, two principals of Upper Canada College, a provincial premier, a president of the Canadian Bank of Commerce, and many others from the scientific branches of the government services, the legal profession, and the clergy.

Over these years the Institute grew in complexity as knowledge branched out into increasingly diverse and more specialized forms. In 1863, the Institute formed a medical and an entomological section. In 1885, it amalgamated with the Natural History Society of Toronto, which became the biological section of the Institute. In 1886, the Institute formed architectural, photographic, philological, historical, geological-mining, and ornithological sections. In 1887, true to the mandate of its Charter, the Institute opened a Museum of Natural History and Archaeology on the third floor of its Richmond Street building, with David Boyle as its curator. The following year the Institute formed the Sociological Committee, dedicated to the collection and dissemination of information about Canada's Native people. Each of the sections had its own executive and ran a lecture series parallel to the regular lectures. All the major activities of the Institute were covered in the Toronto newspapers. International recognition for the Institute's achievements came in 1889, when it hosted the meetings of the American Association for the Advancement of Science, and in 1897, when it hosted the meeting of the British Association for the Advancement of Science in Toronto. Following the 1897 meeting those who were interested had the opportunity to travel to Vancouver by special train, with frequent stops to view aspects of Canadian nature, ranging from insects to the fossils of the Burgess Shale. In 1921 and 1924, the Institute again helped to host both associations and also in 1924, the International Mathematical Congress.

Even before the new century arrived, the executive of the Institute began to realize that they could no longer serve all the growing segments of knowledge. Some of the Institute's sections were encouraged to hive off and form their own associations. Thus, for example, the photographic section became the Toronto Camera Club and the Sociological Committee became the Canadian Indian Research and Aid Society. The considerable holdings of the Institute's museum, particularly in provincial archaeology and ornithology, were handed over to the museum of the Toronto Normal School, Ontario Department of Education, in 1896 and were eventually incorporated into the Royal Ontario Museum in 1924. By 1910 the Institute's library had grown to 34,000 volumes with an annual increment of over 2,000 works. Unable to house and manage such a library the Institute moved it to the University of Toronto in 1911 and in 1948 finalized the move by selling the collection to that institution for a nominal sum, thus ensuring better preservation and greater accessibility.

The Institute's periodical continued to attract the best scientists in Canada. Its pages revealed the most current research in the pure and applied sciences as well as history, archaeology, and some of the other social sciences. But here too a problem developed in that the Institute, true to its original aims, tried to cover everything in a scientific world that was moving toward greater specialization and away from the descriptive and taxonomic studies of the nineteenth century. The journal articles became increasingly unreadable to the majority of the members.

Over the years the Institute had served an extremely valuable function in the intellectual life of Toronto and as an incubator for new intellectual endeavours. But the time had come for change. No association run entirely by volunteers could do all the things the Institute was doing. In 1913, the council of the Institute decided to abandon its esoteric lectures to small bodies of dedicated specialists and continue to encourage these to form their own societies. Instead of a wide range of intellectual pursuits the council of the Institute decided to focus on science and instead of promoting science to the scientist they decided to embark on a program of popular lectures that brought current scientific research and thinking to an educated public. With its access to the most prominent scientists in Canada and its international reputation, the Institute hoped to bring the best people with the latest ideas and research results to a public lecture series. These new aims moved the Institute into the field of public education, a role it fills to this day. The response to the change in the lecture series was dramatic. Within a few years, the Saturday evening lectures of the Institute were drawing over a thousand people.

In its periodical, the *Transactions,* the Institute continued to publish scientific papers but increasingly these were devoted almost purely to the biological sciences. The *Transactions* finally succumbed to growing printing costs in 1969. In order to provide a printed record of its popular lecture series, the Institute had begun a new *Proceedings* in 1936. With this publication the public could follow the latest advances in the natural and social sciences in a readable form. Unfortunately, it too followed the *Transactions* as growing costs forced the Institute to abandon it.

In 1913, the year the Institute decided to recast itself into a more popular form, it had hosted the Twelfth International Geological Congress. To honour the event, the Institute decided to draw on its membership of prominent scientists in the Toronto area for the publication of a guidebook. The result was *The Natural History of the Toronto Region, Ontario, Canada,* edited by Professor J.H. Faull with a committee headed by the Institute's president, Joseph Burr Tyrell, the well-known and respected explorer and mining engineer. The volume was intended not only as a guide to the natural history of the Toronto area but also as an up-to-date compendium of what was known about the various parts of nature. The volume begins with two highly descriptive, almost anecdotal chapters on the Native and European history of the region, typical of the semi-popular writing of the day. There is little attempt at explanation, ecological or historical. Three chapters on geology, climate, and life zones attempt to give a broad background to the specific chapters that follow on the various life forms. These three chapters are probably the most satisfactory in the book because the authors attempted a readable synthesis that went beyond the purely descriptive. The seventeen chapters on the various life forms are exhaustive checklists introduced with brief comments on the origin of the data, followed by more or less elaborate bibliographies. Some, such as the chapter on fish, include brief descriptions of habitat and life cycle, but few deal with problems of decline (no mention of the once-abundant salmon in the fish chapter) or alterations of habitat or future prospects, themes that are so prominent today. A brief chapter on excursions finishes the book. The volume met with great critical acclaim when it appeared and is still in demand through antiquarian bookstores. It was the first summary of its kind for any city region in Canada, and it displayed what the various natural sciences had accomplished to date.

The following year, on 2 April 1914, on the eve of the First World War, in a meeting rife with patriotic fervour and congratulations pouring in for its considerable achievements, the Canadian Institute was given permission by the king to add the title 'Royal' to its name. It was henceforth known officially as The Royal Canadian Institute.

Afterword

Rainer Maria Rilke was a poet attuned not just to the power of words but also to the majesty of nature. He wrote in his letters to a young poet that 'the future enters into us, in order to transform itself in us, long before it happens.'

Rilke is saying that rather than being something that will come about, the future is being nurtured today in the connections we detect, the relationships we develop, and in the choices we make.

The beauty of a book like *Special Places* lies in how it carefully outlines those connections, relationships, and choices. Based on a comprehensive study of the Toronto region, this document honours the Royal Canadian Institute's one-hundred-and-fifty-year-old commitment to scientific inquiry. The book also links that commitment to the Institute's equally vital tradition of taking knowledge out of the laboratory and making it exciting and relevant to the people in the streets.

The interesting thing for me is that one of the keys to the Institute's longevity has been its ability to mimic nature by constantly adapting to changing times and needs. As noted in 'The History of the Royal Canadian Institute,' the Institute has 'served as an incubator for new intellectual endeavours,' but like a good mother hen it also lets its chicks fly away when they are strong enough. That's a testament to the success of spin-offs like the Toronto Camera Club and the Canadian Indian Research and Aid Society. It also highlights the desire to leave a legacy of learning that inspired the RCI when it was formed in 1849.

I like to think that many of the Institute's legacies are reflected in the work and evolution of the Waterfront Regeneration Trust. Since its creation in 1992 the Trust has brought people, ideas, and resources to invest in the revitalization of the Lake Ontario waterfront. Now we're going through our own transition, moving into a non-profit environment – but in the process, like the RCI, never forgetting what we were created to do.

For example, just as the Royal Canadian Institute's public lectures have spread the word of science to laypeople, so has the Waterfront Trail served as a ribbon of progress and partnership along the shores of Lake Ontario, tying together exciting waterfront projects while connecting communities and individuals to their natural world.

The Trail is the link between the new pike habitats in the Royal Botanical Gardens and the remediated industrial soils on Cobourg's waterfront. It ties the award-winning Humber River bicycle-pedestrian bridge to the vibrant Whitby waterfront. Thus, in communities large and

Afterword

small along the length of the Lake Ontario waterfront, the Trail manifests the application of the scientific principles that guide ecosystem planning to real regeneration projects, and by extension to the future of our ecosystem.

Special Places makes those connections, builds relationships, and helps us to make choices like the Waterfront Trail while it asks us to think about what needs to happen as we strive to care for our ecosystem. Simply put, the task of the future, which is even now transforming our present, is to meld the art of inclusion, the processes of communication, and the science of ecology. We need to continue to discover new ways of including more people and organizations in stewardship of the Toronto region. We need to create networks of committed folks. And we need to excite people as we do it.

Only by creating new networks can we bring science and people together. Only by challenging the makers of both policy and poetry to turn their minds and hearts to the ecosystem can we make the changes that *Special Places* so eloquently puts forward.

Returning to my friend Mr Rilke, he also said, 'the most visible joy can only reveal itself to us when we've transformed it, within.' May this book help uncloak the joy that lies within our ecosystems and expose it to a whole new generation of readers.

David Crombie

References and Additional Reading

CHAPTER 1: THE PHYSICAL SETTING

Clayton, J.S., Ehrlich, W.A., Cann, D.B., Day, J.H., and Marshall, I.B. 1977. *Soils of Canada.* Vol. 1, *Soil Report.* Canada Department of Agriculture, Ottawa.

Eyles, N. (ed.). 1997. *Environmental Geology of Urban Areas.* GEOtext 3, Geological Association of Canada, St. John's, NF.

Eyles, N., and Williams, N.E. 1992. The sedimentary and biological record of the last interglacial-glacial transition at Toronto, Canada. Pp. 119-37 in P.U. Clark and P.D. Lea (eds.), *The Last Interglacial-Glacial Transition in North America.* Geological Society of America, Special Paper 270.

Freeman, E.B. (ed.). 1978. Geological highway map, southern Ontario. Ontario Geological Survey, Map 2418.

Karrow, P.F. 1967. *Pleistocene Geology of the Scarborough Area.* Geological Report 46. Ontario Department of Mines, Toronto.

–. 1990. Interglacial beds at Toronto, Ontario. *Géographie, Physique et Quaternaire* 44:289-97.

McAndrews, J.H. 1970. Fossil pollen and our changing landscape and climate. *Rotunda: Bulletin of the Royal Ontario Museum* 3(2):30-7.

Soil Classification Working Group. 1998. *The Canadian System of Soil Classification.* 3rd ed. NRC/CNRC 41647. National Research Council Press, Ottawa.

Thurston, P.C., Williams, H.R., Sutcliffe, R.H., and Stott, G.M. (eds.). 1991. *Geology of Ontario.* Ontario Geological Survey, special vol. 4(1). Queen's Printer for Ontario, Sudbury, ON.

–. 1992. *Geology of Ontario.* Ontario Geological Survey, special vol. 4(2). Queen's Printer for Ontario, Sudbury, ON.

CHAPTER 2: CLIMATE

Findlay, B.F., and Hirt, M.S. 1969. An urban-induced mesocirculation. *Atmospheric Environment* 3:537-42.

Higgins, A.G., and Findlay, B.F. 1969. Local climate of the Lower Rouge valley. Meteorological Branch, Climatological Division Project Report no. 1 (unpublished).

Hilborn, J., and Still, M. 1990. *Canadian Perspectives in Air Pollution.* State of the Environment Report no. 90-1, Environment Canada, Ottawa.

Mateer, C.L. 1961. Note on the effect of the weekly cycle of air pollution on solar radiation at Toronto. *International Journal of Air and Water Pollution* 4(1/2):52-4.

Middleton, W.E.K., and Millar, F.G. 1936. Temperature profiles in Toronto. *Journal of the Royal Astronomical Society, Canada* 30(7):265-72.

Munn, R.E., Hirt, M.S., and Findlay, B.F. 1969. A climatological study of the urban temperature anomaly in the lakeshore environment at Toronto. *Journal of Applied Meteorology* 8:411-22.

Ontario Ministry of the Environment. 1973. *Controlling Air Pollution in Metropolitan Toronto.* Ontario Ministry of the Environment, Toronto.

Palmer-Benson, T. 1987. Wind breakers. *Canadian Geographic* 107:34-41.

Patterson, J. 1917. Long-range weather forecasting and other weather illusions. *Journal of the Royal Astronomical Society, Canada* 11(7):261-80.

Skinner, W.R., and Gullett, D.W. 1993. Trends of daily maximum and minimum temperatures in Canada during the past century. *Climatology Bulletin* 27(2):63-77.

Stupart, R.F. 1913. The climate of Toronto. Pp. 82-90 in *The Natural History of the Toronto Region, Ontario, Canada.* Edited by J.H. Faull. Canadian Institute, Toronto.

–. 1917. Is the climate changing? *Journal of the Royal Astronomical Society, Canada* 11(6):197-207.

Toronto Atmospheric Fund. 1997. *Annual Report for 1996.* Toronto City Hall, Toronto.

Verseghy, D., and Munro, D.S. 1989. Sensitivity studies on the calculation of the radiation balance of urban surfaces: a) Shortwave radiation, *Boundary-Layer Meteorol-*

ogy 46:309-31; b) Longwave radiation, *Boundary-Layer Meteorology* 48:1-18.

Williams, C.J., and Hunter, M.A. 1985. Pedestrian level wind study, Bloor Street, Toronto, Ont., Report submitted to City of Toronto Planning and Development Department, Morrison Hershfield Ltd. (now RWDI) Rep. 48411393, 23 July 1985.

CHAPTER 3: WATERSHEDS

Gordon, N.D., McMahon, T.A., and Finlayson, B.L. 1992. *Stream Hydrology: An Introduction for Ecologists*. John Wiley and Sons, Chichester, UK.

Hallam, J.C. 1959. Habitat and associated fauna of four species of fish in Ontario streams. *Journal of the Fisheries Research Board of Canada* 16:147-73.

Huntsman, A.G. 1944. Why did the Lake Ontario salmon disappear? *Transactions of the Royal Society of Canada*, Section 5, Series 3, 38:83-102.

Hynes, H.B.N. 1975. The stream and its valley. *Verhandlungen Internationale Vereinigung für Theoretische und Angewandte Limnologie* 19:1-15.

Imhof, J.G., Kaushik, N.K., Bowlby, J.B., Gordon, A.M., and Hall, R. 1990. Natural river ecosystems: the ultimate integrator. Pp. 114-27 in *Managing Ontario's Streams*. Edited by J. Fitzgibbon and P. Mason. Canadian Water Resources Association, Cambridge, ON.

Kaufman, J., Rennick, P., Regier, H., Holmes, J., and Wichert, G. 1992. *Metro Waterfront Environmental Study*. Metropolitan Toronto Planning Department, Toronto.

Kaushik, N.K., and Hynes, H.B.N. 1971. The fate of dead leaves that fall into streams. *Archive für Hydrobiologie* 68:465-515.

Metropolitan Toronto Planning Department. 1995. *The State of the Environment Report: Metropolitan Toronto*. The Department, Toronto.

Morisawa, M. 1968. *Streams: Their Dynamics and Morphology*. McGraw-Hill, New York.

Regier, H.A. 1992. Ecosystem integrity in the Great Lakes Basin: an historical sketch of ideas and actions. *Journal of Aquatic Ecosystem Health* 1:25-38.

Rudolph, D., Goss, M., and Rudy, H. 1992. *Ontario Farm Groundwater Quality Survey*. Report prepared for Agriculture Canada under the Federal-Provincial Environmental Sustainability Initiative. Waterloo Centre for Groundwater Research, Waterloo, ON.

Sprules, W.M. 1941. The effect of a beaver dam on the insect fauna of a trout stream. *Transactions of the American Fisheries Society* 70:236-48.

Tavares-Cromar, A.F., and Williams, D.D. 1996. The importance of temporal resolution in food web analysis: evidence from a detritus-based stream. *Ecological Monographs* 66:91-113.

Wichert, G.A. 1994. Fish as indicators of ecological sustainability: historical sequences in Toronto Area streams. *Water Pollution Research Journal of Canada* 29:599-617.

Wiggins, G.B., Mackay, R.J., and Smith, I.M. 1980. Evolutionary and ecological strategies of animals in annual temporary pools. *Archive für Hydrobiologie,* supplement 58:97-206.

Williams, D.D., and Feltmate, B.W. 1992. *Aquatic Insects*. C.A.B. International, Wallingford, Oxford, UK.

Williams, D.D., and Hogg, I.D. 1999. Ecological disruption resulting from the creation and modification of agricultural and urban drainage channels in Canada and Australia – can such habitats contribute to invertebrate conservation? In *The Ecology and Management of Drainage Channels*. Edited by M. Wade and A.J.W. Harpley. John Wiley and Sons, London.

Williams, N.E., and Williams, D.D. 1997. Palaeoecological reconstruction of natural and human influences of groundwater outflows. Pp. 172-80 in *Freshwater Quality: Defining the Indefinable?* Edited by P.J. Boon and S.L. Howell. Scottish Natural Heritage, Edinburgh.

CHAPTER 4: NATIVE SETTLEMENT

Brandão, J.A. 1997. *'Your Fyre Shall Burn No More': Iroquois Policy toward New France and Its Native Allies to 1701*. University of Nebraska Press, Lincoln.

Brodhead, J.R., and O'Callaghan, E.B. (eds.). 1855. *Documents Relative to the Colonial History of New York*. Vols. 4 and 9. Weed, Parsons, Albany, NY.

Canada. 1891. *Indian Treaties and Surrenders from 1680 to 1890*. Brown and Chamberlin, Queen's Printer, Ottawa.

Ellis, C.J., and Ferris, N. 1990. *The Archaeology of Southern Ontario to A.D. 1650*. Occasional Papers of the London Chapter of the Ontario Archaeological Society, no. 5. London, ON.

Fecteau, R. 1985. The introduction and diffusion of cultivated plants in southern Ontario. MA thesis, York University, North York, ON.

Ferris, N., and Spence, M.J. 1995. The Woodland traditions in southern Ontario. *Revista Arqueologia Americana* 9:83-138.

Harris, R.C. (ed.), and Matthews, G.J. (cart.). 1987. *Historical Atlas of Canada*. Vol. 1, *From the Beginning to 1800*. University of Toronto Press, Toronto.

Innis, M.Q. (ed.). 1965. *Mrs. Simcoe's Diary*. Macmillan, Toronto.

McAndrews, J.H., and Jackson, L.J. 1988. Age and environment of late Pleistocene mastodon and mammoth in southern Ontario. *Bulletin of the Buffalo Society of Natural Sciences* 33:161-72.

Robinson, P.J. 1965. *Toronto during the French Régime*. 2nd ed. University of Toronto Press, Toronto.

Rogers, E.S., and Smith, D.B. 1994. *Aboriginal Ontario: Historical Perspectives on the First Nations*. Dundurn Press, Toronto.

Severance, F.H. 1917. *An Old Frontier of New France: The Niagara Region and Adjacent Lakes under French Control.* 2 vols. Dodd, Mead, New York.

Smith, D.B. 1987. *Sacred Feathers: The Reverend Peter Jones (Kahkewaquonaby) and the Mississauga Indians.* University of Toronto Press, Toronto.

Thwaites, R.G. (ed.). 1905. *New Voyages to North America by the Baron De Lahontan.* Burt Franklin, New York.

Trigger, B.G. (ed.). 1987. *Handbook of North American Indians.* Vol. 15, *Northeast.* Smithsonian Institution, Washington, DC.

Weninger, J.M., and McAndrews, J.H. 1989. Late Holocene aggradation in the lower Humber River valley, Toronto, Ontario. *Canadian Journal of Earth Sciences* 26:1842-9.

Wright, J.V. 1972. *Ontario Prehistory.* National Museum of Man, Ottawa.

CHAPTER 5: SPATIAL GROWTH

Austen, M.J.W., Cadman, M.E., and James, R.J. 1994. *Ontario Birds at Risk.* Federation of Ontario Naturalists, Toronto.

Berchem, F.R. 1996. *The Yonge Street Story 1793-1860.* Natural Heritage Books, Toronto.

Board of Registration and Statistics. 1853. *Census of the Canadas, 1851-52.* John Lovell, Quebec.

Boothroyd, Jim. 1998. Driving ourselves sane. *Canadian Geographic* (May-June):55-62.

Brace, Catherine. 1995. Public works in the Canadian city: provision of sewers in Toronto, 1870-1913. *Urban History Review* 23(2):33-43.

Canada. 1855. *1851-52 Census of Canada.* Ottawa.

Firth, Edith G. 1983. *Toronto in Art: 150 Years through Artists' Eyes.* Fitzhenry and Whiteside and the City of Toronto, Toronto.

Kerr, Donald, and Spelt, Jacob. 1965. *The Changing Face of Toronto.* Queen's Printer, Ottawa.

Mandel, Charles. 1998. Shaping new sanctuaries. *Canadian Geographic* (May-June):31-40.

Masters, D.C. 1947. *The Rise of Toronto, 1850-1890.* University of Toronto Press, Toronto.

Metro Planning Commission. 1995. *The Municipality of Metropolitan Toronto: Key Facts.* Metro Toronto Planning Department, Research and Special Studies Division, Toronto.

Mulvany, E. Pelham. 1884. *Toronto: Past and Present.* W.E. Caiger. Reprint 1970, Ontario Reprint Press, Toronto.

Nader, George A. 1976. *Cities of Canada: Profiles of Fifteen Metropolitan Centres.* Vol. 2. Macmillan of Canada/Maclean-Hunter Press, Toronto.

Statistics Canada. 1992. *Census Metropolitan Areas and Census Agglomerations.* Cat. 93-303. 1991 Census of Canada. Supply and Services Canada, Ottawa.

–. 1993. *Ethnic Origin.* Cat. 93-315. 1991 Census of Canada. Supply and Services Canada, Ottawa.

–. 1997. *A National Overview.* Cat. 93-357-XPE. 1996 Census of Canada. Industry Canada, Ottawa.

Vincent, Mary. 1998. GeoMap: the whole Hog(town). *Canadian Geographic* (May-June):28-9.

Wood, David J. 1988. Population change in an agricultural frontier: Upper Canada, 1796 to 1841. In *Patterns of the Past: Interpreting Ontario's History.* Edited by Roger Hall, W. Westfall, and L.S. MacDowell. Dundurn Press, Toronto.

Wynn, Graeme. 1987. On the margins of empire, 1760-1840. Pp. 189-278 in *The Illustrated History of Canada.* Edited by Craig Brown. Lester and Orpen Dennys, Toronto.

CHAPTER 6: ECOLOGY, ECOSYSTEMS, AND THE GREATER TORONTO REGION

Andrewartha, H.G., and Birch, L.C. 1954. *The Distribution and Abundance of Animals.* University of Chicago Press, Chicago.

Begon, M., Harper, J.L., and Townsend, C.R. 1996. *Ecology: Individuals, Populations, and Communities.* Blackwell Scientific, Oxford.

Botkin, D.B., and Keller, E.A. 1998. *Environmental Science: Earth as a Living Planet.* John Wiley and Sons, New York.

Clements, F.E. 1916. *Plant Succession: Analysis of the Development of Vegetation.* Carnegie Institute of Washington Publication 242, Washington, DC.

Costanza, R., d'Argo, R., de Groot, R., Farber, S., Grasso, M., Hannon, B., Limburg, K., Naeem, S., O'Neill, R., Paruelo, J., Raskin, R., Sutton, P., and van den Belt, M. 1997. The value of the world's ecosystem services and natural capital. *Nature* 498:252-60.

Dodson, S.I., Allen, T.R.H., Carpenter, S.R., Ives, A.R., Jeanne, R.L., Kitchell, J.F., Langston, N.E., and Turner, M.G. 1998. *Ecology.* Oxford University Press, New York.

Elton, C.S. 1927. *Animal Ecology.* Sidgwick and Jackson, London.

Foster, D.R. 1992. Land-use history (1730-1990) and vegetation dynamics in central New England, USA. *Ecology* 80:753-72.

Levins, R. 1970. Extinction. *Lectures on Mathematical Analysis of Biological Phenomena. Annals of the New York Academy of Science* 231:123-38.

Lindemann, R.L. 1942. The trophic-dynamic aspect of ecology. *Ecology* 23:399-418.

MacArthur, R.H. 1958. Population ecology of some warblers of northeastern coniferous forests. *Ecology* 39:599-619.

Raven, P. 1995. Tomorrow's world. P. 564 in P. Raven, L. Berg, and G. Johnson. *Environment.* Saunders College Publishing, Orlando, FL.

Turner, M. 1998. Landscape ecology. Pp. 77-122 in S.I. Dodson et al. *Ecology.* Oxford University Press, New York.

Watt, A.S. 1947. Pattern and process in the plant community. *Ecology* 35:1-22.

Wegner, J., and Merriam, G. 1979. Movements by birds and small mammals between a wood and adjoining farm habitats. *Journal of Applied Ecology* 16:349-57.

Wilson, E.O. 1992. *The Diversity of Life*. W.W. Norton, New York.

CHAPTER 7: VASCULAR PLANTS

Ivey, T.J. 1913. Ferns and fern allies. Pp. 141-9 in *The Natural History of the Toronto Region, Ontario, Canada*. Edited by J.H. Faull. Canadian Institute, Toronto.

Riley, J.L. 1989. *Distribution and Status of the Vascular Plants of Central Region*. Ontario Ministry of Natural Resources, Richmond Hill, ON.

Scott, W. 1913. The seed plants of Toronto and vicinity. Pp. 100-40 in *The Natural History of the Toronto Region, Ontario, Canada*. Edited by J.H. Faull. Canadian Institute, Toronto.

CHAPTER 8: MOSSES, LIVERWORTS, HORNWORTS, AND LICHENS

Alexopoulos, C.J., Mims, C.W., and Blackwell, M. 1996. *Introductory Mycology*. 4th ed. John Wiley and Sons, New York.

Barkman, J.J. 1958. *Phytosociology and Ecology of Cryptogamic Epiphytes*. Van Gorcum and Comp. N.V., Assen, Netherlands.

Braithwaite, R. 1880. *The Sphagnaceae or Peat Mosses of Europe and North America*. David Brogue, London.

Brown, D.H. 1992. Impact of agriculture on bryophytes and lichens. Pp. 259-83 in *Bryophytes and Lichens in a Changing Environment*. Edited by J.W. Bates and A.M. Farmer. Oxford University Press, Oxford.

Büdel, B., and Scheidegger, C. 1996. Thallus morphology and anatomy. Pp. 37-64 in *Lichen Biology*. Edited by T.H. Nash III. Cambridge University Press, Cambridge.

Farmer, A.M., Bates, J.W., and Bell, J.N.B. 1992. Ecophysiological effects of acid rain on bryophytes and lichens. Pp. 284-313 in *Bryophytes and Lichens in a Changing Environment*. Edited by J.W. Bates and A.M. Farmer. Oxford University Press, Oxford.

Faull, J.H. 1913. Lichens. Pp. 180-7 in *The Natural History of the Toronto Region, Ontario, Canada*. Edited by J.H. Faull. Canadian Institute, Toronto.

Friedl, T., and Büdel, B. 1996. Photobionts. Pp. 8-23 in *Lichen Biology*. Edited by T.H. Nash III. Cambridge University Press, Cambridge.

Glime, J.M. 1978. Insect utilization of bryophytes. *The Bryologist* 81:186-7.

–. 1992. Effects of pollutants on aquatic species. Pp. 333-61 in *Bryophytes and Lichens in a Changing Environment*. Edited by J.W. Bates and A.M. Farmer. Oxford University Press, Oxford.

Graham, G.H. 1913. Mosses and liverworts. Pp. 150-7 in *The Natural History of the Toronto Region, Ontario, Canada*. Edited by J.H. Faull. Canadian Institute, Toronto.

Grout, A.J. 1903. *Mosses with Hand-lens and Microscope*. Published by the author, New York.

Hale, M.E., Jr. 1983. *The Biology of Lichens*. 3rd ed. Edward Arnold, London.

Hawksworth, D.L., and Rose, F. 1976. *Lichens as Pollution Monitors*. Studies in Biology 66. Edward Arnold, London.

Honegger, R. 1996. Mycobionts. Pp. 24-36 in *Lichen Biology*. Edited by T.H. Nash III. Cambridge University Press, Cambridge.

Ireland, R.R., and Bellolio-Trucco, G. 1987. Illustrated guide to some hornworts, liverworts, and mosses, with glossary and structure illustrations by Linda M. Ley. *Syllogeus* 62:1-205.

Ireland, R.R., and Cain, R.F. 1975. *Checklist of the Mosses of Ontario*. Publications in Botany no. 5. National Museum of Natural Sciences, Ottawa.

Ireland, R.R., and Ley, L.M. 1992. Atlas of Ontario mosses. *Syllogeus* 70:1-138.

James, P.W., Hawksworth, D.L., and Rose, F. 1977. Lichen communities in the British Isles: a preliminary conspectus. Pp. 295-413 in *Lichen Ecology*. Edited by M.R.D. Seaward. Academic Press, London.

LeBlanc, F., and De Sloover, J. 1970. Relation between industrialization and the distribution and growth of epiphytic lichens and mosses in Montreal. *Canadian Journal of Botany* 48:1485-96.

Ley, L.M., and Crowe, J.M. 1999. *An Enthusiast's Guide to the Liverworts and Hornworts of Ontario*. Lakehead University Herbarium, Thunder Bay, ON.

Moxley, E.A. 1939. Mosses of the Toronto region of Ontario. Unpublished manuscript distributed by the Moss Club of Toronto.

–. 1940. Mosses of the Toronto region of Ontario. *The Bryologist* 43:158.

Richardson, D.H.S. 1992. *Pollution Monitoring with Lichens*. Colour plates by Claire Dalby. Naturalists' Handbooks 19. Richmond Publishing Company, Slough, UK.

Rose, F. 1976. Lichenological indicators of age and environmental continuity in woodlands. Pp. 279-307 in *Lichenology: Progress and Problems*. Edited by D.H. Brown, D.L. Hawksworth, and R.H. Bailey. Academic Press, London.

–. 1992. Temperate forest management: its effects on bryophyte and lichen floras and habitats. Pp. 211-333 in *Bryophytes and Lichens in a Changing Environment*. Edited by J.W. Bates and A.M. Farmer. Oxford University Press, Oxford.

Schofield, W.B. 1985. *Introduction to Bryology*. Macmillan, New York.

Watson, E.V. 1971. *The Structure and Life of Bryophytes*. Hutchinson, London.

Wong, P.Y., and Brodo, I.M. 1992. The lichens of southern Ontario, Canada. *Syllogeus* 69:1-79.

CHAPTER 9: FUNGI

Alexopoulos, C.J., Mims, C.W., and Blackwell, M. 1996. *Introductory Mycology*. 4th ed. John Wiley and Sons, New York.

Barron, G.L. 1968. *The Genera of Hyphomycetes from Soil*. Williams & Wilkins, Baltimore.

Bell, G.S. 1933. List of the larger fungi, Toronto region. *Transactions of the Royal Canadian Institute* 19(2):275-99.

Estey, R.H. 1994. *Essays on the Early History of Plant Pathology and Mycology in Canada*. McGill-Queen's University Press, Montreal and Kingston.

Hawksworth, D.L. 1991. The fungal dimension of biodiversity: magnitude, significance, and conservation. *Mycology Research* 95:641-55.

Hiratsuka, Y. 1987. *Forest Tree Diseases of the Prairie Provinces*. Information Report NOR-X-286. Canadian Forestry Service, Edmonton.

Langton, T. 1911. Partial list of Canadian fungi. *Transactions of the Canadian Institute* 9:69-81.

–. 1913. Mushrooms and other fungi. Pp. 158-73 in *The Natural History of the Toronto Region, Ontario, Canada*. Edited by J.H. Faull. Canadian Institute, Toronto.

Lindau, G. 1897. Hypocreales. Pp. 343-72 in *Die natürlichen Pflanzenfamilien*. Teil. 1, Abt. 1 by A. Engler and K. Prantl. Wilhelm Engelmann, Leipzig.

Littlefield, L.J. 1981. *Biology of the Plant Rusts*. Iowa State University Press, Ames.

Malloch, D. 1981. *Moulds: Their Isolation, Cultivation, and Identification*. University of Toronto Press, Toronto.

Manion, P.D. 1991. *Tree Disease Concepts*. 2nd ed. Prentice-Hall, Englewood Cliffs, NJ.

Miller, S.L. 1995. Functional diversity in fungi. *Canadian Journal of Botany* 73 (supplement 1):S50-7.

Parmelee, J.A., and Malloch, D. 1972. *Puccinia hysterium* on *Tragopogon*: a new North American rust record. *Mycologia* 64:922-4.

Schröter, J. 1897a. Peronosporineae. Pp. 108-19 in *Die natürlichen Pflanzenfamilien*. Teil. 1, Abt. 1 by A. Engler and K. Prantl. Wilhelm Engelmann, Leipzig.

–. 1897b. Mucorineae. Pp. 119-34 in *Die natürlichen Pflanzenfamilien*. Teil. 1, Abt. 1 by A. Engler and K. Prantl. Wilhelm Engelmann, Leipzig.

Ziller, W.G. 1974. *The Tree Rusts of Western Canada*. Canadian Forestry Service Publication 1329. Environment Canada, Ottawa.

CHAPTER 10: INVERTEBRATES

Anderson, D.V., and Clayton, D. 1959. *Plankton in Lake Ontario*. Great Lakes Geophysical Research Group, Ontario Department of Lands and Forests, Physical Research Note no. 1. Ontario Department of Lands and Forests, Toronto.

Ashworth, W. 1986. *The Late, Great Lakes*. Alfred Knopf, New York.

Bailey, B.D. 1973. A history of the Toronto waterfront from Etobicoke to Pickering. Paper prepared for the Metropolitan Toronto and Region Conservation Authority, Toronto.

Baker, F.C. 1939. *Fieldbook of Illinois Land Snails*. Illinois Natural History Survey (INHS) Division, Manual 2. INHS, Urbana, IL.

Brewer, R. 1979. *Principles of Ecology*. Saunders College, Philadelphia, PA.

Brinkhurst, R.O. 1970. Distribution and abundance of tubificid (Oligochaeta) species in Toronto Harbour, Lake Ontario. *Journal of the Fisheries Research Board of Canada* 27:1961-9.

Burch, J.B. 1965. *How to Know the Eastern Land Snails*. Wm. C. Brown, Dubuque, IA.

–. 1989. *North American Freshwater Snails*. Malacological Publications, Hamburg, MI.

Chichester, L.F., and Getz, L.L. 1969. The zoogeography and ecology of arionid and limacid slugs introduced into northeastern North America. *Malacologia* 7:313-46.

Clarke, A.H. 1981. *Corbicula fluminea* in Lake Erie. *Nautilus* 95:83-4.

Counts, C.L., III. 1986. The zoogeography and history of the invasion of the United States by *Corbicula fluminea* (Bivalvia: Corbiculidae). *American Malacological Bulletin,* special ed. 2:7-39.

Cressey, R.F. 1978. Marine flora and fauna of the northeastern United States. Crustacea: Branchiura. *NOAA Tech. Rep. Circular* 413:1-10.

Crofton, H.D. 1968. *Nematodes*. Hutchinson University Library, London.

Curds, C.R. 1992. *Protozoa in the Water Industry*. Biology in Focus. Cambridge University Press, Cambridge, UK.

De Jong, D., Morse, R.A., and Eickwort, G.C. 1982. Mite pests of honeybees. *Annual Review of Entomology* 27:229-52.

DePinto, J.V. 1991. State of the Lake Ontario ecosystem: introduction to an ecosystem perspective on a vital resource. *Canadian Journal of Fisheries and Aquatic Sciences* 48:1500-2.

Dickman, M.D., Yang, J.R., and Brindle, I.D. 1990. Impacts of heavy metals on higher aquatic plant, diatom and benthic invertebrate communities in the Niagara River watershed near Welland, Ontario. *Water Pollution Research Journal of Canada* 25:131-59.

Dillon, P.J., Jeffries, D.S., Scheider, W.A., and Yan, N.D. 1980. Some aspects of acidification in southern Ontario. Pp. 212-3 in *Ecological Impact of Acid Precipitation*. Edited by D. Drablos and A. Tollan. Proceedings of International Conference, Sandefjord, Norway. Oslo-Aas. SNSF Project, Oslo.

Don Watershed Task Force. 1994. *Forty Steps to a New Don.* Metropolitan Toronto and Region Conservation Authority, Downsview, ON.

Dondale, C.D., and Redner, J.H. 1982. *The Insects and Arachnids of Canada.* Part 9, *The Sac Spiders of Canada and Alaska, Araneae: Clubionidae and Anyphaenidae.* Agriculture Canada, Ottawa.

Garvey, J.E., and Stein, R.A. 1993. Evaluating how chela size influences the invasion potential of an introduced crayfish (*Orconectes rusticus*). *American Midland Naturalist* 129:172-81.

Griffiths, R.W., Schloesser, D.W., Leach, J.H., and Kovalak, W.P. 1991. Distribution and dispersal of the zebra mussel (*Dreissena polymorpha*) in the Great Lakes Region. *Canadian Journal of Fisheries and Aquatic Sciences* 48:1381-8.

Gunn, J., and Keller, W. 1990. Biological recovery of an acid lake after reductions in industrial emissions of sulphur. *Nature* 345:431-3.

Hall, R.J., and Ide, F.P. 1987. Evidence of acidification effects on stream insect communities in central Ontario between 1937 and 1985. *Canadian Journal of Fisheries and Aquatic Sciences* 44:1652-7.

Hengeveld, H.G. 1990. Global climate change: implications for air temperature and water supply in Canada. *Transactions of the American Fisheries Society* 119:176-82.

Hiltunen, J.K. 1969. The benthic macrofauna of Lake Ontario. *Great Lakes Fisheries Commission Technical Report* 14:39-50.

Hogg, I.D., and Williams, D.D. 1996. Response of stream invertebrates to a global-warming thermal regime: an ecosystem-level manipulation. *Ecology* 77:395-407.

Howard, K.W.F., Boyce, J.I., Livingstone, S., and Salvatori, S.L. 1993. Road salt impacts on groundwater quality – the worst is yet to come. *Geological Society of America, Today* 3:301-21.

IPCC. 1990. Intergovernmental Panel on Climate Change. *Scientific Assessment of Climate Change. Report of Working Group 1.* World Meteorological Organisation, Geneva.

Jones, E. 1966. *Towns and Cities.* Oxford University Press, London.

Judd, W.W. 1965. Terrestrial sowbugs (Crustacea: Isopoda) in the vicinity of London, Ontario. *Canadian Field-Naturalist* 79:197-202.

Kalisz, P.J., and Dotson, D.B. 1989. Land-use history and the occurrence of exotic earthworms in the mountains of eastern Kentucky. *American Midland Naturalist* 122:288-97.

Kaston, B.J. 1978. *How to Know the Spiders.* 3rd ed. Wm. C. Brown, Dubuque, IA.

Kevan, D.K. McE. 1983a. A preliminary survey of the known and potentially Canadian and Alaskan centipedes (Chilopoda). *Canadian Journal of Zoology* 61:2938-55.

–. 1983b. A preliminary survey of the known and potentially Canadian millipedes (Diplopoda). *Canadian Journal of Zoology* 61:2956-75.

Latchford, F.R. 1930. Some introduced molluscs. *Canadian Field-Naturalist* 44:33-4.

Lehman, J.T., and Caceres, C.E. 1993. Food-web responses to species invasion by a predatory invertebrate: *Bythotrephes* in Lake Michigan. *Limnology and Oceanography* 38:879-91.

Lewis, J.G.E. 1981. *The Biology of Centipedes.* Cambridge University Press, Cambridge.

Lodge, D.M. 1993. Biological invasions: lessons for ecology. *Trends in Ecology and Evolution* 8:133-7.

Lowe, R.L., and Pillsbury, R.W. 1995. Shifts in benthic algal community structure and function following the appearance of zebra mussels (*Dreissena polymorpha*) in Saginaw Bay, Lake Huron. *Journal of Great Lakes Research* 21:558-66.

McKay, S.M., and Carling, P.M. 1979. *Trees, Shrubs and Flowers to Know in Ontario.* J.M. Dent and Sons, Oshawa, ON.

Matthews, J.V. 1979. Tertiary and quaternary environments: historical background for an analysis of the Canadian insect fauna. Pp. 31-86 in *Canada and Its Insect Fauna.* Edited by H.V. Danks. Memoirs of the Entomological Society of Canada, no. 108. Entomological Society of Canada, Ottawa.

Maude, S.H. 1988. A new Ontario locality record for the crayfish *Orconectes rusticus* from West Duffins Creek, Durham Regional Municipality. *Canadian Field-Naturalist* 102:66-7.

May, B., and Marsden, J.E. 1992. Genetic identification and implications of another invasive species of Dreissenid mussel in the Great Lakes. *Canadian Journal of Fisheries and Aquatic Sciences* 49:1501-6.

Mills, E.L., Leach, J.H., Carlton, J.T., and Secor, C.L. 1993. Exotic species in the Great Lakes: a history of biotic crises and anthropogenic introductions. *Journal of Great Lakes Research* 19:1-54.

Mott, J.B., and Harrison, A.D. 1983. Nematodes from river drift and surface drinking water supplies in southern Ontario. *Hydrobiologia* 102:27-38.

Nalepa, T.F. 1991. Status and trends of the Lake Ontario macrobenthos. *Canadian Journal of Fisheries and Aquatic Sciences* 48:1558-67.

Oughton, J. 1948. *A Zoogeographical Study of the Land Snails of Ontario.* University of Toronto Studies, Biology Series. University of Toronto Press, Toronto.

Patalas, K. 1969. Composition and horizontal distribution of crustacean plankton in Lake Ontario. *Journal of the Fisheries Research Board of Canada* 26:2135-64.

Patrick, R., and Williams, D.D. 1990. Aquatic biota in North America. Pp. 233-54 in *The Geology of North America,* Vol. O-1, *Surface Water Hydrology.* Edited by M.G.

Wolman and H.C. Riggs. Geological Society of America, Boulder, CO.

Phelps, H.L. 1994. The Asiatic clam (*Corbicula fluminea*) invasion and system-level ecological change in the Potomac River Estuary near Washington, D.C. *Estuaries* 17:614-21.

Pilsbry, H.A. 1939. *Land mollusca of North America (north of Mexico)*. 3 vols. The Academy of Natural Sciences of Philadelphia Monographs. The Academy, Philadelphia, PA.

Reynolds, J.W. 1977. *The Earthworms (Lumbricidae and Sparganophilidae) of Ontario*. Royal Ontario Museum, Life Sciences Miscellaneous Publication. Royal Ontario Museum, Toronto.

Scoggan, H.J. 1978. *The Flora of Canada*. National Museum of Natural Sciences, Ottawa.

Scott-Dupree, C.D., and Otis, G.W. 1989. Parasitic mites of honeybees, to bee or not to bee. Ontario Ministry of Agriculture and Food, *Apiculture Newsletter* no. 01/89:24.

Shaw, M.A., and Mackie, G.L. 1989. Reproductive success of *Amnicola limosa* (Gastropoda) in low alkalinity lakes in south-central Ontario. *Canadian Journal of Fisheries and Aquatic Sciences* 46:863-9.

Shelley, R.M. 1988. The millipedes of eastern Canada (Arthropoda: Diplopoda). *Canadian Journal of Zoology* 66:1638-63.

Soper, J.H., and Heimburger, M.L. 1982. *Shrubs of Ontario*. Royal Ontario Museum, Life Sciences Miscellaneous Publication. Royal Ontario Museum, Toronto.

Spielman, A., and Levi, H.W. 1970. Probable envenomation by *Chiracanthium mildei*, a spider found in houses. *American Journal of Tropical Medicine and Hygiene* 19:729-32.

Sprules, W.G., Reissen, H.P., and Jin, E.H. 1990. Dynamics of the *Bythotrephes* invasion of the St. Lawrence Great Lakes. *Journal of Great Lakes Research* 16:346-51.

Strayer, D.L. 1991. Projected distribution of the zebra mussel, *Dreissena polymorpha*, in North America. *Canadian Journal of Fisheries and Aquatic Sciences* 48:1389-95.

Walker, E.M. 1927. The woodlice or Oniscoidea of Canada (Crustacea, Isopoda). *Canadian Field-Naturalist* 41:173-9.

Warwick, W.F., Fitchko, J., McKee, P.M., Hart, D.R., and Burt, A.J. 1987. The incidence of deformities in *Chironomus* spp. from Port Hope Harbour, Lake Ontario. *Journal of Great Lakes Research* 13:88-92.

Wells, L. 1970. Effects of alewife predation on zooplankton populations in Lake Michigan. *Limnology and Oceanography* 13:556-65.

Williams, D.D. 1987. *The Ecology of Temporary Waters*. Chapman and Hall, London.

Williams, D.D., Williams, N.E., and Cao, Y. 1997. Spatial differences in macroinvertebrate community structure in springs of southeastern Ontario in relation to their chemical and physical environments. *Canadian Journal of Zoology* 75:1404-14.

CHAPTER 11: INSECTS

Borror, D.J., and White, R.E. 1970. *A Field Guide to the Insects of America North of Mexico*. Peterson Field Guide Series. Houghton Mifflin, Boston.

Covell, C.V. 1984. *A Field Guide to the Moths of Eastern North America*. Peterson Field Guide Series. Houghton Mifflin, Boston.

Dethier, V.G. 1992. *Crickets and Katydids, Concerts and Solos*. Harvard University Press, Cambridge, MA.

Evans, H.E. 1984. *Life on a Little-Known Planet*. University of Chicago Press, Chicago.

Holmes, A.M., Hess, Q.F., Tasker, R.R., and Hanks, A.J. 1991. *The Ontario Butterfly Atlas*. Toronto Entomologists' Association, Toronto.

Layberry, R.A., Hall, P.W., and Lafontaine, J.D. 1998. *The Butterflies of Canada*. University of Toronto Press, Toronto.

McLintock, J. 1976. The arbovirus problem in Canada. *Canadian Journal of Public Health* 67, supplement 1:8-12.

Malcolm, S.B., and Zalucki, M.P. (eds.). 1993. *Biology and Conservation of the Monarch Butterfly*. Science Series no. 38, Natural History Museum of Los Angeles County, CA.

Marshall, S. 1995. Ladybird, fly away home. *Seasons* (spring):30-3.

Oldroyd, H. 1966. *The Natural History of Flies*. W.W. Norton, New York.

Opler, P.A. *A Field Guide to Eastern Butterflies*. Illustrated by V. Malikul. Peterson Field Guide Series. Houghton Mifflin, Boston.

Riotte, J.C.E. 1992. *Annotated List of Ontario Lepidoptera*. Life Sciences Miscellaneous Publications, Royal Ontario Museum, Toronto.

Walker, E.M. 1913. Insects and their allies. Pp. 295-403 in *The Natural History of the Toronto Region, Ontario, Canada*. Edited by J.H. Faull. Canadian Institute, Toronto.

–. 1957. Changes in the insect fauna of Ontario (with special reference to the Orthoptera). Pp. 4-12 in *Changes in the Fauna of Ontario*. Edited by F.A. Urquhart. University of Toronto Press for Royal Ontario Museum, Division of Zoology and Palaeontology, Toronto.

White, R.E. 1983. *A Field Guide to the Beetles of North America*. Peterson Field Guide Series. Houghton Mifflin, Boston.

Wiggins, G.B. (ed.). 1966. *Centennial of Entomology in Canada, 1863-1963: A Tribute to Edmund M. Walker*. Life Sciences Contribution no. 69, Royal Ontario Museum. University of Toronto Press, Toronto.

Wilson, E.O. 1992. *The Diversity of Life*. Belknap Press, Harvard University Press, Cambridge, MA.

—. 1996. *In Search of Nature*. Island Press/Shearwater Books, Washington, DC.

CHAPTER 12: FISH

Rapport, D.J., and Regier, H.A. 1995. Disturbance and stress effects on ecological systems. Pp. 397-414 in *Complex Ecology: The Part-Whole Relation in Ecosystems*. Edited by B. Patten and S.E. Jorgensen. Prentice-Hall, Englewood Cliffs, NJ.

Regier, H.A., and Kay, J.J. 1996. An heuristic model of transformations of the aquatic ecosystems of the Great Lakes-St. Lawrence River Basin. *Journal of Aquatic Ecosystem Health* 5:3-21.

Regier, H.A., Lin, P., Ing, K.K., and Wichert, G.A. 1996. Likely responses to climate change in the Laurentian Great Lakes Basin: concepts, methods and findings. *Boreal Environment Research* 1:1-15.

Wichert, G.A. 1994. Fish as indicators of ecological sustainability: historical sequences in Toronto Area streams. *Water Pollution Research Journal of Canada* 29:599-617.

Wichert, G.A., and Regier, H.A. 1998. Four decades of sustained use, of degradation, and of rehabilitation in various streams of Toronto, Canada. Pp. 189-214 in *Rehabilitation of Rivers: Principles and Practice*. Edited by L. de Waal, A.R.G. Large, and P.M. Wade. John Wiley and Sons, London.

CHAPTER 13: AMPHIBIANS AND REPTILES

Barnett, J.M. 1971. Ashbridge's Bay. *Ontario Naturalist*. N.d., n.p.

Eagles, P., and Adindu, G. 1978. A manual for ESA planning and management. University of Waterloo, ON.

Faull, J.H. 1913. *The Natural History of the Toronto Region, Ontario, Canada*. Canadian Institute, Toronto.

Graham, G.H. 1913. Mosses and liverworts. Pp. 150-7 in *The Natural History of the Toronto Region, Ontario, Canada*. Edited by J.H. Faull. Canadian Institute, Toronto.

Gosselin, H., and Johnson, B. 1995. *The Urban Outback – Wetlands for Wildlife: A Guide to Wetland Restoration and Frog-Friendly Backyards*. Metro Toronto Zoo, Toronto.

Ivey, T.J. 1913. Ferns and fern allies. Pp. 141-9 in *The Natural History of the Toronto Region, Ontario, Canada*. Edited by J.H. Faull. Canadian Institute, Toronto.

Johnson, B. 1982. *Amphibians and Reptiles in Metropolitan Toronto: Inventory and Guide*. Toronto Field Naturalists, Toronto.

—. 1989. *Familiar Amphibians and Reptiles of Ontario*. Natural Heritage Press, Toronto.

Keys, D.R. 1913. Toronto: an historical and descriptive sketch. Pp. 11-43 in *The Natural History of the Toronto Region, Ontario, Canada*. Edited by J.H. Faull. Canadian Institute, Toronto.

Lamond, W.G. 1994. *The Reptiles and Amphibians of the Hamilton Area*. Hamilton Field Naturalists' Club, Hamilton, ON.

Nash, C.W. 1913. Fishes. Pp. 249-71 in *The Natural History of the Toronto Region, Ontario, Canada*. Edited by J.H. Faull. Canadian Institute, Toronto.

Piersol, W.H. 1913. Amphibians. Pp. 242-8 in *The Natural History of the Toronto Region, Ontario, Canada*. Edited by J.H. Faull. Canadian Institute, Toronto.

Plourde, S.A., Szepesi, E.L., Riley, J.L., Oldham, M.J., and Campbell, C. 1989. Distribution and status of the herpetofauna of Central Region, Ontario Ministry of Natural Resources. Parks and Recreational Areas Section, OMNR, Open File Ecological Report SR8903, Central Region, Richmond Hill, ON.

Royal Commission on the Future of the Toronto Waterfront. 1992. *Regeneration: Toronto's Waterfront and the Sustainable City, Final Report*. The Commission, Toronto.

Scott, P.W. 1913. The seed plants of Toronto and vicinity. Pp. 100-40 in *The Natural History of the Toronto Region, Ontario, Canada*. Edited by J.H. Faull. Canadian Institute, Toronto.

Williams, J.B. 1913. Reptiles. Pp. 238-41 in *The Natural History of the Toronto Region, Ontario, Canada*. Edited by J.H. Faull. Canadian Institute, Toronto.

CHAPTER 14: MAMMALS

Aanes, R., and Andersen, R. 1996. The effects of sex, time of birth, and habitat on the vulnerability of roe deer fawns to red fox predation. *Canadian Journal of Zoology* 74:1857-65.

Barnes, D.M., and Mallik, A.U. 1996. Use of woody plants in construction of beaver dams in northern Ontario. *Canadian Journal of Zoology* 74:1781-6.

Brigham, R.M. 1991. Flexibility in foraging and roosting behaviour by the big brown bat (*Eptesicus fuscus*). *Canadian Journal of Zoology* 69:117-21.

Brigham, R.M., and Fenton, M.B. 1989. The effect of roost sealing as a method to control maternity colonies of big brown bats. *Canadian Journal of Public Health* 78:47-50.

Buesh, R.R. 1995. Sex differences in behaviour of beavers living in near-arboreal lake habitat. *Canadian Journal of Zoology* 73:2133-43.

Busher, P.E. 1996. Food caching behaviour of beavers (*Castor canadensis*): selection and use of woody species. *American Midland Naturalist* 135:343-8.

Chubbs, T.E., and Phillips, F.R. 1996. Apparent longevity records for red foxes, *Vulpes vulpes*, in Labrador. *Canadian Field-Naturalist* 110:348-9.

Daoust, P.Y., Wandeler, A.I., and Casey, G.A. 1996. Cluster of rabies cases of probable bat origin among red foxes in Prince Edward Island, Canada. *Journal of Wildlife Diseases* 32:403-6.

de la Cueva Salcedo, H., Fenton, M.B., Hickey, M.B.C., and Blake, R.W. 1995. Energetic consequences of flight speeds of foraging red and hoary bats (*Lasiurus borealis* and *Lasiurus cinereus*; Chiroptera: Vespertilionidae). *Journal of Experimental Biology* 198:2245-51.

Dobson, A.P., Bradshaw, A.D., and Baker, A.J.M. 1997. Hopes for the future: restoration ecology and conservation biology. *Science* 277:515-22.

Ferron, J. 1996. How do woodchucks (*Marmota monax*) cope with harsh winter conditions? *Journal of Mammalogy* 77:412-6.

Furlonger, C.L., Dewar, H.J., and Fenton, M.B. 1987. Habitat use by foraging insectivorous bats. *Canadian Journal of Zoology* 65:284-8.

Gese, E.M., Ruff, R.L., and Crabtree, R.L. 1996. Intrinsic and extrinsic factors influencing coyote predation of small mammals in Yellowstone National Park. *Canadian Journal of Zoology* 74:784-97.

Gese, E.M., Stotts, T.E., and Grothe, S. 1996. Interactions between coyotes and red foxes in Yellowstone National Park, Wyoming. *Journal of Mammalogy* 77:377-82.

Harington, C.R. 1978. Quaternary vertebrate faunas of Canada and Alaska and their suggested chronological sequence. *Syllogeus* 15:1-105.

Henry, J.D. 1986. *Red Fox: The Cat-like Canine.* Smithsonian Institution Press, Washington, DC.

Herero, S. 1985. *Bear Attacks: Their Causes and Avoidance.* Winchester Press, Piscataway, NJ.

Hickey, M.B.C., and Fenton, M.B. 1996. Behavioural and thermoregulatory responses of female hoary bats, *Lasiurus cinereus* Beauvois (Chiroptera: Vespertilionidae) to variations in prey availability. *Ecoscience* 3:414-22.

Key, G.E., and Woods, R.D. 1996. Spool-and-line studies on the behavioural ecology of rats (*Rattus* spp.) in the Galapagos Islands. *Canadian Journal of Zoology* 74:733-7.

Lacher, T.E., and Mares, M.A. 1996. Availability of resources and use of space in eastern chipmunks, *Tamias striatus. Journal of Mammalogy* 77:833-49

Lampman, K.P., Taylor, M.E., and Blokpoel, H. 1996. Caspian terns (*Sterna caspia*) breed successfully on a nesting raft. *Colonial Waterbirds* 19:135-8.

Lariviere S., and Messier, F. 1996. Aposematic behaviour in the striped skunk, *Mephitis mephitis. Ethology* 102:986-92.

Marks, C.A., Nijk, M., Gigliotti, F., Busana, F., and Short, R.V. 1996. Preliminary field assessment of cabergoline baiting campaign for reproductive control of the red fox (*Vulpes vulpes*). *Wildlife Research* 23:161-8.

Peterson, R.L. 1966. *The Mammals of Eastern Canada.* Oxford University Press, Toronto.

Rosatte, R.C., MacInnes, C.D., Williams, R.T., and Williams, O. 1997. A proactive prevention strategy for raccoon rabies in Ontario, Canada. *Wildlife Society Bulletin* 25:110-6.

Steele, M.A., Hadj-Chikh, L.Z., and Hazeltine, J. 1996. Caching and feeding decisions by *Sciurus carolinensis*: responses to weevil-infested acorns. *Journal of Mammalogy* 77:305-14.

Vitousek, P.M., Mooney, H.A., Lubchenco, J., and Melillo, J.M. 1997. Human domination of earth's ecosystems. *Science* 277:494-9.

Williams, L., and Brittingham, M.C. 1997. Selection of maternity roosts by big brown bats. *Journal of Wildlife Management* 61:359-68.

Wolff, J.O., Schauber, E.M., and Edge, W.D. 1996. Can dispersal barriers really be used to depict emigrating small mammals? *Canadian Journal of Zoology* 74:1826-30.

CHAPTER 15: BIRDS

Abraham, Carolyn. 1997. Feathered friends get a helping hand: birdman. *Calgary Herald.* 14 June, p. 1.

Austen, M., Cadman, M.D., and James, R.D. 1994. *Ontario Birds at Risk.* Federation of Ontario Naturalists and Long Point Bird Observatory, Toronto.

Cadman, M.D., Eagles, P.F., and Helleiner, F.M. 1994. *Atlas of the Breeding Birds of Ontario.* University of Waterloo Press, Waterloo, ON.

Dales, J. 1961. The lazy broadwings. *South Peel Naturalists' Newsletter* 3(7):11-5.

Ehrlich, P.R., Dobkin, D.S., and Wheye, D. 1988. *The Birders Handbook.* Simon and Schuster, Toronto.

Fleming, J.H. 1913. Birds. Pp. 212-37 in *The Natural History of the Toronto Region, Ontario, Canada.* Edited by J.H. Faull. Canadian Institute, Toronto.

Foot, D.K., and Stoffman, D. 1996. *Boom, Bust, and Echo.* Macfarlane Walter and Ross, Toronto.

Goodwin, C.E. 1995. *A Bird Finding Guide to Ontario.* University of Toronto Press, Toronto.

McKeating, G.B., and Creighton, W.A. 1974. *Backyard Habitat.* Ontario Ministry of Natural Resources, Toronto.

CHAPTER 16: FROM ACQUISITION TO RESTORATION

Anderson, R. 1993. Garbage disposal in the Greater Toronto Area: a preliminary historical geography. *Operational Geographer* 11(1):7-13.

Chambers, C.E. 1912. *Report upon Park Distribution, Parkways, and Main Boulevard System.* Toronto Parks Department, Toronto.

Chester Springs Marsh. 1997. *Bring Back the Don: Seasonal Update* (spring/summer):1.

Colton, T.J. 1980. *Big Daddy: Frederick G. Gardiner and the Building of Metropolitan Toronto.* University of Toronto Press, Toronto.

Conservation Council of Ontario. 1971. *The Urban Landscape.* The Council, Toronto.

Courval, J. 1990. Friends of the Spit: the greening of a community-based environmental group. Pp. 243-54 in *Green*

Cities: Ecologically Sound Approaches to Urban Space. Edited by D. Gordon. Black Rose Books, Montreal.

Desfor, G., Morley, B., Stromberg, R., and Carr, A. 1990. *Environmental Audit of the East Bayfront/Port Industrial Area, Phase 1: Built Heritage*. Technical Paper no. 2. Royal Commission on the Future of the Toronto Waterfront, Toronto.

Don Watershed Task Force. 1994. *Forty Steps to a New Don*. Metropolitan Toronto and Region Conservation Authority, Downsview, ON.

Friends of the Spit. 1987. *Tommy Thompson Park, Phase III: A Better Concept Plan*. The Friends, Toronto.

Gardiner, F.G. 1948. Regional planning. Pp. 96-107 in *Conservation in South Central Ontario*. Edited by Ontario Department of Planning and Development. King's Printer, Toronto.

Gemmil, A. 1978. *Toronto's Outer Harbour Eastern Headland: The Changing Role of a Transportation Facility*. Research Report no. 55. Joint Program in Transportation, University of Toronto/York University, Toronto.

Geomantics International. 1992. *Natural Areas and Environmentally Significant Areas in the City of Toronto*. Geomantics, Toronto.

Gibson, S. 1984. *More than an Island: A History of the Toronto Island*. Irwin, Toronto.

Grady, W. 1995. *Toronto the Wild: Field Notes of an Urban Naturalist*. Macfarlane Walter and Ross, Toronto.

Granger, W. 1990. Naturalizing existing parklands. Pp. 99-112 in *Green Cities: Ecologically Sound Approaches to Urban Space*. Edited by D. Gordon. Black Rose Books, Montreal.

Hanna, R. 1984. *Life Science Areas of Natural and Scientific Interest in Site District 7-4*. Ontario Ministry of Natural Resources, Richmond Hill, ON.

Hough, M. 1990. *Out of Place: Restoring Identity to the Regional Landscape*. Yale University Press, New Haven, CT.

Hough Woodland Naylor Dance Ltd./Gore and Storrie Ltd. 1994. *Ecological Restoration Opportunities for the Lake Ontario Greenway*. Waterfront Regeneration Trust, Toronto.

–. 1995. *Restoring Natural Habitats: A Manual for Habitat Restoration in the Greater Toronto Bioregion*. Waterfront Regeneration Trust, Toronto.

Humber Watershed Task Force. 1997. *Legacy: A Strategy for a Healthy Humber*. Metropolitan Toronto and Region Conservation Authority, Downsview, ON.

Kanter, R. 1990. *Space for All: Options for a Greater Toronto Area Greenlands Strategy*. N.p., Toronto.

Kaplan, H. 1967. *Urban Political Systems: A Functional Analysis of Metro Toronto*. Columbia University Press, New York.

McFarland, E.M. 1970. *The Development of Public Recreation in Canada*. Canadian Parks/Recreation Association, Vanier, ON.

Maciejewski, A., and Lebrecht, S. 1997. *Toronto Parks*. Klotzek Press, Toronto.

Martin, L., and Seagrave, K. 1983. *City Parks of Canada*. Mosaic Press, Oakville, ON.

Metropolitan Toronto and Region Conservation Authority (MTRCA). 1959. *Plan for Flood Control and Water Conservation*. The Authority, Woodbridge, ON.

–. 1965. *A Compendium of Information*. The Authority, Woodbridge, ON.

–. 1967. *Conservation, 1957-1967*. The Authority, Woodbridge, ON.

–. 1982. *Environmentally Significant Areas Study*. The Authority, Downsview, ON.

–. 1989. *Tommy Thompson Park Master Plan and Environmental Assessment*. The Authority, Downsview, ON.

–. 1991. *Municipality of Metropolitan Toronto: Valley and Shoreline Regeneration Project, 1992-1996*. The Authority, Downsview, ON.

–. 1992. *Addendum: Tommy Thompson Park Master Plan and Environmental Assessment*. The Authority, Downsview, ON.

–. 1993a. *Annual Report 1992*. The Authority, Downsview, ON.

–. 1993b. *Environmentally Significant Areas Criteria*. The Authority, Downsview, ON.

–. 1993c. *Toronto Waterfront Habitat Rehabilitation Pilot Projects, 1992: Technical Report*. The Authority, Downsview, ON.

Metropolitan Toronto and Region Remedial Action Plan. 1994. *Clean Waters, Clear Choices: Recommendations for Action*. The Plan, Toronto.

Metropolitan Toronto Council. 1953-66. *Minutes*. The Council, Toronto.

–. 1980. *Official Plan for the Urban Structure*. Toronto.

Metropolitan Toronto Parks and Culture Department (Metro Parks and Culture). 1996a. *Metro Parkland Naturalization Compendium*. The Department, Toronto.

–. 1996b. *The Metro Parks: Fact Sheets and Information*. The Department, Toronto.

Metropolitan Toronto Parks Department. 1973. *Metropolitan Toronto Parks: A Compendium*. The Department, Toronto.

–. 1977. *Metropolitan Toronto Parks: A Compendium*. The Department, Toronto.

Metropolitan Toronto Planning Board (MTPB). 1954. *Metropolitan Toronto 1954*. The Board, Toronto.

–. 1963. *Metropolitan Toronto, 1953-1963: A Decade of Progress*. The Board, Toronto.

–. 1967. *The Waterfront Plan for the Metropolitan Toronto Planning Area*. The Board, Toronto.

Metropolitan Toronto Planning Department (Metro Planning). 1976. *Public Open Space*. Background Studies in the Metropolitan Plan Preparation Programme. The Department, Toronto.

–. 1988. *Parks and Open Space*. Metropolitan Plan Review Report no. 7. The Department, Toronto.
–. 1994. *Metropolitan Waterfront Plan*. The Department, Toronto.
–. 1995. *State of the Environment Report: Metropolitan Toronto*. The Department, Toronto.
North York Parks and Recreation Department. 1984. *Naturalization Areas in North York*. The Department, North York, ON.
Ohlendorf-Moffat, P. 1991. Protecting the Bluffs. *University of Toronto Magazine* (spring):11-4.
Ontario Department of Planning and Development. 1953. *Draft of Proposed Scheme for a Humber Valley and Don Valley Greenbelt*. The Department, Toronto.
–. 1956. *The Rouge-Duffin-Highland-Petticoat Conservation Report*. The Department, Toronto.
Ontario Ministry of Municipal Affairs. 1985. *Niagara Escarpment Plan*. The Ministry, Toronto.
–. 1995. *Comprehensive Set of Policy Statements*. The Ministry, Toronto.
–. 1997. *Provincial Policy Statement*. The Ministry, Toronto.
Ontario Ministry of Natural Resources. 1983. *Maple District Land Use Guidelines*. The Ministry, Toronto.
–. 1991. *Implementation Guidelines for Oak Ridges Moraine within the Greater Toronto Area*. The Ministry, Toronto.
–. 1994. *Rouge Park Management Plan*. The Ministry, Toronto.
The Park System of Toronto. 1941. N.p., Toronto.
Reeves, W.C. 1992. *Visions for the Metropolitan Toronto Waterfront, I: Toward Comprehensive Planning, 1852-1935*. Major Report no. 27. Centre for Urban and Community Studies, University of Toronto, Toronto.
–. 1993. *Visions for the Metropolitan Toronto Waterfront, II: Forging a Regional Identity, 1913-1968*. Major Report no. 28. Centre for Urban and Community Studies, University of Toronto, Toronto.
Richardson, A.H. 1974. *Conservation by the People: The History of the Conservation Movement in Ontario to 1970*. University of Toronto Press/Conservation Authorities of Ontario, Toronto.
Royal Commission on the Future of the Toronto Waterfront. 1992. *Regeneration: Toronto's Waterfront and the Sustainable City, Final Report*. The Commission, Toronto.
Saunders, R.M. 1965. *Toronto Field Naturalists' Club: Its History and Constitution*. The Club, Toronto.
Scarborough City Council. 1991. *Scarborough Greenlands*. The Council, Scarborough, ON.
Scarborough Works Department. 1983. *At Last Count: An Assessment of Natural Areas in Scarborough, Part 2*. The Department, Scarborough, ON.
Suhanic, G. 1997. Things growing 'berry' well in 'Fire Country.' *High Park: A Park Lover's Quarterly* 4(3):7.
Task Force to Bring Back the Don. 1989. *Bringing Back the Don*. The Task Force, Toronto.
Toronto and Suburban Planning Board (TSPB). 1947. *Report*. The Board, Toronto.
Toronto and York Planning Board (TYPB). 1949. *Report*. The Board, Toronto.
–. 1951. *Second Report*. The Board, Toronto.
Toronto City Planning Board (TCPB). 1943a. *Master Plan for the City of Toronto and Environs*. The Board, Toronto.
–. 1943b. *Second Annual Report*. The Board, Toronto.
–. 1960. *Natural Parklands*. The Board, Toronto.
Toronto Field Naturalists' Club. 1976. *Toronto the Green*. The Club, Toronto.
Toronto Guild of Civic Art. 1909. *Report on a Comprehensive Plan for Systematic Civic Improvements in Toronto*. The Guild, Toronto.
Toronto Parks, Recreation and Culture Commissioners. 1997. *Building Blocks: A Liveable City through Parks, Recreation and Culture, a Proposal for the New Toronto*. Toronto Transition Team, Toronto.
Toronto Parks and Recreation Department. 1992. *High Park: Proposals for Restoration and Management and Framework for Implementation*. The Department, Toronto.
Varga, S. 1982. Environmentally significant natural areas in Metro Toronto. *Toronto Field Naturalist* 348:9-18.
Waterfront Regeneration Trust. 1995. *Lake Ontario Greenway Strategy*. The Trust, Toronto.
Wilson, J. 1910. *Report upon a Suitable System of Boulevards and Connecting Driveways for the City of Toronto*. Toronto Parks Department, Toronto.
Wright, J.R. 1983. *Urban Parks in Ontario, Part I: Origins to 1860*. N.p., Ottawa.
–. 1984. *Urban Parks in Ontario, Part II: The Public Park Movement, 1860-1914*. N.p., Ottawa.

CHAPTER 17: SPECIAL PLACES

The City of Toronto has developed a program of self-guided walks and published brochures giving full details. For information, call Parks and Recreation Information at 416/392-1111.

Waterfront Ecosystems: Restoring Is Remembering
Aitken, A. 1793. *Plan of York Harbour*. Colonial Office, London. National Archives of Canada, map VI/440-Toronto 1793.
Bonnycastle, R. 1841. *The Canadas in 1841*. Henry Colburn, London.
Bouchette, J. 1792. *Plan of Toronto Harbour with Rocks, Shoals and Soundings Thereof*. National Archives of Canada, map VI/440-1792.
Brinkhurst, R.O. 1970. Distribution and abundance of tubificid (Oligochaeta) species in Toronto Harbour, Lake Ontario. *Journal of the Fisheries Research Board of Canada* 27:1961-9.
Collins, J. 1788. Description of Toronto Bay. In Memorabilia – being a collection of extracts relating to the origin

of the name Toronto and a history of the waterfront of the city, its harbour and shipping from 1669 to 1912, gathered from standard authorities and newspapers. Edited by V.M. Roberts, 1912. Unpublished manuscript, Metropolitan Toronto Central Library.

Doyle, M. 1893. Testimony. In Dominion Fisheries Commission. Commissioners S. Wilmot and E. Harris. Parliament of Canada, *Sessional Papers* 26(7):10c.

Fleming, S. 1853-4. Toronto Harbour – its formation and preservation. *Canadian Journal* 2:105-7.

Hind, H.Y. 1852-3. Notes on the geology of Toronto. *Canadian Journal* 1:147-51.

Kane, P. 1859. *Wanderings of an Artist among the Indians of North America*. Longman, Brown, Green, Longmans and Roberts, London.

Kerr, J.W., and Kerr, F.W. 1860-98. Their daily diaries as the fishing inspectors. Unpublished manuscript, Royal Ontario Museum, Toronto.

MacCrimmon, H.R. 1977. *Animals, Man, and Change*. McClelland and Stewart, Toronto.

Magrath, T.W. [1832] 1953. Letter to Rev. Thomas Radcliffe, Dublin. Pp. 174-82 in *Authentic Letters from Upper Canada*. Edited by T. Radcliff. Macmillan, Toronto.

Nash, C.W. 1913. Fishes. Pp. 249-71 in *The Natural History of the Toronto Region, Ontario, Canada*. Edited by J.H. Faull. Canadian Institute, Toronto.

Nicholson, H.A. 1872. Preliminary report on dredgings in Lake Ontario. *Annals and Magazine of Natural History*, 4th series, 10:276-85.

Nicolls, G. 1833. *Plan of the Town and Harbour of York, Upper Canada*. Royal Engineer's Office, Quebec, map. National Archives of Canada NMC16818.

Scadding, H. 1873. *Toronto of Old*. Oxford University Press, Toronto.

Simcoe, E.P.G. 1934. *The Diary of Mrs. John Graves Simcoe*. Edited by J.R. Robertson. Ontario Publishing, Toronto.

Snell, E. 1987. *Wetland Distribution and Conversion in Southern Ontario*. Working Paper 48, Canada Land Use Monitoring Program, Inland Waters and Lands Directorate, Environment Canada, Ottawa.

Steedman, R.J. 1986. Historical streams of Toronto. *Toronto Field Naturalist* 382:14-8.

Stephenson, T.D. 1990. Fish reproductive utilization of coastal marshes of Lake Ontario near Toronto. *Journal of Great Lakes Research* 16(1):71-81.

Ure, G.P. 1858. *The Handbook of Toronto by a Member of the Press*. Lovell and Gibson, Toronto.

Vidal, E.E. 1815. *Map from old French Fort to marsh*. National Archives of Canada, map, V30/440.

Whillans, T.H. 1979. Historic transformations of fish communities in three Great Lakes bays. *Journal of Great Lakes Research* 5:195-215.

–. 1980. *Feasibility of Rehabilitating the Shorezone Fishery of Lake Ontario*. Lake Ontario Tactical Fisheries Plan, Ontario Ministry of Natural Resources, Richmond Hill, ON.

–. 1982. Changes in marsh area along the Canadian shore of Lake Ontario. *Journal of Great Lakes Research* 8:570-7.

Williamson, A.E. 1861. Notes on land and freshwater shells collected in the environs of Toronto. *Canadian Journal of Industry, Science, and Art,* series 2, 6:327-9.

Toronto's Port Lands

Hough, M., Benson, B., and Evenson, J. 1997. *Greening the Toronto Port Lands*. Waterfront Regeneration Trust, Toronto.

Hough Stansbury and Woodland Ltd. 1987. An illustrative proposal for overall landscape principles. In *Harbourfront 2000: A Report to the Futures Committee of Harbourfront*. Prepared by Baird Sampson, Architects, in association with Jones & Kirkland, Architects, Hotson Bakker, Architects, Ziedler Roberts Partnership/Architects. October.

Kalff, S., McPherson, G., and Miller, G. 1991. Environmental audit of the East Bayfront/Port Industrial Area, Phase II natural heritage. Report of Royal Commission on the Future of the Toronto Waterfront, 1991. Toronto.

Royal Commission on the Future of the Toronto Waterfront. 1990. *Environments in Transition*. April. Toronto.

–. 1991. *Pathways: Towards an Ecosystem Approach*. April. Toronto.

Stinson, J. 1990. *The Heritage of the Port Industrial District*. Vol. 1. Toronto Harbour Commission, Toronto.

Task Force to Bring Back the Don. 1991. *Bringing Back the Don*. Hough Stansbury Woodland Ltd. et al. consulting team. Report to the City of Toronto, July. City of Toronto, Toronto.

Scarborough Bluffs

Coleman, A.P. 1933. The Pleistocene of the Toronto region. v. XLI, Part VII, 1-69. Ontario Department of Mines, Toronto.

Eyles, N. 1997. Environmental geology of a supercity: the Greater Toronto area. Pp. 7-80 in *Environmental Geology of Urban Areas*. Edited by N. Eyles. GeoText 3. Geological Association of Canada, St. John's, NF.

Hinde, G.J. 1878. The glacial and interglacial strata of Scarboro' Heights and other localities near Toronto, Ontario. *Canadian Journal of Science*, new series 15:388-413.

Karrow, P.F. 1967. *Pleistocene Geology of the Scarborough Area*. Ontario Department of Mines, Geological Report 46. Ontario Department of Mines, Toronto.

The Savannahs of High Park

Bakowksy, W.D. 1993. *A Review and Assessment of Prairie, Oak Savannah and Woodland in Site Regions 7 and 6 (Southern Region)*. Ontario Ministry of Natural Resources, Southern Region, Peterborough, ON.

Catling, P.M., Catling, V.R., and McKay-Kuja, S.M. 1992. The extent, floristic composition, and maintenance of the Rice Lake Plains, Ontario, based on historical records. *Canadian Field-Naturalist* 106(1):73-86.

Goldie, J. 1819. *Diary of a Journey through Upper Canada and some of the New England States.* Privately published by his granddaughter Theresa Goldie Falkner. Original manuscript in the Baldwin Room of the Toronto Public Library.

Lizars, K.M. 1913. *The Valley of the Humber, 1615-1913.* William Briggs, Toronto.

O'Brien, M. 1968. *The Journals of Mary O'Brien, 1828-1838.* Edited by Audrey Saunders Miller. Macmillan, Toronto.

Reznicek, A.A. 1983. Association of relict prairie flora with Indian trails in central Ontario. Pp. 33-9 in *Proceedings of the 8th North American Prairie Conference, 1982.* Edited by R. Brewer. Western Michigan University, Kalamazoo.

Rodger, L. 1998. *Tallgrass Communities of Southern Ontario: A Recovery Plan.* World Wildlife Fund Canada and the Ontario Ministry of Natural Resources, Toronto.

Varga, S. 1989. *High Park Oak Woodlands Area of Natural and Scientific Interest.* Ontario Ministry of Natural Resources, Southern Region, Aurora, ON.

Varga, S., Jalava, J., and Riley, J.L. 1991. *Ecological Survey of the Rouge Valley Park.* Ontario Ministry of Natural Resources, Southern Region, Aurora, ON.

Webber, M.J. 1984. *The Vascular Plant Flora of Peel County, Ontario,* Botany Press, Mississauga, ON.

Oak Ridges Moraine

Day, G. 1953. The Indian as an ecological factor in the northeastern forest. *Ecology* 34(2):329-46.

Oak Ridges Moraine Technical Working Committee. 1994. The Oak Ridges Moraine area strategy for the Greater Toronto Area. Draft for public discussion. April. Ministry of Natural Resources, Peterborough, ON.

Oak Ridges Trail Association. 1997. *Oak Ridges Trail Guidebook.* Oak Ridges Trail Association, Aurora, ON.

Royal Commission on the Future of the Toronto Waterfront. 1992. *Regeneration: Toronto's Waterfront and the Sustainable City, Final Report.* The Commission, Toronto.

STORM Coalition. 1997. *Oak Ridges Moraine.* Boston Mills Press, Erin, ON.

Credit River

Beaumont, R. 1974. *Cataract and the Forks of the Credit.* Boston Mills Press, Cheltenham, ON.

Chapman, L.J., and Putnam, D.F. 1984. *The Physiography of Southern Ontario.* 3rd ed. Ministry of Natural Resources, Toronto.

Clarkson, B. 1977. *At the Mouth of the Credit.* Boston Mills Press, Cheltenham, ON.

Corporation of the County of Peel. 1967. *A History of Peel County.* County of Peel, Brampton, ON.

Credit Valley Conservation Authority (CVCA). 1990. Terra Cotta-Silver Creek Conservation Areas master plan and environmental study. Hough Stansbury Woodland Ltd. Unpublished report. Meadowvale, ON.

–.1992. *Credit River Water Management Strategy.* The Authority, Meadowvale, ON.

–. 1994. A conservation areas strategy for the Credit River watershed. Unpublished report. The Authority, Meadowvale, ON.

–. 1996. *Facing Change – A Business Plan for Watershed Management.* The Authority, Meadowvale, ON.

–. 1997. Rattray Marsh Protection Association wins international environmental award. Press release. Meadowvale, ON.

De Visser, J. 1992. *Credit River Valley.* Boston Mills Press, Erin, ON.

Faull, J.H. (ed.). 1913. *The Natural History of the Toronto Region, Ontario, Canada.* Canadian Institute, Toronto.

Flynn, G.D., and Mersey, J.E. 1997. Change in forest cover and fragmentation in the Forks of the Credit area of the Niagara Escarpment, 1977-1996. Unpublished ms.

Gertler, L.O. 1968. *Niagara Escarpment Study – Conservation and Recreation Report.* Niagara Escarpment Study Group, Treasury Department Finance and Economics, Toronto.

Gould, J. 1984. *A Reconnaissance Biological Inventory of Forks of the Credit Provincial Park.* Ministry of Natural Resources, Central Region, Aurora, ON.

Greenland, G. 1972. Credit Forks historical research summary. Unpublished report. Ministry of Natural Resources, Alliston, ON.

Hussey, R. 1990. *Rattray Marsh Then and Now.* Judith M. Goulin, co-ordinating author. Rattray Marsh Protection Association, Ajax, ON.

Kaiser, J. 1990. *A Biological Inventory and Evaluation of the Credit Forks Area of Natural and Scientific Interest.* Ontario Ministry of Natural Resources, Aurora, ON.

Ontario Department of Planning and Development. 1956. *Credit Valley Conservation Report.* Queen's Printer, Toronto.

Ontario Ministry of Environment and Energy. 1990. *The Niagara Escarpment Plan.* Queen's Printer for Ontario, Toronto.

Ontario Ministry of Natural Resources. 1990. *Atlantic Salmon in Ontario.* Queen's Printer for Ontario, Toronto.

Puddister, M. 1994. Preface. Pp. 6-7 in *Ecosystem Protection in an Urbanizing Environment.* Edited by M.J. Puddister and M.P. Nelischer. Credit Valley Conservation Authority, Meadowvale, ON.

Riley, J.L. 1989. *Distribution and Status of the Vascular Plants of Central Region.* Ministry of Natural Resources, Richmond Hill, ON.

Roulston, P.J. 1978. *Place Names of Peel: Past and Present.* Boston Mills Press, Cheltenham, ON.

Royal Commission on the Future of the Toronto Waterfront. 1992. *Regeneration: Toronto's Waterfront and the Sustainable City, Final Report.* The Commission, Toronto.

Trimble, B. 1975. *'Belfountain' Caves Castles and Quarries in the Caledon Hills.* Herrington Printing and Publishing, Erin, ON.

Varga, S., Jalava, J.V., and Larson, B. 1994. *Biological Inventory and Evaluation of the Silver Creek Valley Area of Natural and Scientific Interest.* Ontario Ministry of Natural Resources, Aurora, ON.

Zatyko, M. 1979. *Terra Cotta – A Capsule History.* Boston Mills Press, Cheltenham, ON.

Humber Valley

Anderson, T.W., and Lewis, C.F.M. 1985. Postglacial water level history of the Lake Ontario basin. Pp. 231-53 in *Quaternary Evolution of the Great Lakes.* Edited by P.F. Karrow and P.E. Calkin. Special paper 30. Geological Association of Canada, St. John's, NF.

Banville, D. (comp.). 1994. *The Vascular Plants of Metropolitan Toronto.* 2nd ed. Toronto Field Naturalists, Toronto.

Hawkins, W. 1834. Plan of the King's Mill reserve. Modified from original map, Records of the Ministry of Natural Resources, RG1-470-0-0-305, Archives of Ontario, OTAR (SR988).

Kennedy, B. 1979. *Hurricane Hazel.* Macmillan, Toronto.

Lewis, C.F.M., Cameron, G.D.M., King, E.L., Todd, B.J., and Blasco, S.M. 1995. Structural contour, isopach and feature maps of Quaternary sediments in western Lake Ontario. Atomic Energy Control Board of Canada Report on Project no. 2.243.1.

Lisars, K. 1913. *The Valley of the Humber, 1615-1913.* William Briggs, Toronto.

Robinson, P.J. 1965. *Toronto during the French Regime.* 2nd ed. University of Toronto Press, Toronto.

Sharpe, D.R. 1980. Quaternary geology of Toronto and surrounding area. Ontario Geological Survey Preliminary Map P. 2204.

Weninger, J.M., and McAndrews, J.H. 1989. Late Holocene aggradation in the lower Humber valley, Toronto, Ontario. *Canadian Journal of Earth Sciences* 26:1842-9.

Don Valley

Avigdor, J. 1994. *The Scadding Cabin, 1794: Toronto's Oldest House.* York Pioneer Society, Toronto.

Banville, D. (comp.). 1994. *The Vascular Plants of Metropolitan Toronto.* 2nd ed. Toronto Field Naturalists, Toronto.

Banville, D., and Cardini, L. 1978. *West Don Ravine.* Toronto Field Naturalists, Toronto.

The Cardinal (newsletter of the Don Valley Conservation Association). 1951-6. Edited by C. Sauriol.

Cardini, L., and Juhola, H. 1977. *Taylor Creek/Woodbine Bridge Ravines.* Toronto Field Naturalists, Toronto.

Carling, P.M., and McKay, S.M. 1975. Associations of halophytic plants in the Toronto region. *Ontario Field Biologist* 29:75.

Cranmer-Byng, J., Cunningham, R., and Hamilton, E. 1976. *Burke Ravine.* Toronto Field Naturalists, Toronto.

Cranmer-Byng, J., Hamilton, E., and Hilts, S. 1973. *Chatsworth Ravine.* Toronto Field Naturalists, Toronto.

Crawford, P., Dorret, G., Frost, P., Kuutan, E., Lang, J., Laywine, P., MacHattie, G., McGuigan, S., and Waverman, D. 1978. River mosaic: a study of the landscape quality and visual character of the lower Don valley. Department of Landscape Architecture, University of Toronto.

Cruickshank, B., and Parker, B. 1974. *Brookbanks Ravine.* Toronto Field Naturalists, Toronto.

Darke, E. 1995. *A Mill Should Be Build* [sic] *Thereon: An Early History of Todmorden Mills.* Natural History/Natural Heritage, Toronto.

Don Watershed Task Force. 1994. *Forty Steps to a New Don.* Metropolitan Toronto and Region Conservation Authority, Downsview, ON.

Faull, J.H. (ed.). 1913. *The Natural History of the Toronto Region, Ontario, Canada.* Canadian Institute, Toronto.

Freeman, E. 1975. *Toronto's Geological Past: An Introduction.* Ontario Ministry of Natural Resources, Division of Mines, Toronto.

Friends of the Don East York. 1993- . Newsletters.

Guthrie, A. 1986. *A Don Legacy: A Pioneer History.* Boston Mills Press, Erin, ON.

Herzberg, L., and Juhola, H. 1987. *Todmorden Mills: A Human and Natural History.* Toronto Field Naturalists, Toronto.

Kelly, C. 1984. *Cabbagetown in Pictures.* Toronto Public Library Board Local History Handbook no. 4. The Board, Toronto.

Kelly, D., and Greenbaum, A. 1975. *Wigmore Ravine.* Toronto Field Naturalists, Toronto.

Metropolitan Toronto Parks Department. 1978. *Metropolitan Toronto Parks: A Compendium.* The Municipality, Toronto.

Myrvold, B. 1979. *The Danforth in Pictures.* Toronto Public Library Board Local History Handbook no. 3. The Board, Toronto.

Ohlendorf-Moffat, P. 1989. Rebirth of a river. *Toronto Magazine,* April, 24-9, 85-6.

–. 1990. Back to the Don. *Toronto Magazine,* April, 26-9, 60-2.

Ontario Department of Planning and Development. 1950. *Don Valley Conservation Report.* The Province, Toronto.

Richardson, A.H. 1974. *Conservation by the People: The History of the Conservation Movement in Ontario to 1970.* University of Toronto Press, Toronto.

Royal Commission on the Future of the Toronto Waterfront. 1992. *Regeneration: Toronto's Waterfront and the Sustainable City, Final Report.* The Commission, Toronto.

Sauriol, C. 1981. *Remembering the Don: A Rare Record of Earlier Times within the Don River Valley.* Consolidated Amethyst Communications, Scarborough, ON.

–. 1984. *Tales of the Don.* Natural Heritage/Natural History, Toronto.

–. 1991. *Green Footsteps: Recollections of a Grassroots Conservationist.* Hemlock Press, Toronto.

–. 1992. *Trails of the Don.* Hemlock Press, Orillia, ON.

–. 1995. *Pioneers of the Don.* Author, Toronto.

Scadding, H. [1873] 1987. *Toronto of Old.* Edited by F. Armstrong. Dundurn Press, Toronto.

Seton, E.T. [1878] 1977. *Wild Animals I Have Known.* McClelland and Stewart, Toronto.

Simcoe, E.P. [1911] 1983. *Mrs. Simcoe's Diary.* Edited by M.Q. Innis. Macmillan, Toronto.

Smith, D. 1992. Restoration project: Todmorden Mills – heritage restoration and wildflower preserve. *Wildflower* 8(3):40-1.

Task Force to Bring Back the Don. 1991. *Bringing Back the Don.* Metropolitan Toronto and Region Conservation Authority, Toronto.

–. 1995- . Bring Back the Don Seasonal Updates. Metropolitan Toronto and Region Conservation Authority, Toronto.

Taylor, D., and Scrivener, P. 1976. *Park Drive Ravine.* Toronto Field Naturalists, Toronto.

Toronto Field Naturalists. 1938- . Newsletters.

–. 1976. *Toronto the Green.*

Duffins Creek

Ajax Historical Board. 1972. *The Pictorial History of Ajax, 1941-1972.* The Board, Ajax, ON.

Bowen, G.S., Lam, D.C.L., Kay, D., and Booty, W.G. 1995. An introduction to the Duffins Creek Project. Pp. 17-21 in *Proceedings of the Watershed Management Symposium, December 6-8, 1995.* Canada Centre for Inland Waters, Burlington, ON.

Emery, L. 1985. *Review of Fish Species Introduced into the Great Lakes, 1819-1975.* Great Lakes Fishery Commission Technical Report no. 45.

Greenwald, M. 1973. *The Historical Complexities of Pickering-Markham-Scarborough-Uxbridge.* Report. North Pickering Community Development Project, Toronto, ON.

Harkness, W.J.K. 1941. Catches of speckled trout from the plantings of hatchery-raised fish in private waters of Ontario. *Transactions of the American Fisheries Society* 70:410-13.

Holm, E., and Crossman, E.J. 1986. A report on a 1985 attempt to resurvey some areas within the Ontario distribution of *Clinostomus elongatus,* the redside dace, and to summarize previous records. Unpublished report, on file, Ontario Ministry of Natural Resources, Toronto.

Huntsman, A.G. 1944. Why did Ontario salmon disappear? *Transactions of the Royal Society of Canada* Series 3, 38:83-102.

Lockrey, D. 1934. *Checklist of Vascular Plants of Pickering and Ajax.* D. Lockrey, Pickering, ON.

McCrimmon, H.R. 1954. Stream studies on planted Atlantic salmon. *Journal of the Fisheries Research Board of Canada* 11:362-403.

McCrimmon, H.R., and Berst, A.H. 1961. An analysis of sixty-five years of fishing in a trout pond unit. *Journal of Wildlife Management* 25:168-78.

Mackay, H.H. 1963. *Fishes of Ontario.* Ontario Department of Lands and Forests, Bryant Press, Toronto.

McKay, W.A. 1961. *The Pickering Story.* Township of Pickering Historical Society, Pickering, ON.

Morley, L.M. (ed.). 1970. *The Village of Pickering, 1800-1970.* Corporation of the Village of Pickering, Pickering, ON.

Ontario Department of Planning and Development. 1956. *The Rouge-Duffin-Highland-Petticoat Conservation Report.* The Department, Toronto.

Ontario Ministry of Culture and Recreation. 1980. *Three Heritage Studies on the History of the HBC Michipicoten Post and on the Archeology of the North Pickering Area.* The Ministry, Toronto.

Scott, W.B., and Crossman, E.J. 1973. *Freshwater Fishes of Canada.* Bulletin of the Fisheries Research Board of Canada no. 184.

Sly, P.G. 1991. The effects of land use and cultural development on the Lake Ontario ecosystem since 1750. *Hydrobiologia* 213:1-75.

Speirs, J.M., and Falls, J.B. 1992. Fishes of Duffins Creek. Pp. 42-57 in *Fishes of Duffins Creek.* Pickering Naturalist, Pickering, ON.

Walden, F.A. 1974. *A Biological Survey of the North Pickering Project Site and Toronto II Airport Site.* Ontario Ministry of Housing, Toronto.

Weaver, L.A., and Garman, G.C. 1994. Urbanization of a watershed and historical changes in a stream fish assemblage. *Transactions of the American Fisheries Society* 123:162-72.

Whitcher, W.F., and Venning, W.H. 1870. Special report on fish breeding at Newcastle, Ontario. Pp. 66-9 in *Annual Report of the Department of Marine and Fisheries,* Appendix 2. Hunter, Rose and Company, Ottawa, ON.

Wilmot, S. 1869. *Annual Report of the Department of Marine and Fisheries for the Year 1868.* Hunter, Rose and Company, Ottawa, ON.

–. 1870. *Annual Report of the Department of Marine and Fisheries for the Year 1869.* Pp. 59-65. Hunter, Rose and Company, Ottawa, ON.

Wood, W.R. 1911. *Past Years in Pickering: Sketches of the History of the Community.* W. Briggs, Toronto.

Rouge Valley

Galbraith, J. 1833. Field notes, broken fronts in Pickering, 14th day of June and 4th day of July 1833, notes of the line in front of the 3rd Range from Lot 1 to 35. Survey Records Branch. Ontario Ministry of Natural Resources, Toronto.

Goldie, J. 1819. Diary of a Journey through Upper Canada and some of the New England States. Original ms. Baldwin Room, Toronto Public Library, Toronto.

Ontario Department of Planning and Development. 1956. *The Rouge-Duffin-Highland-Petticoat Conservation Report.* The Department, Toronto.

Riley, J.L. 1978. Guide to the vascular plants and wildlife of the Rouge River valley in Metropolitan Toronto and Durham Region. *Ontario Field Biologist,* special publication 1:1-53.

–. 1980. The recognition of environmentally significant areas in Metropolitan Toronto – the lower Rouge River valley. Pp. 163-85 in *Protection of Natural Areas in Ontario: Conference Proceedings.* Edited by S.W. Barrett and J.L. Riley. Working Paper no. 3, Faculty of Environmental Studies, York University, North York, ON.

Smith, D.W. 1793. A sketch showing the situation of 230 acres of land, etc. Map Library, University of Toronto Library, Toronto.

Varga, S., Jalava, J., and Riley, J.L. 1991. *Ecological Survey of the Rouge Valley Park.* Open File Ecological Report 9104. Parks and Recreational Areas Section, Ontario Ministry of Natural Resources, Central Region, Aurora, ON.

ADDITIONAL READING

Chapter 4: Native Settlement

The standard work on the archaeology of Ontario for the general public is the well-illustrated *Ontario Prehistory* by Wright (1972). The most recent and comprehensive compendium on the prehistory of Ontario is *The Archaeology of Southern Ontario* (Ellis and Ferris 1990). This massive work also contains a chapter on environmental change. Further information on changing vegetation patterns can be found on plates constructed by McAndrews for the *Historical Atlas of Canada* (Harris and Matthews 1987), and on changing lake levels in Weninger and McAndrews (1989). An excellent case study showing the relationship between the environment and the early spear hunters is by McAndrews and Jackson (1988). For a good overview of the complex Woodland period, there is no better paper than that by Ferris and Spence (1995). The diffusion of cultivated plants into Ontario and their gradual acceptance by the Native population has been researched in detail by Fecteau (1985). For the best general account of the Natives of northeastern North America, see the Smithsonian *Handbook of North American Indians,* edited by Trigger (1987), and for Ontario *Aboriginal Ontario,* edited by Rogers and Smith (1994).

First published in 1933, but still the best book on the Toronto area in the seventeenth and eighteenth centuries is Robinson's *Toronto during the French Régime, 1615 to 1793* (1965). A similar work on the Niagara area that contains good information on Toronto is by Severance (1917). Many of the documents used by Robinson and Severance are in Brodhead and O'Callaghan (1855). Lahontan (Thwaites 1905) gives an eyewitness account of Native methods of beaver hunting in the seventeenth century. The most recent and authoritative work on the seventeenth century is by Brandão (1997). Other information on Ontario during the French period is in the *Historical Atlas* and a chapter in Ellis and Ferris (1990).

A good description of the natural environment of the Toronto area just before it was transformed by European settlement is the diary of Mrs. Elizabeth Simcoe. This wonderful edition by Mary Quayle Innis (1965) also contains copious references to other early accounts of the area. By far the best account of the Mississauga is the sensitive book *Sacred Feathers* (1987), by Donald Smith on the great Mississauga Methodist preacher of the Credit River Reserve, the Reverend Peter Jones.

For a visual and textual overview of the entire period under study, see the *Historical Atlas of Canada.*

Chapter 6: Ecology, Ecosystems, and the Greater Toronto Region

Begon, Harper, and Townsend's *Ecology: Individuals, Populations and Communities* (1996) is a second- or third-year university text. It assumes the reader has had an introductory university course in biology. This is a tome of accumulated ecological knowledge to the level of the community but is weak at the levels of ecosystem and landscape ecology and has little focus on ecological implications of human disturbance. Dodson et al., *Ecology* (1998) is an overview of the subdisciplines relevant to modern ecology with emphasis on how ecologists approach questions rather than on what they have found. Humans and their relationship to nature form an implicit and explicit subtext. An excellent annotated reading list closes each chapter. Wilson's *The Diversity of Life* (1992) is a compilation of essays on evolution, ecology, and biodiversity. Suited for general readership.

Chapter 8: Mosses, Liverworts, Hornworts, and Lichens

Bates, J.W., and Farmer, A.M. (eds.). 1992. *Bryophytes and Lichens in a Changing Environment.* Oxford University Press, Oxford.

Brodo, I.M. 1988. *Lichens of the Ottawa Region.* 2nd ed. Special publication 3, Ottawa Field-Naturalists' Club, Ottawa.

Conrad, H.S. 1979. *How to Know the Mosses and Liverworts*. 2nd ed. revised by P.L. Redfearn, Jr. W.C. Brown, Dubuque, IA.

Crum, H. 1976. *Mosses of the Great Lakes Forest*. Rev. ed. University of Michigan Herbarium, Ann Arbor.

–. 1991. *Liverworts and Hornworts of Southern Michigan*. University of Michigan Herbarium, Ann Arbor.

Hale, M.E., Jr. 1979. *How to Know the Lichens*. 2nd ed. W.C. Brown, Dubuque, IA.

Muma, R. 1985. *A Graphic Guide to Ontario Mosses*. Rev. ed. Published by the author, Toronto.

Nash, T.H., III (ed.). 1996. *Lichen Biology*. Cambridge University Press, Cambridge.

Seaward, M.R.D. (ed.). 1977. *Lichen Ecology*. Academic Press, London.

Chapter 9: Fungi

Boyce, J.S. 1948. *Forest Pathology*. 2nd ed. McGraw-Hill, New York.

Cummins, G.B., and Hiratsuka, Y. 1983. *Illustrated Genera of Rust Fungi*. Rev. ed. APS Press, St. Paul, MN.

Ellis, M.B., and Ellis, J.P. 1997. *Microfungi on Land Plants: An Identification Handbook*. New enlarged ed. Richmond Publishing, Slough, UK.

Perone, P.P. 1978. *Diseases and Pests of Ornamental Plants*. 5th ed. John Wiley and Sons, New York.

Sinclair, W.A., Lyon, H.H., and Johnson, W.T. 1987. *Diseases of Trees and Shrubs*. Cornell University Press, Ithaca, NY.

Walker, J.C. 1969. *Plant Pathology*. 3rd ed. McGraw-Hill, New York.

Contributors' Acknowledgments

CHAPTER 1: PHYSICAL SETTING

We thank all those scientists whose work we have consulted during the course of writing this paper but have not been able to cite because of the nature of this publication. A major contribution was made by Mike Doughty (University of Toronto at Scarborough), who did most of the illustrative work. We acknowledge and thank David Rudkin, Vincent Vertolli, Joan Burke, Kathy David, Christopher Rancourt, and Nancy Gahm (all of the Royal Ontario Museum) and Edward Freeman (consultant) for their generous and capable assistance. Finally, we appreciate the contributions of Sian Williams and Conrad Heidenreich for supplying the information on A.P. Coleman.

CHAPTER 4: NATIVE SETTLEMENT

We acknowledge our debt to the many professional and amateur archaeologists whose labours form the synthesis of the archaeological section of this chapter.

CHAPTER 8: MOSSES, LIVERWORTS, HORNWORTS, AND LICHENS

I wish to express my deep appreciation to Dr. I.M. Brodo (Ottawa) and Ms. Joan Crowe (Owen Sound) for reading and commenting on an earlier version of this chapter. Their comments and suggestions have enhanced considerably the quality of the text, for which I am extremely thankful. I am also grateful to Ms Jenny Bull for assistance with preparation of the line drawings. Research support from the Natural Sciences and Engineering Research Council of Canada is acknowledged with thanks.

CHAPTER 9: FUNGI

Research support from the Natural Sciences and Engineering Research Council of Canada is acknowledged with thanks. John Krug expresses appreciation to Dr. Martin Hubbes for reading the manuscript, helpful discussion, and providing a photograph of the beetle chambers associated with the Dutch elm disease fungus. He is grateful to Ms Jenny Bull, who assisted him with the preparation of the line drawings.

CHAPTER 11: INSECTS

Comments on the manuscript were provided by D.C. Currie and D.C. Darling, both of the Centre for Biodiversity and Conservation Biology, Royal Ontario Museum, and by W.K. Gall, Division of Invertebrate Zoology, Buffalo Museum of Science. Some information was provided by S.A. Marshall, Department of Environmental Biology, University of Guelph. Word processing was done by Diana Powell, Department of Zoology, University of Toronto.

CHAPTER 13: AMPHIBIANS AND REPTILES

The distribution data were collected for and augmented by members of the Toronto Field Naturalists. TFN member Helen Juhola has provided support for this project for over 15 years. Andrew Peart updated the maps, and Mark Wright assisted in their production. Heather Passmore and Barbara Tessman provided editorial assistance. I would like to thank Sherry Pettigrew and Betty Roots for assistance in the preparation of this chapter.

CHAPTER 15: BIRDS

I am indebted to Karen Murphy for her invaluable research assistance. We thank Dr. Bruce Falls for providing material for our sidebar on Dick Saunders.

Contributors

Roberta L. Bondar, Canada's first woman in space, is Adjunct Professor, Faculty of Kinesiology, and Distinguished Professor, Ryerson Polytechnic University. She is Chair of the Science Advisory Board for Health Protection Branch, Health Canada.

Robert C. Burgar is Project Archaeologist at the Toronto and Region Conservation Authority (TRCA) as well as Research Associate at the Royal Ontario Museum.

Donald A. Chant is Professor Emeritus in and former Chairman of the Department of Zoology, University of Toronto; former Chairman and President of the Ontario Waste Management Corporation; Chairman of the Board, World Wildlife Fund Canada; and has been Vice-President of the RCI since 1997. In 1989 Dr. Chant was awarded the Order of Canada.

Terence Dickinson is an astronomer and editor of *SkyNews* magazine.

Jenna Dunlop is a research associate in the Department of Biology, York University, and the Centre for Biodiversity and Conservation, Royal Ontario Museum.

James E. Eckenwalder is Associate Professor of Botany at the University of Toronto and Research Associate in the Centre for Biodiversity at the Royal Ontario Museum.

Nicholas Eyles is Professor of Geology at the University of Toronto at Scarborough. He has worked in South America, North and South Africa, Europe, and Australia.

M. Brock Fenton is Associate Vice President (Research) and Professor of Biology at York University. He has been awarded the Distinguished Canadian Biologist Award from the Canadian Council of University Biology Chairs.

Harold Harvey is Professor Emeritus in the Department of Zoology, University of Toronto. In addition to his work in Ontario, Dr. Harvey has also worked with Natural Resources, Manitoba; the International Pacific Salmon Fisheries Commission; the Fisheries Research Board of Canada; Fisheries and Oceans, Canada; the Canadian International Development Agency in the South Pacific; and the International Development Research Council in Turkey.

Conrad E. Heidenreich is Professor of Geography at York University. Dr. Heidenreich served on the editorial board of *The Historical Atlas of Canada,* vol. 1, *From the Beginning to 1800,* and was President of the RCI in 1991.

Louise Herzberg is a teacher and counsellor and has worked in educational research and as a consultant for exceptional children.

Michael Hough is a practising landscape architect and founder of Hough Woodland Naylor Dance Leinster who has taught at the Faculty of Environmental Studies, York University, since 1971. He taught at the School of Architecture and initiated the program in Landscape Architecture at the University of Toronto between 1962 and 1970. He is author of *Cities and Natural Process* (Routledge 1995) and *Out of Place* (Yale University Press 1990). In association with the Waterfront Regeneration Trust, he published a paper in 1997, *Greening of the Toronto Port Lands.*

Ken Howard is a hydrogeologist, certified by the American Institute of Hydrology and chartered by the British Geological Society. Dr. Howard is the Director of the Groundwater Research Group at the University of Toronto.

Robert (Bob) Johnson is Curator of Amphibians and Reptiles at the Metro Toronto Zoo and the author of *Familiar Amphibians and Reptiles of Ontario.*

Contributors

Helen Juhola is the editor of the Toronto Field Naturalists' newsletter and a past president of the group.

Madeline Kalbach is Director of the Research Unit for Ethnic Studies, Chair of Ethnic Studies, Assistant Professor of Sociology and Director of the Ethnic Studies and Population Research Lab at the University of Calgary. She is a past president of the South Peel Naturalists' Club and was editor of its newsletter for many years. An enthusiastic bird watcher, she participated in the Ontario Breeding Bird Atlas project from 1980 to 1985.

Warren Kalbach is Canada's foremost demographer and expert in population, urban studies, and ethnic populations. He is a Fellow of the Royal Society of Canada (FRSC); Professor Emeritus of Sociology at the University of Toronto; and Adjunct Professor at the University of Calgary.

John C. Krug is Lecturer and Curator of the herbarium, Department of Botany, University of Toronto.

John (Jock) McAndrews is Curator Emeritus at the Royal Ontario Museum and Professor Emeritus of Botany and Geology at the University of Toronto.

David McQueen is Professor Emeritus of Economics at Glendon College of York University.

David W. Malloch is Professor of Botany at the University of Toronto. Before joining the university, he worked at the Biosystematics Research Institute of Agriculture Canada on the Central Experimental Farm in Ottawa.

Tom Mason is Curator of Invertebrates at the Toronto Zoo.

R.E. (Ted) Munn is Professor Emeritus at the Institute for Environmental Studies, University of Toronto. Dr. Munn is editor-in-chief of the forthcoming *Encyclopaedia of Global Environmental Change* and former Chief Scientist for Environment Canada's Air Quality Research Branch.

Michael J. Puddister is Senior Planner with the Credit Valley Conservation Authority.

Wayne C. Reeves is Policy Officer with the Policy and Development Division of the City of Toronto Department of Economic Development, Culture and Tourism. His publications include *Visions for the Metropolitan Toronto Waterfront*.

Henry A. Regier is Professor Emeritus of Zoology at the University of Toronto. He has been Director of the Institute for Environmental Studies at the University of Toronto; Commissioner of the Great Lakes Fishery Commission; an executive committee member of the Great Lakes Science Advisory Board, International Joint Commission; and President of the American Fishery Society.

John Riley is Director of Conservation Science for the Nature Conservancy of Canada, Toronto, Ontario. He is co-author of *Natural Heritage of Southern Ontario's Settled Landscapes* and *Ecological Survey of the Rouge Valley Park*, as well as the forthcoming *Ecological Survey of the Niagara Escarpment Biosphere Reserve*.

Betty I. Roots is Professor Emeritus and former Chair, Department of Zoology, University of Toronto. Dr. Roots has also been Director, Life Sciences Division of the Academy of Science, Royal Society of Canada; Dean of Sciences, Erindale College, University of Toronto; and President of the Royal Canadian Institute from 1994 to 1995.

Morley Thomas was Director General of the Canadian Climate Centre from 1979 to 1983. Mr. Thomas has represented Canada on the World Meteorological Organisation Commission for Climatology for over 25 years and is a past president of the Commission and of the Canadian Meteorological and Oceanographic Society. He has been awarded the Patterson Medal (1980) and the Massey Medal (1985).

Vic Timmer is Professor of Forestry at the University of Toronto.

Steve Varga is Inventory Biologist with the Ontario Ministry of Natural Resources, Aurora District.

Peter von Bitter is Curator and Head of the Department of Palaeobiology at the Royal Ontario Museum and Professor of Geology at the University of Toronto.

John Westgate is Professor of Geology in the Physical Sciences Division at the University of Toronto at Scarborough.

Tom Whillans is Professor of Environmental and Resources Studies at Trent University.

Gordon Wichert received his doctorate from the University of Toronto and has worked for the Ontario Ministry of Natural Resources. He is currently in Bangladesh working with the Mennonite Central Committee as a fisheries adviser on an agricultural project.

Glenn Wiggins is Curator Emeritus of Entomology at the Royal Ontario Museum and author of numerous publications, including the book *Larvae of the North American Caddisfly General (Trichoptera)*. He was awarded the 1992 Gold Medal of the Entomological Society of Canada for outstanding achievement.

D. Dudley Williams is Professor of Zoology and Professor of Environmental Sciences at the University of Toronto at Scarborough. He is also Honorary Professor at the University of Wales, Bangor.

David Yap is supervisor in the Environmental Monitoring and Reporting Branch (Specialized Monitoring and Air Quality) with the Ontario Ministry of Environment. He has been Chair of the Ontario Climate Advisory Committee since 1992.

Ann P. Zimmerman is Professor of Zoology and Director, Division of the Environment, at the University of Toronto.

Index

Note: All references are to Toronto unless otherwise stated; (t) = table; (f) = figure.

Acid rain, effect, 124-5, 152-3
Admiralty, Lake, 63, 64-5
Agaricus spp. (summer and fall mushrooms), 134, 136
Agriculture: bryophytes, effect on, 125-6; fungi expansion, 142-3; habitat continuity, 189; and Native peoples, 65-6, 70; seed remnants/ banks, 106, 111-2; water, effect on, 178-9
Agrocybe spp. (spring and fall mushrooms), 132-3, 136
Air Pollution Control Act (1967), 41
Air Pollution Index, 42
Air quality, 41-2, 45
Air Quality Advisories and Index, 42
Algonquian language group, 67
Algonquin Arch, 11
Amabel Formation, 14(t), 15
Amanita virosa (destroying angel mushroom), 138
Amatoxins (mushroom toxin), 138
Ambystoma spp. (salamanders), 188, 192(t), 193, 194
American toad *(Bufo americanus)*, 192(t), 194
Amnicola limosa (snail), 153
Amphibians: distribution/numbers (1982-97), 191-3; in Don valley, 283; frogs and toads, 194-6; habitats, 187-9, 198-9; in Rouge valley, 293; urbanization, effects, 189-91
Analysis of the Development of Vegetation (Frederic Clements, 1916), 96

Animal Ecology (Charles Elton, 1927), 98
Anoplura (insect order), 169
Antheridium (bryophyte), 117
Aondironnon (Native peoples), 67
Apalone spinifera spinifera (eastern spiny softshell turtles), 197
Aphids (order Homoptera), 170
Apiosporina morbosa (black knot of cherry fungus), 145
Apothecia (fungal reproduction), 140
Apple scab *(Venturia inaequalis)*, 145
Apterygota (wingless insects), 163-4
Aquifers, 27-9, 181, 268
Aquitards, 28
Archaeology, direct historical approach, 68
Archaic period, Native peoples, 64-5
Archegonia (bryophyte), 117
Area of Toronto, 4(f), 77, 83(f), 84-5, 86-7(f)
Argulus japonicus (crustacean), 155
Armillariella mellea (fungus), 143
Ascocarps/ascospores (fungal reproduction), 140
Ascomycetes or sac fungi, 119, 140, 141(f)
Ascus (fungal reproduction), 140
Ashbridge's Bay: coastal wetland, 57; ecosystem 'service,' 94; in-filling, 94, 95(f), 231, 246(f), 251; reptiles/ amphibians (1913), 187-8
Asiatic clams *(Corbicula fluminea)*, 154(f), 155

Astronomy, interest in, 49
Atlantic salmon: in Credit River, 271-2, 276(f), 277; in Duffins Creek, 287-8; human interventions, 57; preferred habitat, 179, 181
Automobiles and environment, 43, 45
Autotrophs, 98, 99

Baby Point/Baby Post, 68, 70(f)
Balsam poplar *(Populus balsamifera)*, 109
Bark fungus *(Xylaria polymorpha)*, 147
Bascule bridge, Cherry Street, 253
Basidiomycetes (club fungi), 119, 143, 144(f)
Basidium and basidiospores (fungal reproduction), 140
Bass. *See under* Centrarchids
Bass, Brad, 45
Bats, big brown *(Eptesicus fuscus)*, 208-9, 214
Bats, hoary *(Lasiurus cinereus)*, 209-10
Bean-tree *(Catalpa speciosa)*, 110
Beavers: creation of wetlands, 199; giant *(Castoroides ohioensis)*, 20, 24(f); in Humber River valley, 278; Iroquois hunting grounds, 68, 69(f), 70; in old-growth ecosystems, 55; present-day *(Castor canadensis)*, 213, 214
Bedrock geology of Toronto, 12(f), 13(f), 20(f)
Beech *(Fagus grandifolia)*, 109

Beetles (order Coleoptera), 171
Belfountain, 272-3
Bethune, C.J.S., 162
'Big Bang,' 48-9
Bigtooth aspen *(P. grandidentata)*, 110
Biodiversity, 100-2, 111-2
Bioindicators, of environmental changes, 165, 179, 181, 203, 205-6
Birds: feeders, impact of, 222-3; in Forks of Credit Park, 275; habitats, 217-9; in High Park savannahs (1819), 261; in Humber River valley, 278; migration, 220-1, 222; in Rouge valley, 293; seasonal species, 219-21; status designations, recommended, 224(t); in Toronto Port Lands, 251, 253; urbanization, effects of, 221-2, 224
Bird's nest fungus *(Crucibulum laeve)*, 148, 149(f)
Bitternut and shagbark hickories *(Carya ovata)*, 108
Black ash *(Fraxinus nigra)*, 109
Black knot of cherry fungus *(Apiosporina morbosa)*, 145
Black locust *(Robinia pseudoacacia)*, 109
Black maple *(Acer nigrum)*, 108
Black newts, 188
Black oak *(Quercus velutina)*, 108, 261
Black spot (roses), 145-6
Black spruce *(Picea mariana)*, 109
Black walnut *(Juglans nigra)*, 108
Black-capped chickadee, 222
Blanding's turtles *(Emydoidea blandingi)*, 192(t), 196
Blattodea (insect order), cockroaches, 167
Bloor Street Wind Study, 41(f)
Blowing adder, 189
Blue Mountain Formation, 14(t), 15
Blue-beech *(Carpinus caroliniana)*, 108
Blue-spotted salamander *(Ambystoma laterale)*, 192(t), 194
Bluffers Park, 57, 259
Bobcaygeon Formation, 14(t), 16
Bond Head, Francis (lieutenant governor of Ontario), 75
Boreal forests, 106, 108-9
Botrytis cinerea (fungus), 146
Boyle, David, 306
Broad-winged hawks, 220-1

Brodie, William, 162
Brook trout, 57, 179, 181-3, 184(f), 275
Brown rot of peaches *(Monilinia fructicola)*, 145
Brown snake, northern *(Storeria d. dekayi)*, 188-9, 192(t), 197
Brown spot needle blight *(Scirrhia acicola)*, 147
Brown trout, 179, 181, 182-3, 184(f), 275
Bruce Trail, Niagara Escarpment, 274
Brundtland, Grö Harlem, 300
Brunisol (soil), 29-30
Bryophytes, 117, 121-3, 124-7. *See also* Hornworts; Lichens; Liverworts; Mosses
Bufflehead (duck), 220
Bufo americanus (American toad), 192(t), 194
Bugs (order Hemiptera), 170
Bullfrogs *(Rana catesbeiana)*, 188, 195-6
Bur oak *(Quercus macrocarpa)*, 109
Butter and eggs *(Linaria vulgaris)*, 111
Butterflies (order Lepidoptera): description of order, 172, 174; karner blue, 261, 265; Monarch and Viceroy butterflies, 172-3; *Ontario Butterfly Atlas*, 162
Butternut *(Juglans cinerea)*, 145
Bythotrephes cederstroemi (spiny water flea), 155

Cabot Head Formation, 14(t), 15
Caddisflies (order Trichoptera), 171-2
Calocybe carnea (summer mushroom), 134
Calvatia gigantea (puffball), 148, 149(f)
Canada, origin of name, 74
Canada geese, 219, 223(f)
Canada-France-Hawaii Telescope, 48
Canadian Entomological Club, 162
Canadian Environmental Law Association, 6
Canadian Indian Research and Aid Society, 306
Canadian Institute. *See* Royal Canadian Institute
Canadian Journal (journal of Royal Canadian Institute), 305-6
Canadian Wildflower Society, 112
Canis latrans (coyotes), 207, 251

Canker (in fungal asexual reproduction), 141
Cankers (fungal), 147
Carbon dioxide, removal by vascular plants, 114
Cardinal, northern, 222, 223(f)
Carnivores (mammals), 206-8
Carolina Quest (chapter by R.M. Saunders, 1971), 218
Carolinian Zone forests, 106-7, 108
Carrying capacity of ecosystems, 99-100
Castoroides ohioensis (giant beaver), 20, 24(f)
Census Metropolitan Area (CMA) population figures, 77, 85
Centrarchids (sunfishes and basses), 179, 181-3
Cepaea nemoralis (European garden snails), 156
Char. *See under* Salmonids
Cheiracanthium mildei (blackfooted sac spider), 156(f), 157
Chelydra serpentina (snapping turtles), 189, 192(t), 196
Chester Springs Marsh, 239
Chestnut, sweet *(Castanea dentata)*, 144-5
Chestnut blight *(Cryphonectria parasitica)*, 144-5
Chinese elm *(Ulmus pumila)*. *See* Siberian elm
Chinook salmon, 179, 181
Chipmunks, eastern *(Tamias striatus)*, 211
Chitin, in fungi, 130
Chlorophyllum molybdites (fall mushroom), 138-9
Chondrostereum purpureum (fungus), 143
Chryomyxa ledi (rust fungi), 144
Chrysemys picta marginata (Midland painted turtles), 192(t), 196
Cicadas (order Homoptera), 170
Claremont Conservation Area, 289
Claviceps purpurea (fungus), 146, 147(f)
Clean Air Campaign, 46
Clements, Frederic, 96-8
Clemmys insculpta (wood turtles), 197
Climate: in book *Natural History* (1913), 33; historical trends, 43-4; macroclimate, 34-6; mesoclimate (*see* Mesoclimate); microclimates,

39-41; observing locations, 33, 37(f); projection for next century, 44-5, 46. *See also* Precipitation; Temperatures
'Climate of Toronto' (R.F. Stupart, 1913), 33
Clitocybe dealbata (summer mushroom), 134
Cloud Gardens, 85
Club mosses (lycopod group), 107
CMA (Census Metropolitan Area) population figures, 77, 85
Cockroaches (order Blattodea), 167
Coexistence, in ecosystems, 98, 100-2
Coho salmon, 179, 181
Coleman, Arthur Philemon (1852-1939), 17
Coleoptera (insect order), 171
Collembola (insect order), 164
Colonel Samuel Smith Park, 57
Common buckthorn *(Rhamnus cathartica)*, 109
Common goldeneye (duck), 220
Commute project, 46
Competitive exclusion in ecosystems, 98, 100
Conidia (asexual spores), 141
'Connectivity' of habitats, 102, 188, 189, 199-201
Conocybe spp. (summer mushrooms), 134-5
'Conservancy district,' 232(f), 233
Conservation: 'conservancy district,' 232(f), 233; early parks (1873-1942), 229-31; green belt *(Master Plan)* proposals (1943), 232(f), 233, 234(f); landscape conservation, 200-1; planning authorities, 233-4, 297; recreation purposes (1873-1942), 231; restoration (1976 to present), 238-41; water rehabilitation (1940s to present), 58-9. *See also* Naturalization; Parks of Toronto; Regional planning
Conservation Authorities Act (1946), 233
Conservation ecology, 102-3
Contaminants in groundwater, 29, 56-7. *See also* Pollution
Cooksville Creek watershed, 52(f)
Coprinus spp. (spring, summer, and fall mushrooms), 133, 135, 136, 137(f)
Corbicula fluminea (Asiatic clams), 154(f), 155

Cordyceps capitata (fungus), 148
Corkscrew willow *(Salix matsudana* Tortuosa), 110
Cottonwood, eastern *(Populus deltoides)*, 108
Coventry, Alan F., 275
Coyotes *(Canis latrans)*, 207, 251
Crabapples *(Malus* spp.), 110
Crack willow *(Salix X rubens)*, 109
Crayfish *(Orconectes rusticus)*, 155-6
Credit River watershed: bird-watching, 218(t); conservation/ management of area, 52(f), 271, 272(f), 273-4, 277; early settlement, 271-3; landscape diversity, 271; quarry mining, 272-3; Rattray Marsh, 275-7; salmon reintroduction, 271-2, 276(f), 277; a 'special place,' 271-7; warming of water, effect, 183
Credit Valley Conservation Authority, 6, 59, 273
Cree (Native peoples) language group, 67
'Creeping incrementalism' (urbanization), 302
Crickets (order Orthoptera), 166-7
Croft, Henry, 162
Crombie, David, 295, 308-9
Crucibulum laeve (bird's nest fungus), 148, 149(f)
Crustaceans, non-native, 155-6
Crustal rebound, postglacial, 280-1
Cryphonectria parasitica (chestnut blight), 144-5

Dacromyces palmatus* (a jelly fungus), 148, 149(f)
Damselflies (order Odonata), 164-5
Daphnia, 155
Darters, 179, 181, 182, 183(f)
David Dunlap Observatory, 48
'Dead man's fingers' (bark fungus), 147
Deer mouse *(Peromyscus maniculatus)*, 212
Deforestation, effects: amphibians/ reptiles, 189; environment, 27, 29; increase in nest predators, 221; Oak Ridges Moraine, 30-1, 267
Denton, Susan Marie, 231
Dermaptera (insect order), 169
Dermea balsamea (fungus), 147
Desmognathus fuscus (dusky salamanders), 188

Destroying angel mushroom *(Amanita virosa)*, 138
Diadophis punctatus edwardsi (northern ringback snake), 189, 198
Diamicts (stony muds), Scarborough Bluffs, 23, 25(f), 257-8
Die-back diseases (fungal), 147
Diplocarpon rosae (fungus), 146
Diplodia pinea (fungus), 147
Diptera (insect order), 174
Dog lichen *(Peltigera canina)*, 124
Don Formation, 19(f), 20-1, 23(f), 24(f), 258(f)
Don River, 52(f), 58, 182-3, 188
Don valley: bird-watching, 218(t); efforts of Charles Sauriol, 284-5; green corridor, 151; Hurricane Hazel, effect, 284; plant species, 109, 110, 111; pollution, 283, 284, 297; rehabilitation, 239, 283-4, 297; a 'special place,' 283-5
Don Valley Brickyard, Pleistocene sediments, 19, 23(f)
Don Valley Conservation Association, 285
Don Watershed Task Force, 284
Dorchester, Lord (governor general), 73-4
Dothistroma pini (fungus), 147
Dragonflies (order Odonata), 164-5, 166(f)
Dreissina polymorpha (zebra mussels), 154-5, 220
Ducks, wintering, 219-20
Duffins Creek: fish species, 287-8; forest cover, 286-7; headwaters, 270(f), 286; Hurricane Hazel, effect, 286, 287(f); sea lamprey, first account, 180; settlement, 287; a 'special place,' 286-9; watershed, 52(f), 56(f), 286-7
Dusky salamanders *(Desmognathus fuscus)*, 188
Dutch elm disease *(Ophiostoma* spp.), 145

Early (Lower) Silurian period, sedimentation, 14(t), 15
Early Palaeozoic period, marine communities, 16
Earwigs (order Dermaptera), 169
'Ecological memory,' and waterfront ecosystems, 248
Ecology, conservation, 102-3

Ecology, landscape, 102, 200-1
Ecology, population, 98, 99-102, 103
Ecology, urban, 298-9
Ecosystems: ancient, marine life, 15-6; definitions, 3-4, 5; development models (Clements and Gleason), 96-8; 'eco-approach,' 6-7; 'ecological memory,' 248; ecosystem 'services,' 94; energetics approach (Odum, Lindeman), 98-9; fundamental unit of life, 93-4; habitats and species connected, 297-8; improvement, 297; niches/pyramids (Elton), 98; rate of change (mid-1800s on), 301-3; 'special places' (*see* Special places); successional recovery, 247-8; urban landscape (1913), 187-8. *See also* Conservation; Environment; Forests; Naturalization; Plants, vascular; *names of individual flora and fauna*
Ecosystems, abuse-tolerant: adaptive species, 193, 194; effect of development, 57; 'generalist' species (*see* 'Generalist' or opportunistic species); Toronto, an ecological slum, 58, 101-2, 178-9. *See also* Forests; Plants, vascular
Ecosystems, aquatic: degradation of water, 178-9; impaired, effect on amphibians/reptiles, 189-91; pollution, 56-7, 125, 177-8; reform after 1940s, 58-9, 184; water quality/temperature and fish, 57, 181-3. *See also* Groundwater; Streams
Ecosystems, old-growth: features, 53-4; habitat and species connected, 297-8; human and natural disturbances, 54-5, 57, 229; pristine stream courses, 54, 177-8; selective sequential resource 'mining,' 55-6
Ectomycorrhizae, 131, 136-7
Edge habitats, 102-3
Edwards Gardens, 85
Eels, 180
Elaters (in liverworts), 118
Elton, Charles, 98
Emydoidea blandingi (Blanding's turtles), 192(t), 196
Endopterygota (insects with metamorphosis), 170-2, 174-5
English elm (*Ulmus procera*), 110

Entoloma sericeum (summer mushroom), 135
Entomological Society of Canada, 162
Entomology, 162. *See also* Insects
Entropy in ecosystems, 99
Environment: effects of European settlement, 27-9, 29-31, 57; intense urbanization, 151-2, 229, 295-7, 301-3; and natural flora, 106-7; natural setting (late 1700s), 77-8, 121-2, 188; regional policy lacking, 300-1; Toronto, an ecological slum, 58, 101-2, 178-9
Environment Protection Act (1971), 41
Environmentally Significant Areas Study (MTRCA, 1982), 238
Ephemeroptera (insect order), 164
Eptesicus fuscus (big brown bats), 208-9, 214
Erie (Native peoples), 68
Etobicoke Creek watershed, 52(f)
Eubosmina coregoni (water flea), 155
European garden snails (*Cepaea nemoralis*), 156
European striped slugs (*Limax maximus*), 157
European valve snails (*Valvata piscinalis*), 155
Eurytemora affinis (crustacean), 155

Fairy ring mushroom (*Marasmius oreades*), 135-6
Farms, reduction of acreage, 84
Fatal Light Awareness Program (FLAP), for birds, 224
Faull, J.H., 3, 307
Ferguson, Mary, 218
Ferns, 107
Filaments (hyphae) in fungi, 130, 141
First Nations. *See* Native peoples
Fish species: bioindicators, 179, 181; centrarchids, 179, 181-3; in Duffins Creek, 287-8; habitat preferences, 179, 181; human interventions, effects, 57; in Humber marshes and wetlands, 247; percids, 179; in Port Lands, 251; in Rouge valley, 293; salmonids, 179, 181-3, 184(f); warming of water, effects, 181-3
Fishflies (order Neuroptera), 171

Five Nations language, 67
Flashing Wings (R.M. Saunders, 1947), 218
Fleas (order Siphonaptera), 174-5
Fleming, Sandford, 305
Flies (order Diptera), 174
Flooding. *See* Streams
Flour beetles, 101
Flowers, 113, 261
Fluorides and bryophytes, 124-5
Forbes, Stephen: *The Lake as a Microcosm* (address, 1887), 3-4
Forest fires, 54-5
Forests: along Niagara Escarpment, 275; Archaic period, 64-5; in Duffins Creek, 286-7; fungi in original forests, 131, 142, 148; in Humber River valley, 278; improved state, 297; managed forests, 123; natural, complexity of, 105; Palaeo-Indian period, 63; pre-settlement, 77-8, 105, 121-2; in ravines of High Park (1819), 261; remnant forests, 106; in Rouge valley in 1800s, 291; types of, 105-6, 108-9; understorey plants, 111. *See also* Deforestation; Plants, vascular; Trees
Forks of Credit Provincial Park, 272(f), 273, 275
Fort Frontenac (French, 1673), 71
Fort Niagara (French, 1668), 71
Fort Rouillé, 70(f), 71-2
Fort Toronto (later Fort Rouillé), 70(f), 71-2
Fort York, 78, 79(f)
Fossil Hill Formation, 14(t), 16
Fossils, 16, 20-1, 24(f), 26-7, 281-2
Foster, D.R., 102
Fothergill, Charles, 180
Framework Convention on Climate Change (UN), 45
French and Native peoples, 71-2
Friends of the Don, 284
Friends of the Spit (Leslie Street), 238-9
Friends of the Valley [Don valley], 284
Frogs, 188, 191-3, 194-6
Fungi, 129-32, 142, 148. *See also* Mushrooms
Fungi, pathogenic: agriculture, with introduction, 142-3; Ascomycetes (sac fungi), 119, 140, 141(f); asexual reproduction, 141;

Index

Basidiomycetes (club fungi), 140, 143, 144(f); on deciduous trees and conifers, 144-5, 147; in gardens and orchards, 145-6; habitat, 142; leaf spots, 147; mildews, 146; Oomycetes, 141; in original forests, 142; parasitic, 131, 140, 147-8; rusts, 143-4; saprotrophic, 131, 140; in wild grasses, 146; Zygomycetes, 141, 142(f). *See also* Lichens

Fur trade, English and French, 71-2

Gabion baskets (riverbank reinforcement), 285
Gametophytes, bryophytes, and ferns, 107, 117
Gammarus fasciatus (crustacean), 155
Gammarus pseudolimnaeus (amphipod), 153
Ganaraska Forest, 267, 269
Ganaraska watershed, rehabilitation, 267
Gardiner, Frederick G., 233-4
Garter snake, eastern (*Thamnophis s. sirtalis*), 189, 192(t), 197-8
Gasteromycete (mushroom), 135
Gastrocybe lateritia (summer mushroom), 135
Gause, G.F., 100
'Generalist' or opportunistic species: amphibians/reptiles, 193; birds, 217, 221, 223; in degraded aquatic ecosystems, 177-8; description, 57, 97, 298, 299; freshwater invertebrates, 152; mammals, 214; and plant species, 115
Geology: bedrock geology, 12(f), 13(f), 20(f); Laurentian Channel, 16-7, 20(f); Middle Devonian period, 15; Ordovician periods, 12-3, 14(t), 15; postglacial period, 24-7; Sauk Sea sequence, 12, 14(t); sedimentary rock units, cycles, 13, 14(t); Silurian periods, 14(t), 15; surficial geology, 17-8; Tippecanoe Sea sequence, 13, 14(t). *See also* Glaciers; Sediments
Georgian Bay Formation, 14(t), 15, 16
Georgian Bay ice lobe, 25(f)
Gertler Report on Niagara Escarpment (1968), 274-5
Giardia lamblia invertebrate, 159

Glaciers: effect on landforms, 17-8, 63-4; ice lobes, 17, 18(f), 25(f); ice-flow direction, 23-4, 25(f); ice-wedge casts in Toronto region, 19, 22-3; Laurentide Ice Sheet, 23-4, 25(f), 258; Little Ice Age, 26-7
Gleason, Henry S., 96-8
Glen Major Angling Club, 288
Global warming and Toronto temperatures, 153
Goat willow (*Salix caprea*), 110
Goldie, John, 260
Gondwanaland, 15
Gowganda tillite in Northern till, 23-4, 25(f)
Grapes, fungal diseases, 146
Graptemys geographica (map turtles), 192(t), 196
Grasshoppers (order Orthoptera), 166-7
Gray treefrogs (*Hyla versicolor*), 188, 192(t), 195
Great Lakes, effect on weather, 34
Great Lakes-St. Lawrence Zone forests, 106-7
Green frogs (*Rana clamitans*), 188, 192(t), 194-5
Greenhouse gas emissions and climate, 45, 181
Greenwood Conservation Area, 289
Gremmeniella abietina (fungus), 147
Grenadier Pond (High Park), 260(f), 261, 264, 265
Grenville Province (Precambrian geological belt), 11
Grifolia frondosa (fungus), 142
Groundhogs (*Marmota monax*), 210-1
Groundwater, 27-9, 54, 181, 268. *See also* Ecosystems, aquatic; Streams
Grylloblatta campodeiformis, 168
Grylloblattodea (insect order), 167-8
Guelph Formation, 14(t)
Gull River Formation, 14(t)

Habitats: amphibians/reptiles (1913), 187-9, 189-91, 198-9; birds, 217-9, 221-2; edge habitats, 102-3; fish species, and human activities, 57, 181-3; in forest ecosystem, pre-settlement, 119-20; insect, effect of changes, 161; linking by corridors, 102-3, 189-90, 199-201; loss of, 122-3, 189-91, 199-200, 221-2; mammals, 205-6; naturalization program, 239-40; new, 123, 199, 200, 295; non-native species, invasion by, 152, 159; reduced complexity, due to humans, 101, 112; specialized, for shrubs, 111; successional stages, seral, 97, 199. *See also* Ecosystems; Forests; Wetlands

Hale-Bopp Comet, 49
Halley's Comet, 49
Halophytes (salt-tolerant plants), 106
Halton Till, 17, 18(f), 19(f), 28, 258
Hansell, Roger, 45
Hawks, sighting, 220-1
'Heat island' (Toronto), 36, 45, 153
Hebeloma crustuliniforme (fall mushroom), 136-7
Hedges and shrubs (distribution of vegetation in Toronto), 112-3
Hemiptera (insect order), 170
Herbivores, shift to generalist, 115
Heterobasidion annosum (fungus), 143
Heterotrophs, 98, 130
High Park: amphibians/reptiles, 189, 198-9; bird-watching site, 218(t); creation (1873), 229, 261, 264; Grenadier Pond, 260(f), 261, 264, 265; naturalization/restoration, 265; plants, 262(f), 263(f), 264; savannahs, 260-1, 264-5; snapping turtles, 196; a 'special place,' 260-5; trees, 108, 111, 261; wild lupines, 111, 230(f), 265
Highland Creek watershed, 52(f), 56(f)
Hind, Henry Youle, 305
Hogg, Helen Sawyer, 48
Hogg's Hollow, temperature fluctuations, 33, 36
Hognose snake, 189
Holling, C.S., 102
Homoptera (insect order), 170
Honey bees and mites, 156(f), 157-9
Hornworts: life cycle, 117-8
Horse chestnut (*Aesculus hippocastanum*), 108, 109
Horsetails, 107
House mouse (*Mus musculus*), 212
Howard, John G., 229, 261, 264
Hubble, Edwin, and the 'Big Bang,' 48-9
Humber Bay Park, bird-watching site, 218(t), 220

Humber Plains (savannahs), 260, 264
Humber Portage Trail, 278(f), 279-80
Humber River, 182-3
Humber valley: amphibians/reptiles, 188, 193, 194, 198; bird-watching site, 218; flora and fauna, 278-9; future of marshes, 282; geological formation, 280-2; Humber Portage Trail, 278(f), 279-80; Hurricane Hazel, effect, 282; meanders and pond sites, 278-9, 280(f), 281-2; settlement history, 279-80; a 'special place,' 278-82; vascular plants, disappearance, 106; watershed, 52(f)
Humidity, 35, 39
Hunter, Peter (lieutenant governor of Ontario), 79
Hunting by Native peoples, 70, 72
Huron ice lobe, 25(f)
Huron (Native peoples) language group, 67-8
Huronia, 67-8
Hurricane Hazel (1954), effects: in Duffins Creek, 286, 287(f); on flood control in valleys, 284, 285; Humber River flood, 282; on parkland acquisition, 236; in Rouge valley, 291-2
Hussey, Ruth, 276
Hyalella azteca (crustacean), 153
Hybrids of native and exotic trees, 110-1
Hydrodynamic forces, 51
Hydrological cycle, 51-2, 53(f)
Hyla versicolor (gray treefrogs), 192(t), 195
Hymenoptera (insect order), 175
Hyphae, fungal (filaments), 130, 141
Hypomyces lactifluorum (fungus), 148
Hypoxylon mammatum (canker), 147

Ice lobes, 17, 18(f), 25(f)
Immigration, Asian, 88
Inocybe decipientoides (fall mushroom), 137
Inonotus dryadeus (fungus), 142
Insects: aphids and cicadas, 170; Apterygota (wingless), 163-4; beetles, 171; bioindicators, 165; bugs, 170; butterflies and moths, 172-4; caddisflies, 171-2; classification, 163; cockroaches, 167; dragonflies/damselflies, 164-5, 166(f); earwigs, 169; ecological roles, 161-3; Endopterygota (metamorphosis), 170-2, 174-5; fishflies and lacewings, 171; fleas, 174-5; flies and mosquitoes, 174; grasshoppers and crickets, 166-7; grylloblattids, 167-8; lice, 169; mayflies, 164; in *Natural History* (1913), 161; Neoptera (new wings), 165-70; Palaeoptera (primitive wings), 164-5; praying mantid, 167; scorpionflies, 171; silverfish, 163-4; stoneflies, 169; termites, 167; thrips, 169-70; walkingsticks, 168; wasps and ants, 175
Intermediate Aquifer, 28-9
Invertebrates, freshwater, 152-4
Invertebrates, non-native: Asiatic clams, 154(f), 155; crustaceans, 155-6; European valve snails, 155; Isopoda and spiders, 157; mites, 156(f), 157-9; snails and slugs, 156-7; terrestrial invaders, 156-9; zebra mussels, 154-5, 220
Iroquoian Native peoples, 65, 67
Iroquois, Lake: effect on landforms, 19, 22(f), 258; and Humber River valley, 280-1; and Ontario ice lobe, 18, 24-5; Pleistocene sediments, 19(f); and Scarborough Bluffs, 22(f), 256, 258
Iroquois Confederacy (League), 67-8
Iroquois (Native peoples): beaver hunting, 68, 69(f), 70; emergence, 65; language groups, 67; and Mississauga peoples, 71, 75; social organization, 66-7; subsistence activities, 70; village sites, 68-70
Isidia (of lichens), 119-20
Isoptera (insect order), 167

Jones, Peter (Reverend), 74-5

Karner blue butterfly (in savannahs of High Park), 261, 265
Kaskaskia Sea, 15
Katydids (order Orthoptera), 166
Kilgour, Alice, 231
Killaly, H.H., 305
King's Mill Reserve (Humber River valley), 278(f), 280(f)

Lac aux Claies (later Lake Simcoe), 74
Lacewings (order Neuroptera), 171
Lactarius sp., 148
The Lake as a Microcosm (address by Stephen Forbes, 1887), 3-4
Lake Iroquois. *See* Iroquois, Lake
Lakefront. *See* Waterfront
Lamprey, sea, 180
Lampropeltis t. triangulum (eastern milk snake), 189, 192(t), 198
Landscape ecology, 102, 200-1
'Land-use history (1730-1990) and vegetation dynamics in central New England' (D.R. Foster, 1992), 102
Largemouth bass, 179, 181, 182(f)
Lasiurus cinereus (hoary bats), 209-10
Late Ordovician period sedimentation, 13, 14(t), 15-6
Laurentian Channel, 16-7, 20(f), 257
Laurentide Ice Sheet, 23-4, 25(f), 258
Lawns, 113
Lawrence Park, 231
Leaf spots (fungal diseases), 147
Lefroy, Henry, 33, 35
Leopard frogs (*Rana pipiens*), 188, 192(t), 195
Lepidoptera (insect order), 172, 174
Lepiota spp. (fall mushrooms), 138
Lepista spp. (fall mushrooms), 137
Leptodora kindti (crustacean), 155
Les Piquets, Lake (later Lake Simcoe), 74
Leslie Street Spit: bird-watching site, 218, 220; created wetland, 57, 295; ecosystem succession, 97; Friends of the Spit, 238-9; naturalization, 239, 295; stabilizing Toronto Islands, 247; turtles and snakes, 196, 197; vascular plants, 107, 108
Leucoagaricus naucinus (fall mushroom), 138
Levins, Richard, 103
Lice (orders Anoplura, Mallophaga, Psocoptera), 169
Lichens: description, 119-21; habitats, 122-3, 126-7; importance of, 122, 127; pollution and human activities, 122, 124-5
Limax maximus (European striped slugs), 157
Limnodrilus hoffmeisteri (worm), 153

Index

Lindeman, Raymond, 98
Lindsay Formation (Collingwood Member), 14(t), 15
Linné, Carl (Carolus Linneus), 118
Linneus, Carolus (*Species Plantarum*, 1753), 118
Little Ice Age, 26-7
Little-leaf linden (*Tilia cordata*), 110
Liverworts, 118, 119(f), 126-7
Locusts (order Orthoptera), 166
Logan, William E., 305
Lophodermium pinastri (needle cast fungus), 147
Lower Aquifer, 28-9
Lumsden, George F., 47
Lung lichen (*Lobaria pulmonaria*), 124
Lupines, wild (*Lupinus perennis*), resurgence, 111, 230(f), 265
Luvisol (soil), 29-30
Lycopods (mosses), 107

MacArthur, Robert, 100
Macdonald, J.E.H. (*Spring Breezes, High Park*, painting), 264(f), 265
Mackinaw Interstadial period, 19(f), 24
Macroclimate weather systems, 34-6
'Magic mushrooms' (*Psilocybe physaloides*), 133-4
Maidenhair tree (*Ginkgo biloba*), 110
Mallophaga (insect order), 169
Mammals: bats, 208-10, 214; beavers (*see* Beavers); big brown bats, 208-9, 214; as bioindicators, 203, 205-6; carnivores, 206-8; characteristics, 203; coyotes, 207, 251; in Don valley, 283-4; eastern chipmunks, 211; in Forks of the Credit Provincial Park, 275; future for, 214-5; grey squirrels, 211-2; habitats, 205-6; hoary bats, 209-10; meadow voles, 212-3; mice, 212; minks, 208; muskrats, 213, 214; Norway rats, 213-4; rabbits, 210; raccoons, 206; red fox, 206-7; rodents, 210-4; in Rouge valley, 293; species (1800 to present), 204-5(t), 214-5; striped skunks, 207-8; viewing sites, 214; woodchuck (groundhogs), 210-1
Manitoba maple (*Acer negundo*), 108
Manitoulin Formation, 14(t), 15
Mantids (order Mantodea), 167

Mantodea (insect order), 167
Map turtles (*Graptemys geographica*), 192(t), 196
Marasmius spp. (summer mushrooms), 135-6
Marine life, ancient ecosystems, 15-6
Marmota monax (groundhogs or woodchuck), 210-1
Master Plan for the City of Toronto and Environs (1943), 232(f), 233
Mayflies (order Ephemeroptera), 164
Meadow vole (*Microtus pennsylvanicus*), 212-3
Meadowcliffe Till, 19(f), 258
Meadowood Native peoples, 65
Mecoptera (insect order), 171
Mephitis mephitis (striped skunks), 207-8
Merriam, Gray, 102
Mesic optimum, 111
Mesoclimates: 1936 measurements, 33, 34(f); factors affecting, 36; 'heat island,' 36, 45, 153; siting of animals in zoo, 40
Metal contamination and bryophytes, 124-5
Metamorphosis, insect, 170-2, 174-5
'Meta-populations,' 103
Metropolitan Toronto and Region Conservation Authority (MTRCA, now Toronto and Region Conservation Authority), 6, 59, 236, 238, 285
Metropolitan Toronto Parks Department (1955), 234
Metropolitan Toronto Planning Board (MTPB) (1954), 234
Mice, 212
Micmac (Native peoples) language group, 67
Microtus pennsylvanicus (meadow vole), 212-3
Middle Devonian period, 15
Middle Ordovician period, 12-3, 14(t), 15-6
Middle Silurian period, 14(t), 15-6
Midland painted turtles (*Chrysemys picta marginata*), 192(t), 196
Mildews, downy and powdery, 146
Milk snake, eastern (*Lampropeltis t. triangulum*), 189, 192(t), 198
Milkweed plant and Monarch butterfly, 172-3
Milky Way, 48

Millipede, flat-backed (*Oxidus gracilis*), 157
Mimico Creek watershed, 52(f)
Minks (*Mustela vison*), 208
Mississauga (Native peoples), 70(f), 71-5
Mites, varroa and honey bee tracheal, 156(f), 157-9
Mitten-tree (*Sassafras albidum*), 108
Mohawk (Native peoples) language group, 67
Monarch butterfly, 172-3
Monilinia fructicola (brown rot of peaches), 145
Monterey pine fungus (*Dothistroma pini*), 147
Morchella esculenta (morel), 148, 149(f)
Morel (*Morchella esculenta*), 148, 149(f)
Mosquitoes (order Diptera), 174
Mosses, 107, 118-9, 120(f), 124-5, 126
Moths (order Lepidoptera), 172, 174
Mount Pleasant Cemetery, 112, 217, 218(t), 220
Mourning doves, 222, 223(f)
MTRCA. *See* Metropolitan Toronto and Region Conservation Authority
Mudpuppies (*Necturus maculosus*), 188, 193-4
Mus musculus (house mouse), 212
Muscarine (toxin in mushrooms), 134
Mushrooms: fall varieties, 136-9; spring varieties, 132-4; summer varieties, 134-6. *See also* Fungi
Musk turtles (*Sternothurus odoratus*), 197
Muskrats (*Ondatra zibethicus*), 213, 214
Mustela vison (minks), 208
Mutualists, 131
Mycelium (in Zygomycetes), 141
Mycena alcalina (fall mushroom), 138(f), 139
Mycobionts (fungal partner of lichens), 119
Mycorrhizae, 131

Native peoples: agriculture, introduction, 65-6; Archaic period (8000-1000 BC), 64-5; fishing, 70, 72; French, relations with, 71-2; habitation, lakeside orientation,

Index

65; Historic period (1650-1880), 68-75; language groups, 67; in Oak Ridges Moraine, 267; Palaeo-Indian settlement (9000-7000 BC), 63-4; social complexity, 66-7; socio-political organization, 67-8; tools, 64, 65; treaties, misperception of, 73-4; Woodland period, (1000 BC-AD 1650), 65-8. *See also names of individual nations, tribes*

The Natural History of the Toronto Region, Ontario, Canada (Royal Canadian Institute, 1913): amphibian/reptile survey, 187-8; bird species, 224-5; climate description, 33; fish species, 288; insect numbers, 161; published by Royal Canadian Institute, 3, 307

Naturalization: Don valley, 284; High Park, 265; Leslie Street Spit, 239, 295; Metro Toronto's parks, 239-40; waterfront ecosystems, 248

Nectria galligena (canker), 147

Necturus maculosus (mudpuppies), 188, 193-4

Needle cast fungus (*Lophodermium pinastri*), 147

Nematodes, 131, 159

Neoptera (insects with new wings), 165-70

Nerodia s. sipedon (northern water snake), 189, 192(t), 197

Neuroptera (insect order), 171

Neutral (Native peoples) language group, 67

Newts, 191-3

Niagara Escarpment, 271, 273(f), 274-5

'Niches' in ecosystems, 98, 101

Nipissing Flood (rise of Lake Ontario), 281-2

North Toronto Green Community, 284

Northern till, 18(f), 19(f), 23-4, 25(f), 28

Norway maple (*Acer platanoides*), 109

Norway rats (*Rattus norvegicus*), 213-4

Notophthalmus v. viridescens (red-spotted newt), 192(t), 193

Nutrient cycling, by vascular plants, 114-5

Oak Ridges Moraine: amphibian and reptile habitats (1913), 187; aquifer, 28, 266, 268; deforestation and reforestation, 30-1, 267, 268; glacial sediments, 17, 18(f), 19(f), 21(f); gravel and sand extraction, 267, 269(f); groundwater, 27, 268; importance, 267-9; lack of protection, 269; soil changes, 30-1; 'special place,' 266-70; watershed, 52(f), 266, 267, 268

Oak Ridges Moraine (STORM Coalition, 1997), 270

Oak Ridges Trail, 269-70

Oak Ridges Trail Guidebook (Oak Ridges Trail Association, 1997), 270

Odonata (insect order), 164-5, 166(f)

Odonata of Canada and Alaska (E. Walker), 168

Odum, Eugene and Howard, 98-9

Ojibwa (Native peoples) language group, 67

Oldsquaw (duck), 220

Oligochaetes, 153

Ondatra zibethicus (muskrats), 213, 214

Onondaga (Native peoples) language group, 67

Ontario, Lake, 36, 280-2

Ontario, origin of name, 74

Ontario Butterfly Atlas (Holmes et al., 1991), 162

Ontario ice lobe, 17-8, 24-5, 258

Ontario Rock Garden Society, 112

Oomycetes (fungi), 141

Oospores (fungal reproduction), 141

Opheodrys vernalis (smooth green snake), 189, 192(t), 198

Ophiostoma spp. (fungi), 145

Orconectes rusticus (crayfish), 155-6

Orthoptera (insect order), 166-7

Our Common Future (UN, 1986), 300

Oxeye daisy (*Leucanthemum vulgare*), 111

Oxidus gracilis (flat-backed millipede), 157

Ozone and air quality, 42

Palaeo-Indians, 63-4

Palaeoptera (insects with primitive wings), 164-5

Panaeolus foenisecii (spring mushroom), 133

Parasites (insect and fungal), 131, 140, 147-8, 162

Park, Thomas, 101

Parks of Toronto: for bird watching, 217, 218(t); early parks (1873-1942), 229-31; green belt/conservancy district (Putnam), 232(f), 233; Hurricane Hazel and land acquisition, 236; metropolitan, 234-5; naturalization, 239-40; neighbourhood, 231; ravines, 231, 240; Toronto Guild of Civic Art, proposals, 230-1; types of gardens, 85, 86-7(f). *See also* Conservation; Regional planning

Passmore, Frederick, 305

Pathogenic fungi. *See* Fungi, pathogenic

'Pattern and process in the plant community' (A.S. Watt, in the *Journal of Ecology*, 1947), 97

PCBs (polychlorinated biphenyls), 124-5

Peach leaf curl (*Taphrina deformans*), 145

Percids (yellow perch, walleye, and darters), 179, 181-3

Peregrine falcon, 217, 221, 224(t)

Perithecia (fungal reproduction), 140, 141(f)

Peromyscus spp. (mice), 212

Pesticides, DDT, 221

Petticoat Creek watershed, 52(f)

Petun (Native peoples) language group, 67-8

Phaeolus schweinitzii (fungus), 142, 143(f)

Phallus ravenelii (stinkhorns), 148, 149(f)

Phasmatodea (insect order), 168

Photobionts (chlorophyll-bearing partner of lichens), 119

Photosynthetic conversion (by vascular plants), 113-4

The Physiography of Southern Ontario (D.F. Putnam, 1951), 232

Pickerel frogs (*Rana palustris*), 188, 195-6

Pickering Museum Village, 289

Plan for Flood Control and Water Conservation (1959), 236, 238

Plant succession, 122

Plants, binomial naming system, 118

Plants, non-vascular, 117-8. *See also* Bryophytes; Lichens; Mosses

Plants, vascular: agricultural remnants, 106, 111-2; chemical structuring of animal communities, 115; club mosses, 107; contributions to ecosystems, 113-5; cultivated, 106, 112, 113; description, 105-7; diversity, 111-2; effect of human activities, 106-7; ferns and horsetails, 107; fertilization, 161; in Forks of Credit Provincial Park, 275; future, predictions, 115; growth forms, 112-3; native, and the environment, 106-7, 111; in Rouge valley, 291, 293; salt-tolerant, 106; trees (see Trees)
Plasmopara viticola (downy mildew), 146
Plecoptera (insect order), 169
Pleistocene period, 16-8, 19(f), 23(f)
Plethodon cinereus (eastern redback salamander), 192(t), 193
Pleurotus dryinus (fungus), 142
Pollen grains, fossilized, 16, 20-1, 24(f), 26-7, 281-2
Pollution, effects, 56-7, 124-5, 153, 177-8, 181
Pollution Probe, 6, 46
Polychlorinated biphenyls (PCBs), 124-5
Polyporus squamosus (fungus), 142-3
Pontoporeia hoyi (amphipod), 153
Population: census data (1991, 1996), 77, 85; demographic changes, 88; growth and expansion (1793-1961), 83(f), 84-5, 86-7(f); Toronto (1881), 82; York (1834), 79
Population ecology, 98, 99-102, 103
Port Lands, 249-51, 252(f), 253-4, 255
Pottery as archaeological marker, 64
Precambrian period, 11-2, 13(f), 15-6
Precipitation: changes possible in next 100 years, 44, 46; mean annual, 38, 39(f), 44(f); rain and snow shadows, 38; seasonal variations, 34-6
Proceedings (formerly *Canadian Journal*), 305-6
Procyon lotor (raccoons), 206
Psathyrella spp. (summer and fall mushrooms), 136, 138(f), 139
Pseudacris spp. (frogs), 188, 192(t), 195
Psilocybe physaloides (spring mushroom), 133-4

Psocoptera (insect order), 169
Puccinia spp. (rust fungi), 144
Puffball *(Calvatia gigantea),* 148, 149(f)
Putnam, Donald F. (1903-77), 232, 233
Pycnidia (in fungal asexual reproduction), 141
'Pyramid' ecosystem approach, 98

Quaking aspen *(P. tremuloides),* 110
Quaternary sediments, 27-8
Queen Anne's lace *(Daucus carota),* 111
Queenston Delta, 15
Queenston Formation, 14(t), 15

Rabbits, cottontail *(Sylvilagus floridanus),* 210
Rabies and raccoons, 206
Raccoons *(Procyon lotor),* 206
Ragweed *(Ambrosia artemisiifolia),* 106
Rainbow trout, 179, 181, 182-3, 184(f)
Rainfall, 35, 36, 44, 44(f). *See also* Precipitation
Rana spp. (frogs), 188, 192(t), 194-6
Rats, Norway *(Rattus norvegicus),* 213-4
Rattray, James Halliday, 276
Rattray Marsh, 218, 220, 272(f), 275-7
Rattray Marsh Protection Association, 276-7
Rattus norvegicus (Norway rats), 213-4
RCI. *See* Royal Canadian Institute (RCI)
Red clover *(Trifolium pratense),* 111
Red fox *(Vulpes vulpes),* 206-7
Red maple, 109
Red mulberry*(Morus rubra),* 108
Red oak *(Quercus rubra),* 109
Red pine *(Pinus resinosa),* 109
Redbelly snake, northern *(Storeria o. occipitomaculata),* 188, 192(t), 197
Red-spotted newt *(Notophthalmus v. viridescens),* 188, 192(t), 193
Reductionist science, 98
Red-winged blackbird, 221
Regional planning, parks and land use: acquisition of parks, 235-6; Gardiner, support for, 233-4;

Master Plan (1943), 233; metropolitan planning bodies, 233, 234; naturalization program, 239-40; park system, establishment, 231-4, 240; water conservation/flood control under MTRCA, 236, 238
Regosol (soil), 31
Reptiles: distribution and numbers (1982-97), 190(f), 191-3; in Don valley, 283; habitats in 1913, 187-9, 198-9; in Rouge valley, 293; snakes, 197-8; turtles, 196-7; urbanization, effects, 189-91
Rhytisma acerinum (fungal leaf disease), 147
Ribbon snake, northern *(Thamnophis sauritus septentironalis),* 189, 198
rideTOgether program, 46
Ringneck snake, northern *(Diadophis punctatus edwardsi),* 189, 198
Riverdale Park, 229
Rivers. *See* Streams; *names of individual rivers*
Road salt and plants, 106, 284
Rosedale Ravines, 231
Roses, fungal diseases, 145-6
Rouge valley: amphibian and reptile habitat, 192, 198; bird-watching site, 218(t); diversity of flora and fauna, 107, 291, 293; history of area, 290-1; Hurricane Hazel, impact, 291-2; problems facing park, 293; a 'special place,' 290-3; urban wilderness park (1990), 240, 290(f), 292; watershed, 52(f), 240
Rousseau Post, 70(f)
Rowan-tree *(Sorbus aucuparia),* 110
Royal Canadian Institute (RCI): aim of this book, 303; history of, 305-7; mandate, 305; *Natural History, 1913* (see Natural History of the Toronto Region)
Royal Commission on the Future of the Toronto Waterfront, 6, 59
Royal Ontario Museum of Zoology, 162
Royal York Hotel, 84
Russian-olive *(Elaeagnus angustifolia),* 110
Rust fungi *(Puccinia hysterium),* 143-4
Rye, fungal diseases, 146, 147(f)

Sac spider *(Cheiracanthium mildei)*, 156(f), 157
Salamander, eastern redback *(Plethodon cinereus)*, 192(t), 193
Salamanders, 191-3, 194
Salmonids (char, trout, salmon), 179, 181-3, 184(f)
Salt, road, 106, 153-4, 284
Saprotrophs, 131, 140
Saucer magnolia *(Magnolia X soulangeana)*, 110
Sauk Sea, 12, 14(t), 16
Saunders, Richard (Dick) (1904-98), 218
Saunders, William, 162
Sauriol, Charles (1904-95), 284, 285
Charles Sauriol Environmental Trust Fund, 285
Savannahs, in High Park, 260-5
Save the Rouge, conservation organization, 292
Scarborough Bluffs, 21, 23, 25(f), 256-9
Scarborough Formation: description, 21-2; fossilized tree pollen, 22, 24(f); Pleistocene sediments, 19(f), 257, 258(f); showing in Don Valley Brickyard, 23(f)
Scarborough Heights Park, 57
Scirrhia acicola (brown spot needle blight), 147
Sciurus carolinensis (grey squirrels), 211-2
Scleroderris canker disease, 147
Sclerotia (ergots) of fungus, 146, 147(f)
Scopinella sphaerophila (fungus), 147
Scorpionflies (order Mecoptera), 171
Scotch pine *(Pinus sylvestris)*, 144
Sea lamprey, 180
Seaton Hiking Trail, 288(f), 289
Sediments: cycles of sedimentary rock units, 13, 14(t); Pleistocene, 16-8, 19(f); Sauk Sea deposits, 12; types in Toronto region, 21(f)
Seminary Till, 19(f), 258
Seneca (Native peoples) language group, 67
'Sense of place' (Toronto), 299-300
Seral successional stages, 97, 199
Seven Years War, 72
Sewage, 58-9, 82-4, 153

'Shaggy mane' mushroom *(Coprinus comatus)*, 136, 137(f)
Shallow Lake Formation, 14(t)
Sherwood Park, 231
Siberian elm *(Ulmus pumila)*, 108, 109, 110
Significant natural areas in Toronto, 237(f), 238
Silver Creek Conservation Area, 272(f), 274
Silver maple *(Acer saccharinum)*, 108, 109
Silverfish (order Thysanura), 163-4
Simcoe, Elizabeth, 72, 73, 77-8, 256
Simcoe, John Graves (lieutenant governor of Ontario), 74
Simcoe, Lake, early names, 74
Simcoe ice lobe, 17, 18(f), 25(f)
Siphonaptera (insect order), 174-5
Sirococcus clavigignenti-juglandacearum (conidial fungus), 145
Six Nations Iroquois and Mississauga, 75
Skistodiaptomus pallidus (crustacean), 155
Skunks, striped *(Mephitis mephitis)*, 207-8
Slippery elm *(Ulmus rubra)*, 108, 110
Slugs, 156-7
Smallmouth bass, 179, 181, 182
Smith, Robert Home, 231
Smooth green snake *(Opheodrys vernalis)*, 189, 192(t), 198
Snakes: descriptions, 197-8; distribution and numbers (1913; 1982-97), 188-9, 191-3; in Don valley, 283; hibernaculum, 200; in Port Lands, 251
Snapping turtles *(Chelydra serpentina)*, 189, 192(t), 196
Snowfall, 36, 39(f), 44. *See also* Precipitation
Soils, 29-31
Solar radiation, 38
Soredia (of lichens), 119-20
South Peel Naturalists' Club, 275-6
'Special places': with ecological integrity, 93, 103, 243, 244(f). *See also* Credit River watershed; Don valley; Duffins Creek; High Park; Humber valley; Oak Ridges Moraine; Port Lands; Rouge valley; Scarborough Bluffs; Waterfront

'Specialist' species: with abuse of aquatic ecosystems, 177-8; description, 57, 97, 298; freshwater invertebrates, 152; and plant species, 115
Species extinctions, 152. *See also* 'Generalist' species; 'Specialist' species
Species Plantarum (Carolus Linneus, 1753), 118
Spiders, 157
Spiny softshell turtles, eastern *(Apalone spinifera spinifera)*, 197
Spiny water flea *(Bythotrephes cederstroemi)*, 155
Sporangia (in Zygomycetes), 141, 142(f)
Sporophytes, 107, 117
Spotted salamander *(Ambystoma maculatum)*, 188, 192(t), 193
Spring Breezes, High Park (painting by J.E.H. Macdonald), 264(f), 265
Spring peeper *(Pseudacris crucifer)*, 188, 192(t), 195
Springtails (order Collembola), 164
Spruce rust, 144
Squirrels, grey *(Sciurus carolinensis)*, 211-2
St. Lawrence valley, 24-5
Stargazing, changes in last 100 years, 47-9
Sternotherus odoratus (musk turtles), 197
Stinkhorns *(Phallus ravenelii)*, 148, 149(f)
Stoneflies (order Plecoptera), 169
Storeria spp. (snakes), 188-9, 192(t), 197
Streams: channelling, 59, 178, 295, 297; flow 'corrections' and flooding, 56, 59, 178; gabion baskets (riverbank reinforcement), 285; hydrodynamic forces, 51-2; old-growth (pristine) stream courses, 54, 177-8; pollution, effect of, 56-7, 125, 177-8; reform activities after 1940s, 58-9, 184; urbanization and ecosystem, 57; warming, due to human influences, 181-3; water conservation/flood control, 236, 238. *See also* Ecosystems, aquatic; Groundwater

Strepsiptera (insect order), 171
Striped chorus frogs *(Pseudacris t. triseriata)*, 188, 192(t), 195
Stromata (fungal reproduction), 140, 141(f)
Stromotolites, 11
Stupart, R.F., 33, 43
Suburbs, growth, 84-5
Successional stages, seral, 97, 199
Sugar maple *(Acer sacchraum)*, 109
Sulphur dioxide, 41-2, 124
Sunfishes, 179, 181, 182(f)
Sunnybrook Diamict or Till, 19(f), 257-8
Sunnybrook Drift, 28, 257-8
Sunnybrook Park, loss of habitat, 123
Sunshine, averages, 38
Surficial geology of Toronto, 17-8
Susquehannock (Native peoples), 67, 68
Sycamore, eastern *(Platanus occidentalis)*, 108
Sylvilagus floridanus (cottontail rabbits), 210

Taddle Creek, 296, 297
Tamarack *(Larix laricina)*, 109
Tamias striatus (eastern chipmunks), 211
Tansley, A.G., 3
Taphrina deformans (fungus of peaches), 145
Tar spot (fungal disease), 147
Tartarian maple *(Acer ginnala)*, 109-10
Task Force to Bring Back the Don, 239, 284
Teiaiagon (Iroquois village), 68, 70(f)
Temperatures: changes possible in next century, 44, 46; historical trends, 43-4; increase, effect on invertebrates, 153; macroclimate seasonal variations, 34-6; mesoclimates patterns, 36-7; microclimatic changes, 40; warming of water, 181
Ten Thousand Trees project, 46
Termites (order Isoptera), 167
Terra Cotta Conservation Area, 272(f), 274
Thallus (vegetative structure of lichens), 119, 120(f), 121

Thamnophis spp. (snakes), 189, 192(t), 197-8
Thelephora terrestris (turkey tail), 148, 149(f)
Thompson, Tommy (1913-85), 234-5
Thorncliffe Formation, 19(f), 258
Thrips (order Thysanoptera), 169-70
Thysanoptera (insect order), 169-70
Thysanura (insect order), 163-4
Tills, 17, 18(f), 19(f), 21(f)
Tippecanoe Sea, 13, 14(t)
Toads, 191-3, 194-6
Tommy Thompson Park. *See* Leslie Street Spit
Toronto: city of York (1793-1834), 78-9, 80(f); 'city region,' 295; financial centre, 82; Fort York (1793), 74, 78, 79(f); growth and expansion, 4(f), 77, 83(f), 84-5, 86-7(f); 'heat island,' 36, 45, 153; incorporation (1834), 79; manufacturing industry by 1911, 82; megacity (1998), 85, 88-9; Metropolitan Toronto (1953), 84; mid-nineteenth century, 81-2; name, origin, 74; networks, cultural and natural, 299-300; population (*see* Population); 'sense of place,' 299-300; settlement by Native peoples (*see* Native peoples); sewage problems (late 1800s), 82-4; Toronto Purchase (1787), 73-4; traffic and transportation, 84
Toronto and Suburban Planning Board (TSPB), 233
Toronto and York Planning Board (TYPB) (1948), 233
Toronto Atmospheric Fund, 45-6
Toronto Camera Club, 306
Toronto Economic Development Corporation: green infrastructure system (port lands), 254-5
Toronto Entomologists' Association, 162
Toronto Field Naturalists, 231, 238, 284
Toronto Guild of Civic Art and parks, 230-1
Toronto Harbour, pre-settlement, 77
Toronto Islands: amphibian and reptile habitats, 188, 189, 194; bird-watching site, 217, 218, 220; Island Park, 229-30, 236; south

shore, 245; Toronto Island Airport, 295; turtles, 196; vascular plants, 107, 108, 110
Toronto Purchase (1787), 73-4
Toronto Stock Exchange, 82
Toronto the Green (Toronto Field Naturalists, 1976), 238
Toronto Zoo, 40, 200, 236
Tracheoniscus rathkei (Isopoda, sowbugs), 157
Transactions (formerly *Canadian Journal)*, 305-6
Treaty of 1701, 71
Treaty of Ryswick (1697), 71
Tree-of-heaven *(Ailanthus altissima)*, 108, 109-10
Trees: and animal biodiversity, 115; boreal, 106, 108-9; dead trees, ecological use, 115; floodplain and upland (Carolinian), 106-7, 108; fungi, pathogenic, 144-5, 147; Great Lakes-St. Lawrence region, 106-7; in Humber River valley, 278, 279; hybrids of native and exotic, 110-1; mixed forest zone trees, 105-6, 109; native, 107-9; naturalized exotic, 109-11; in savannahs of High Park (1819), 261; in urban and disturbed habitats, 109-10, 112. *See also* Plants, vascular
Tremella mycophaga (a jelly fungus), 147-8
Trichoptera (insect order), 171-2
'The trophic-dynamic aspect of ecology' (Raymond Lindeman, 1942), 98
Trout. *See under* Salmonids
Tully, Kivas, 305
Turkey tail *(Thelephora terrestris)*, 148, 149(f)
Turtles, 189, 191-3, 196-7
Tuscarora (Native peoples), 68
Twisted-wing parasites (order Strepsiptera), 171
Tyrell, Joseph Burr, 307

Understorey plants (in forests), 111
United Nations: Commission on Sustainable Development, 301; Framework Convention on Climate Change, 45
Upper Aquifer (Oak Ridges Moraine), 28

Urban ecology, 298-9
Urbanization, effects: amphibians and reptiles populations, 189-91; aquatic ecosystems, 57; birds, 221-4; 'creeping incrementalism,' 302; environment, 151-2, 229, 295-7, 301-3; and fungi, 131-2; Oak Ridges Moraine, 267; old-growth ecosystems, 57, 229; savannahs, 264; trees, 109-10, 112

Valvata piscinalis (European valve snails), 155
Varroa mites *(Varroa jacobsoni)*, 156(f), 157-9
Vascular flora. *See* Plants, vascular
Venturia inaequalis (apple scab), 145
Vertebrate fossils, in Don Formation, 20, 24(f)
Vertical gardens, 45
Verulam Formation, 14(t)
Viceroy butterfly, 173
Vulpes vulpes (red fox), 206-7

Walker, Edmund (1877-1969), 162, 167-8
Walker, James, 267, 268
Walkingsticks (order Phasmatodea), 168
Walleye, 179, 181
Warblers and competitive exclusion, 100
Wasps (order Hymenoptera), 175
'Wastelands' (on Oak Ridges Moraine), 31
Water. *See* Ecosystems, aquatic; Groundwater; Streams
Water fleas *(Eubosmina coregoni)*, 155
Water snake, northern *(Nerodia s. sipedon)*, 189, 192(t), 197
Water table, lowered, 56, 189

Waterfront: birds over-wintering, 220; bird-watching site, 218(t); current flora and fauna, 247; rehabilitation projects, 240-1; reptile habitat, 188-9, 196; a 'special place,' 245-8; from wetlands to port lands, 57, 105, 246-7, 295
Waterfront Regeneration Trust, 6, 254-5, 284, 308
Waterfront Trail, 308-9
Watersheds: baseflow, importance of, 53-4; Credit River, 271, 272(f), 273-4; Don River, 58; Duffins Creek, 52(f), 286-7; Ganaraska, rehabilitation, 267; greater Toronto region, 51, 52(f); Oak Ridges Moraine, 266, 267, 268; 'parking-lot hydrographs,' 56; planning program (Field Naturalists, 1976), 238, 240
Watt, A.S., 97
Weather. *See* Climate; Mesoclimates
Weed species, 111-2
Wetlands, coastal: amphibians and reptiles, 187, 189-90, 198-9; bird-watching site, 218; boundary definition, 52; destruction along waterfront, 57, 94, 105, 231; improved state, 297; original waterfront, 245-6; Rattray Marsh, 218, 220, 272(f), 275-7
Wetlands, upland, 56, 105, 187-8, 198-9
Wheat rust *(Puccinia graminis)*, 144
Whirlpool Formation, 14(t), 15
White ash *(Fraxinus americana)*, 109
White birch *(Betula papyrifera)*, 109
White cedar *(Thuja occidentalis)*, 109
White elm *(Ulmus americana)*, 109

White mulberry *(Morus alba)*, 109
White oak *(Quercus alba)*, 108
White pine blister rust *(Cronartium ribicola)*, 144
White pine *(Pinus strobus)*, 109
White poplar *(Populus alba)*, 110
White-footed mouse *(Peromyscus leucopus)*, 212
Wilcox Lake and pollen fossils, 26-7
Wilson, Daniel, 305
Wind, in Toronto, 39, 40-1
Wood frogs *(Rana sylvatica)*, 188, 192(t), 195
Wood turtles *(Clemmys insculpta)*, 197
Woodchucks *(Marmota monax)*, 210-1
Woodgate gall rust *(Endocronartium harknessii)*, 144
Woodland periods (Native peoples), 65-8
Woodlands. *See* Forests
World Wildlife Fund, 6
Wych elm *(Ulmus glabra)*, 110

Xylaria polymorpha (bark fungus), 147

Yellow perch, 179, 181, 182-3
Yonge Street, road to Lake Simcoe (1794), 78, 80(f)
York (precursor of Toronto), 74, 78-9, 80(f)
York Till, 19, 22(f)
Yorkville, Village of, 82-3

Zebra mussels *(Dreissina polymorpha)*, 154-5, 220
Zoo, Toronto, 40, 200, 236
Zooplankton, decrease, 153
Zygomycetes (fungi), 141, 142(f)